Lecture Notes in Computer Science 2726

Edited by G. Goos, J. Hartmanis, and J. van Leeuwen

Springer
Berlin
Heidelberg
New York
Barcelona
Hong Kong
London
Milan
Paris
Tokyo

Edwin Hancock Mario Vento (Eds.)

Graph Based Representations in Pattern Recognition

4th IAPR International Workshop, GbRPR 2003
York, UK, June 30 – July 2, 2003
Proceedings

Springer

Series Editors

Gerhard Goos, Karlsruhe University, Germany
Juris Hartmanis, Cornell University, NY, USA
Jan van Leeuwen, Utrecht University, The Netherlands

Volume Editors

Edwin Hancock
University of York, Department of Computer Science
York, YO1 5DD, United Kingdom
E-mail: erh@cs.york.ac.uk

Mario Vento
Università degli Studi di Salerno, D.I.I.I.E.
Via Ponte don Melillo, 1, 84084 Fisciano (SA), Italy
E-mail: mvento@unisa.it

Cataloging-in-Publication Data applied for

A catalog record for this book is available from the Library of Congress

Bibliographic information published by Die Deutsche Bibliothek
Die Deutsche Bibliothek lists this publication in the Deutsche Nationalbibliographie;
detailed bibliographic data is available in the Internet at <http://dnb.ddb.de>.

CR Subject Classification (1998): I.5, I.4, I.3, G.2.2, E.1

ISSN 0302-9743
ISBN 3-540-40452-X Springer-Verlag Berlin Heidelberg New York

Springer-Verlag Berlin Heidelberg New York
a member of BertelsmannSpringer Science+Business Media GmbH

http://www.springer.de

© Springer-Verlag Berlin Heidelberg 2003
Printed in Germany

Typesetting: Camera-ready by author, data conversion by PTP-Berlin GmbH
Printed on acid-free paper   SPIN: 10929032   06/3142   5 4 3 2 1 0

Preface

This volume contains the papers presented at the Fourth IAPR Workshop on Graph Based Representations in Pattern Recognition. The workshop was held at the King's Manor in York, England between 30 June and 2nd July 2003. The previous workshops in the series were held in Lyon, France (1997), Haindorf, Austria (1999), and Ischia, Italy (2001).

The city of York provided an interesting venue for the meeting. It has been said that the history of York is the history of England. There have been both Roman and Viking episodes. For instance, Constantine was proclaimed emperor in York. The city has also been a major seat of ecclesiastical power and was also involved in the development of the railways in the nineteenth century. Much of York's history is evidenced by its buildings, and the King's Manor is one of the most important and attractive of these. Originally part of the Abbey, after the dissolution of the monasteries by Henry VIII, the building became a center of government for the Tudors and the Stuarts (who stayed here regularly on their journeys between London and Edinburgh), serving as the headquarters of the Council of the North until it was disbanded in 1561. The building became part of the University of York at its foundation in 1963.

The papers in the workshop span the topics of representation, segmentation, graph-matching, graph edit-distance, matrix and spectral methods, and graph-clustering. They hence span the use of graphs at all levels of representation from low level image segmentation, all the way through to high level symbolic manipulation. There are papers on formalizing the use of graphs for image representation, on the development of new and efficient means for graph matching, on formalizing the notion of distance between graphs, and on learning the class structure of sets of graphs through clustering. The workshop also sees the convergence and synthesis of ideas from several fields. As well as the use of ideas from classical discrete mathematics, there are also papers that draw heavily on newly emerging paradigms such as spectral graph theory and machine learning. An important theme running through many of the papers is the desire to bring principled methods to bear on important problems which have sometimes eluded rigor for too long. Another encouraging sign are the relatively large number of contributions in the area of clustering. Most of the classical work on statistical pattern recognition has focused on developing methods that can be used with feature vectors. Since graphs are not vectorial in nature, many tasks in pattern recognition and machine learning are not easily accomplished with data that requires a graph-based representation. Hence, the development of graph clustering methods holds out the promise of significant advances in unsupervised learning from relational and structural data.

The papers presented here have all undergone review and have been revised by their authors. I would therefore like to thank the program committee for their help in ensuring that the papers have been given a thorough and critical evaluation. I would also like to thank Antonio Robles-Kelly for help in preparing the final volume, and Alex Hughes for maintaining the workshop website.

June 2003   Edwin Hancock
                                                                  Mario Vento

Program Co-chairs

Edwin Hancock   The University of York, York, UK
Mario Vento   Università di Napoli, Naples, Italy

Program Committee

Luc Brun   Université de Reims, Reims, France
Horst Bunke   University of Bern, Bern, Switzerland
Terry Caelli   University of Alberta, Edmonton, Canada
Luigi Pietro Cordella   Università di Napoli, Naples, Italy
Jean-Michel Jolion   Lyon Research Center for Images and Informa-
                                 tion Systems, Lyon, France
Walter Kropatsch   Vienna Univ. of Technology, Vienna, Austria
Marcello Pelillo   Università Ca' Foscari di Venezia, Venice, Italy
Alberto Sanfeliu   Univ. Politècnica de Catalunya, Barcelona,
                                 Spain
Kaleem Siddiqi   McGill University, Montreal, Canada

Table of Contents

Data Structures and Representation

Segmentation

Graph Edit Distance

Graph Matching

Matrix Methods

Graph Clustering

Construction of Combinatorial Pyramids

Luc Brun[1] and Walter G. Kropatsch[2]

[1] Laboratoire d'Études et de Recherche en Informatique(EA 2618)
Université de Reims – France
luc.brun@univ-reims.fr
[2] Institute for Computer-aided Automation
Pattern Recognition and Image Processing Group
Vienna Univ. of Technology – Austria
krw@prip.tuwien.ac.at

Abstract. Irregular pyramids are made of a stack of successively reduced graphs embedded in the plane. Each vertex of a reduced graph corresponds to a connected set of vertices in the level below. One connected set of vertices reduced into a single vertex at the above level is called the reduction window of this vertex. In the same way, a connected set of vertices in the base level graph reduced to a single vertex at a given level is called the receptive field of this vertex. The graphs used in the pyramid may be region adjacency graphs, dual graphs or combinatorial maps. This last type of pyramids are called Combinatorial Pyramids. Compared to usual graph data structures, combinatorial maps encode one graph and its dual within a same formalism and offer an explicit encoding of the orientation of edges around vertices. This paper describes the construction scheme of a Combinatorial Pyramid. We also provide a constructive definition of the notions of reduction windows and receptive fields within the Combinatorial Pyramid framework.

1   Introduction

Graphs play an important role in different fields of computer vision and pattern recognition. However, many graph algorithms still suffer from a high computational complexity. Irregular pyramids allow coarse to fine strategies by encoding a hierarchy of graphs. The efficiency of an irregular data structure is strongly influenced by two closely related features: The decimation process used to build one graph from the graph below and the data structure used to encode each graph of the pyramid. The decimation process determines the height of the pyramid and the properties that may arise from the decimation process. The model used to encode the graphs of the pyramid determines the properties that may be encoded in each graph.

We present in this paper an implementation of Combinatorial Pyramids [2]. Such pyramids are defined as a stack of successively reduced combinatorial maps. A combinatorial map data structure may be intuitively understood as an encoding of one planar graph and its dual into a same data structure. Combinatorial maps encode additionally the orientation of edges around each vertex.

E. Hancock and M. Vento (Eds.): IAPR Workshop GbRPR 2003, LNCS 2726, pp. 1–12, 2003.

The remaining of this paper is organized as follows: We present in section 2 the decimation process used within the irregular pyramid framework and two graph encodings. The combinatorial map model is presented in Section 3. The construction scheme of a Combinatorial Pyramid is then presented in Section 4. We also describe in this section the notions of reduction window and receptive fields within the Combinatorial Pyramid framework. This section concludes with an application of Combinatorial Pyramids to the segmentation framework.

2   Irregular Pyramids

The irregular pyramids are defined as a stack of successively reduced graphs, each graph being built from the graph below by selecting a set of vertices named surviving vertices and mapping each non-surviving vertex to a surviving one [13]. This reduction operation was first introduced by Meer [12,13] as a stochastic process. Using such framework, the graph $G_{l+1} = (V_{l+1}, E_{l+1})$ defined at level $l+1$ is deduced from the graph defined at level l by the following steps:

1. The selection of the vertices of G_{l+1} among V_l. These vertices are the surviving vertices of the decimation process.
2. A link of each non surviving vertex to a surviving one. This step defines a partition of V_l.
3. A definition of the adjacency relationships between the vertices of G_{l+1} in order to define E_{l+1}.

2.1   Simple Graph Pyramids

In order to obtain a fixed decimation ratio between each level, Meer [12] imposes the following constraints on the set of surviving vertices:

$$\forall v \in V_l - V_{l+1} \ \exists v' \in V_{l+1} : (v, v') \in E_l \tag{1}$$
$$\forall (v, v') \in V_{l+1}^2 : (v, v') \notin E_l \tag{2}$$

Constraint (1) insures that each non-surviving vertex is adjacent to at least a surviving one. Constraint (2) insures that two adjacent vertices cannot both survive. These constraints define a *maximal independent set* (MIS).

Meer [12] defines a set of surviving vertices fulfilling the maximal independent set requirements thanks to a stochastic process: A variable x_i is associated to each vertex v_i and each vertex associated to a local maximum of this variable survives. Since two adjacent vertices can not both represent a local maximum of the variable, the constraint (2) is satisfied. However, some vertices may not represent a local maximum of the random variable while having only non surviving neighbors (see vertex ⑥ in Fig. 1). The constraint (1) is in this case violated. The determination of the MIS is thus performed thanks to an iterative process [13,10]. Jolion [9] and Haxhimusa et al. [8] have recently proposed two improvements of this iterative decimation process.

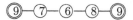

Fig. 1. Decimation of a "1D" graph. The value of the variable x_i is represented inside each vertex. Vertices whose value is locally maximal are represented by double circles.

Meer [12] uses as variable x_i a random variable uniformly distributed between $[0, 1]$. This purely random process has been adapted by Montanvert [13] to the connected component analysis framework by restricting the decimation process to a set of subgraphs of the graph G_l. This restriction of the decimation process is performed by a function $\lambda(v, v')$ defined on each couple of adjacent vertices v and v' and such that $\lambda(v, v') = 1$ if v and v' belong to the same subgraph and 0 otherwise. Jolion [10] proposed to use an interest operator such as the variance computed on a neighborhood of each vertex rather the random variable. Using such a reduction scheme, surviving vertices are located on homogeneous parts of the image.

Given the set of surviving vertices, Meer and Montanvert [12,13] connect each non surviving vertex to its surviving neighbor whose random variable is maximum. Jolion [10] uses a contrast operator such as the difference of gray levels to connect each non surviving vertex to its surviving neighbor with the closest gray level.

Note that within the segmentation framework, the use of the function λ defined by Montanvert [13] supposes to define first an implicit partitioning of the current graph in order to set the value of $\lambda(v, v')$ for each couple of vertices. Using the interest and contrast operators defined by Jolion the partition of the graph is built during the decimation process. The set of non surviving vertices connected to a surviving vertex defines its reduction window and thus the parent child relationship between two consecutive levels.

The final set of surviving vertices defined on V_l corresponds to the set of vertices V_{l+1} of the reduced graph $G_{l+1} = (V_{l+1}, E_{l+1})$. The set of edges E_{l+1} of G_{l+1} is defined by connecting by an edge in G_{l+1} any couple of surviving vertices having adjacent children.

2.2   Dual Graph Pyramids

Using the reduction scheme defined in Section 2.1, two surviving vertices in G_l are connected in G_{l+1} if they have at least two adjacent children. Two reduction windows adjacent by more than a couple of children will thus be connected by a single edge in the reduced graph. The stack of graphs produced by the above decimation process is thus a stack of simple graphs (i.e. graphs without self-loops nor double edges). Such graphs does not allow to encode multiple boundaries between regions nor inclusions relationships. For example, the two boundaries between the left and right region in Fig. 2(a) are encoded by a single edge at the top of a simple graph pyramid (edge (W, B_1) in Fig. 2(b)). In the same way, the inclusion relationship between the center black region and the surrounding one in Fig. 2(a) is encoded by one edge at the top of the simple graph pyramid

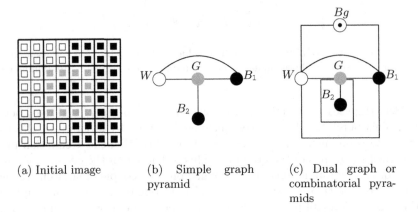

(a) Initial image   (b) Simple graph pyramid   (c) Dual graph or combinatorial pyramids

Fig. 2. Encoding of the connected components of a 8×8 image (a) by a simple graph pyramid (b) and the Dual Graph or Combinatorial Pyramids (c). The vertex B_g in (c) encodes the background of the image.

(edge (G, B_2) in Fig. 2(b)). Such an encoding does not allow to differentiate an inclusion from an adjacency relationship. These drawbacks may be overcome by using the dual graph pyramids introduced by Willersin and Kropatsch [15]. Within the dual graph pyramid framework the reduction process is performed by a set of edge contractions. The edge contraction operation collapses into one vertex two vertices adjacent by an edge and remove the edge. The resulting graph is a minor of the initial one [5]. In order to keep the number of connected components of the initial graph, the set of edges to be contracted in a graph G_l of the pyramid must forms a forest of G_l. The resulting set of edges is called a contraction kernel. Using the decimation process described in Section 2.1, each non surviving vertex v is adjacent to its father v'. If we select one edge between v and v' for each non surviving vertex v, the resulting set of edges K forms a forest of the initial graph composed of trees of depth 1. This set encodes thus a contraction kernel.

The contraction of a graph reduces the number of vertices while maintaining the connections to other vertices. As a consequence some redundant edges such as self-loops or double edges may occur, some encode relevant topological relations (e.g. an island in a lake) others can be removed without any harm to the involved topology. These redundant edges may be characterized in the dual of the contracted graph. The removal of such edges is called a dual decimation step and the set of removed edges is called a removal kernel.

Using the dual graph pyramid framework each receptive field is a connected set of vertices in the base level. Moreover, each edge between two vertices encodes an unique connected boundary between the associated receptive fields. Finally, the use of self-loops within the hierarchy allows to differentiate adjacency relationships between receptive fields from inclusion relations.

3   Combinatorial Maps

Combinatorial maps and generalized combinatorial maps define a general framework which allows to encode any subdivision of nD topological spaces orientable or non-orientable with or without boundaries. An exhaustive comparison of combinatorial maps with other boundary representations such as cell-tuples and quad-edges is presented in [11]. Recent trends in combinatorial maps apply this framework to the segmentation of 3D images [1] and the encoding of hierarchies [3,4].

The remaining of this paper will be based on 2D combinatorial maps which we simply call combinatorial maps. A combinatorial map may be seen as a planar graph encoding explicitly the orientation of edges around a given vertex. It can be shown [7,5] that such maps correspond to a planar graph embedding and encode all the topological properties of the drawing of the graph. Figure 3(a) demonstrates the derivation of a combinatorial map from a plane graph $G = (V, E)$ corresponding to a 3×3 pixel grid. First the edges of E are split into two half edges called *darts*, each dart having its origin at the vertex it is attached to. The fact that two half-edges (darts) stem from the same edge is recorded in the reverse permutation α. A second permutation σ encodes the set of darts encountered when turning counterclockwise around a vertex.

A combinatorial map is thus defined by a triplet $G = (\mathcal{D}, \sigma, \alpha)$, where $\mathcal{D}$ is the set of darts and σ, α are two permutations defined on $\mathcal{D}$ such that α is an involution:

$$\forall d \in \mathcal{D} \quad \alpha^2(d) = d \tag{3}$$

If the darts are encoded by positive and negative integers, the involution α may be implicitly encoded by the sign (Figure 3(a)).

Given a dart d and a permutation π, the π-orbit of d denoted by $\pi^*(d)$ is the series of darts $(\pi^i(d))_{i \in \mathbb{N}}$ defined by the successive applications of π on the dart d. The σ and α orbits of a dart d will be respectively denoted by $\sigma^*(d)$ and $\alpha^*(d)$.

Fig. 3 illustrates the encoding $G = (\{\bullet\}, \{\bullet - \bullet\})$ of a 3×3 4-connected discrete grid by a combinatorial map. Each vertex of the initial combinatorial map (Fig. 3(a)) encodes a pixel of the grid. The σ-orbit of one vertex encodes its adjacency relationships with neighboring vertices. Let us consider the dart 1 in Fig. 3(a). Turning counterclockwise around the central vertex we encounter the darts $1, 13, 24$ and 7. We have thus $\sigma(1) = 13$, $\sigma(13) = 24$, $\sigma(24) = 7$ and $\sigma(7) = 1$. The σ-orbit of 1 is thus defined as $\sigma^*(1) = (1, 13, 24, 7)$.

The permutation α being implicitly encoded by the sign in Fig. 3, we have $\alpha^*(1) = (1, -1)$. The α successors of the darts 13 to 24 are not shown in Fig. 3(a) in order to not overload it. These darts encode the adjacency relationships between the border pixels of the grid and its background. The σ orbit of the background vertex is equal to the sequence of darts from -13 to -24: $(-13, -14, \ldots, -23, -24)$.

Each vertex of a combinatorial map $G = (\mathcal{D}, \sigma, \alpha)$ is implicitly encoded by its σ orbit. This implicit encoding may be transformed into an explicit one by

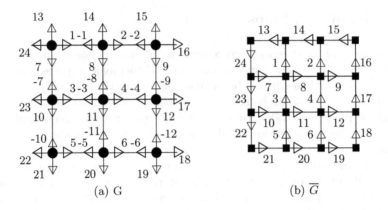

Fig. 3. A 3×3 grid encoded by a combinatorial map

using a vertex's labeling function [2]. Such a function, denoted by μ associates to each dart d a label which identifies its σ-orbit $\sigma^*(d)$. More precisely a vertex's labeling function should satisfy:

$$\forall (d, d')^2 \in \mathcal{D} \ \mu(d) = \mu(d') \Leftrightarrow \sigma^*(d) = \sigma^*(d') \tag{4}$$

For example, if darts are encoded by integers, the function $\mu(d) = min_{d' \in \sigma^*(d)}\{d'\}$ defines a valid vertex's labeling function.

Given a combinatorial map $G = (\mathcal{D}, \sigma, \alpha)$, its dual is defined by $\overline{G} = (\mathcal{D}, \varphi, \alpha)$ with $\varphi = \sigma \circ \alpha$. The orbits of the permutation φ encode the set of darts encountered when turning around a face of G. Note that, using a counter-clockwise orientation for permutation σ, each dart of a φ-orbit has its associated face on its right (see e.g. the φ-orbit $\varphi^*(1) = (1, 8, -3, -7)$ in Figure 3(a)). The dual combinatorial map $\overline{G} = (\{\blacksquare\}, \{\blacksquare - \blacksquare\})$ is shown in Fig. 3(b)

4   Combinatorial Pyramids

As in the dual graph pyramid scheme [15] (Section 2), a combinatorial pyramid is defined by an initial combinatorial map successively reduced by a sequence of contraction or removal operations respectively encoded by contraction and removal kernels.

4.1   Constructing Contraction Kernels Using Union-Find

As mentioned in Section 2, Jolion [10] and Montanvert [13] have proposed two different methods to decimate a graph. These methods may be adapted to build a contraction kernel by selecting one edge between each non surviving vertex and its father(Section 2.2).

We have encoded the decimation process defined by Jolion [10]. Our implementation uses conjointly the interest and contrast operators defined by Jolion,

```
 1    find_compress(array father,vertex v)
 2    {
 3         if (father(v) == v )
 4             return v;
 5         else
 6         {
 7             father(v) = find_compress(father,father(v))
 8             return father(v)
 9         }
10    }
```

Algorithm 1: The find_compress algorithm.

with the function λ defined by Montanvert. This last function is used to forbid the attachment of some non surviving vertices to surviving ones. A similar result may be obtained by setting the contrast operator to an infinite value for such couple of vertices.

The method of Montanvert [13] is based on an implicit partitioning of the graph. Such a partitioning may be obtained by using, for example, a clustering algorithm on the value attached to each vertex. Within this context the design of a contraction kernel is equivalent to the design to a spanning forest of the graph. This problem may be solved efficiently using union and find operations [6] on sequential or on parallel machines with few processors.

Within this technique each tree of a contraction kernel is encoded by storing in each vertex a reference to its father. The father of each vertex is initialized by $father(v) = v$. The find operation applied on a vertex v determines the root of the tree containing v. The union operation merge two trees into one. Given two vertices v_1 and v_2 whose father must be merged ($\lambda(father(v_1), father(v_2)) = 1$) the union of both trees is performed by setting the "grand-father"' of one of the two vertices, say v_1 to the father of v_2. We have thus $father(father(v_1)) = father(v_2)$. Using such a construction scheme for contraction kernels, the time complexity of the algorithm is dominated by the number of indirections required to find the father of each vertex. Indeed, if one vertex v_1 is merged to v_2 and then v_2 to v_3, we have $father(v_1) = v_2$ and $father(v_2) = v_3$. The determination of the father of v_1 requires thus 2 indirections. These indirections may be suppressed by using the algorithm find_compress (Algorithm 1) which connects directly to the root all the nodes in the branch of the tree which connect one vertex to its root.

The construction of the contraction kernel is performed by Algorithm 2. The test performed on line 12 of Algorithm 2 merges two trees if their associated roots are different and if the regions encoded by these trees should be merged according to the function λ. The function μ used on lines 10 and 11 encodes the vertices (equation 4).

Let us consider a combinatorial map $G_l = (\mathcal{SD}_l, \sigma_l, \alpha)$ reduced to $G_{l+1} = (\mathcal{SD}_{l+1} = \mathcal{SD}_l - K_{l+1}, \sigma_{l+1}, \alpha)$ by K_{l+1}. Since an union operation is performed

```
1    define_kernel(pyramide P, function lambda)
2    {
3        contraction kernel K = ∅
4        top of the pyramid G_l = (SD_l, σ_l, α)
5
6        For any vertex v in G_l
7                father(v) = v
8        For any dart b ∈ D_l
9        {
10           f_1 = find_compress(father, μ(b))
11           f_2 = find_compress(father, μ(α(b)))
12           if (f_1 != f_2 and lambda(f_1, f_2))
13           {
14               K = K ∪ {α*(b)}
15               father(f_1) = f_2
16           }
17        }
18        return K
19   }
```

Algorithm 2: Computation of a contraction kernel

for each edge added to the contraction kernel K_{l+1}, the number of union performed by the algorithm is equal to $|K_{l+1}|$. In the same way, we perform two find_compress operations for each dart, the total number of find_compress operations performed by the algorithm is thus equal to $2|SD_l|$. Based on the previous considerations it can be shown [14] that the sequential complexity of Algorithm 2 is equal to $\mathcal{O}(2\beta(2|SD_l|, |K|)|SD_l|))$ where β is a very low growing function.

Tables 1 compares the execution times on a sequential machine of the decimation processes using the iterative definition of survivors (Section 2) and the union and find operations. Both decimation processes reduce a $2^n \times 2^n$ image to a single vertex. Note that the number $|SD_l|$ of initial darts is, in this case a linear function of the size 2^{2n} of the input image. This size is multiplied by 4 between each line of Table 1. The execution time of the Union & Find algorithm is also multiplied by 4 between each line. The complexity of the Union & Find algorithm is thus found to be linear according to the size of the image on this experiment. On the other hand, the execution times of the iterative decimation process is approximately multiplied by 8 between each line. The complexity of this algorithm is thus not linear on this experiment.

4.2   Removals Kernels

Given a combinatorial map $G = (\mathcal{D}, \sigma, \alpha)$ and a contraction kernel K, the contracted combinatorial map $G' = (SD = \mathcal{D} - K, \sigma', \alpha)$ may contain redundant edges corresponding to double edges or empty-self-loops (Section 2.2). An empty

Table 1. Execution times required by the iterative decimation process and the one based on union and find operations.

execution times	Union & Find	Iterative Decimation
64×64	0.01 s	0.39 s
128×128	0.04 s	3.13 s
256×256	0.19 s	24.65 s
512×512	0.78 s	3 min 14s
1024×1024	3.5 s	25 min 33 s

self loop $\alpha^*(d)$ is characterized [2] in G' by $\varphi'(d) = d$. Note that the removal of an edge $\alpha^*(d)$ corresponding to and empty self loop may create a new empty self-loop $\alpha^*(\sigma(d))$ if $\alpha^*(d)$ and $\alpha^*(\sigma(d))$ define nested loops. The removal operation is thus performed by an iterative process which first removes the empty self-loop $\alpha^*(d)$ characterized by $\varphi'(d) = d$ and then removes each dart $\sigma^n(d)$ which become an empty self loop after the removal of $\sigma^{n-1}(d)$. In the same way, double edges are characterized [2] in G' by $\varphi'^2(\sigma(d)) = \alpha(d)$. Such darts are also removed iteratively by checking the double edge condition for the sequence of σ successors of the dart d defining the first double edge.

4.3   Computing the Reduced Combinatorial Maps

The creation of the reduced combinatorial map from a contraction or a removal kernel is performed in parallel by using dart's reduction window [2]. Given a combinatorial map $G = (\mathcal{D}, \sigma, \alpha)$, a kernel K and a surviving dart $d \in \mathcal{SD} = \mathcal{D} - K$, the reduction window of d is either equal to:

$$RW(d) = d, \sigma(d), \ldots, \sigma^{n-1}(d)$$

with $n = Min\{p \in \mathbb{N}^* \mid \sigma^p(d) \in \mathcal{SD}\}$ if K is a removal kernel or

$$RW(d) = d, \varphi(\alpha(d)), \ldots, \varphi^{n-1}(\alpha(d))$$

with $n = Min\{p \in \mathbb{N}^* \mid \varphi^p(\alpha(d)) \in \mathcal{SD}\}$, if K is a contraction kernel.

Given a kernel K and a surviving dart $d \in \mathcal{SD}$, such that $RW(d) = d.d_1 \ldots d_p$, the successor of d within the reduced combinatorial map $G' = G/K = (\mathcal{SD}, \sigma', \alpha)$ is retrieved from $RW(d) = d.d_1 \ldots . d_p$ by [2]:

$$\sigma'(d) = \begin{cases} \sigma(d_p) & \text{if } K \text{ is a removal kernel} \\ \varphi(d_p) & \text{if } K \text{ is a contraction kernel} \end{cases} \tag{5}$$

Note that the reduction window of a surviving dart d connects $d \in G'$ to a sequence of darts in the initial combinatorial map G. The notion of dart's reduction window connects thus two successive levels of the pyramid and corresponds to the usual notion of reduction window [13].

Within the combinatorial pyramid framework each vertex defined at level i is encoded by its σ_i-orbit. The reduction window of such a vertex is defined as the concatenation of the darts belonging to its σ_i-orbit:

$$RW_i(\sigma_i^*(d)) = \bigodot_{j=1}^{n} RW_i(d_j)$$

with $\sigma_i^*(d) = (d_1, \ldots, d_p)$. The symbol $\bigodot$ denotes the concatenation operator.

4.4   Receptive Fields

Dart's reduction windows allow us to reduce a combinatorial map using either contraction or removal kernels. Starting from an initial combinatorial map G_0 and given a sequence of kernels $K_1, \ldots, K_n$ we can thus build the sequence of reduced combinatorial maps $G_0, G_1, \ldots, G_n$.

The transitive closure of the father-child relationship defined by the dart's reduction window corresponds to the notion of dart's receptive field. A dart's receptive field connects one dart in a combinatorial map $G_i = (\mathcal{SD}_i, \sigma_i, \alpha)$ to a sequence of darts defined in the base level combinatorial map $G_0 = (\mathcal{D}, \sigma, \alpha)$. These sequences are defined [2] recursively by $RF_0(d) = d$ for any $d \in \mathcal{D}$ and:

$$\forall i \in \{1, \ldots, n\}, \forall d \in \mathcal{SD}_i$$
$$RF_i(d) = \begin{cases} \bigodot_{j=1}^{p} RF_{i-1}(d_j) & \text{if } K_i \text{ is a removal kernel} \\ \bigodot_{j=1}^{p} d_j RF_{i-1}^*(\alpha(d_j)) & \text{if } K_i \text{ is a contraction kernel} \end{cases} \tag{6}$$

with $RW_i(d) = d_1 \ldots, d_p$. The symbol $RF_{i-1}^*(\alpha(d_j))$, $j \in \{1, \ldots, p\}$ denotes the sequence $RF_{i-1}(\alpha(d_j))$ without its first dart.

The receptive field of a dart encodes its embedding in the base level combinatorial map. If this initial combinatorial map is associated to a planar sampling grid, each initial vertex corresponds to one pixel and a vertex defined at level i of the pyramid encodes a connected set of vertices, i.e. a region. Within the combinatorial map framework the embedding of a vertex $\sigma_i^*(d)$ in the initial combinatorial map is called a vertex's receptive field and is defined using the same construction scheme than the vertex's reduction window. Given one dart $d \in \mathcal{SD}_i$, the receptive field of its associated vertex $\sigma_i^*(d) = (d_1, \ldots, d_p)$ is thus defined as the concatenation of the dart's receptive fields belonging to $\sigma_i^*(d)$:

$$R_{\sigma_i^*(d)} = \bigodot_{j=1}^{p} RF_i(d_j)$$

The receptive field of a vertex $\sigma_i^*(d)$ encodes the darts of the vertices reduced to the vertex $\sigma_i^*(d)$ at level i [2]. Using a labeling function μ (equation 4) in the base level combinatorial map, each label identifies a pixel in the associated image. This connected set of pixels may thus be retrieved by computing the set of reduced vertices: $\cup_{d' \in R_{\sigma_i^*(d)}} \mu(d')$.

4.5   Application to the Segmentation Framework

Fig. 4 shows an application of Combinatorial Pyramids to the segmentation framework. The original image (Fig. 4(a)) is quantized into K gray levels. The

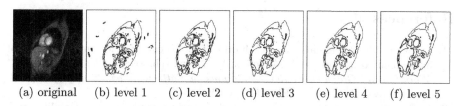

(a) original   (b) level 1   (c) level 2   (d) level 3   (e) level 4   (f) level 5

Fig. 4. One Combinatorial Pyramid encoding a stack of partitions of a heart.

quantization algorithm used in this experiment [16] decomposes the initial range of image's gray levels into K intervals. This set of intervals is designed in order to minimize the sum of the K variances computed on each interval.

Each region in (Fig. 4(b)) encodes a connected component of the pixels whose gray value are mapped onto a same interval. The background of the image is determined by selecting the region with the greatest cardinal adjacent to the exterior of the image. Then, all regions included in the background whose size is lower than a given threshold T are merged with the background (Fig. 4(c)). The mean gray level of each region is then used to initialize a gray level histogram. The frequency $h(i)$ of one entry i of the histogram is set to the sum of region's cardinal whose mean gray level is equal to i. This histogram is then quantized into K values and we merge any couple of adjacent region whose mean gray levels are mapped onto a same interval . This process is iterated form Fig. 4(d) to Fig. 4(f) until no merge occurs. Note that, at each level of the pyramid, the quantization algorithm only provides a partition of the range of grays values. The encoding of the partition and the merge operations are performed using the combinatorial pyramid model. The partition which may be obtained by using only [16] is represented on Fig. 4(b). The segmentation results represented in Fig. 4 have been obtained with $K = 4$ and $T = 10$.

The mean gray level and the cardinal of each region are computed during the construction of the pyramid by updating the parameters of each surviving vertex from the ones of its reduction window. The borders of the partitions represented in Fig. 4 are deduced from the receptive fields (Section 4.4)

5   Conclusion

We have defined in this article the construction scheme of a Combinatorial Pyramid. Two methods to build a contraction kernel have been proposed: one based on the method initially defined by Meer, Montanvert and Jolion [12,13,10] and a new one based on Union and Find operations. Our experiments show that this last method is more efficient than the first one when the contraction criteria may be determined before the construction of the contraction kernel. We have also presented the notions of Reduction Window and Receptive fields within the Combinatorial Pyramid framework. The receptive fields have been used in our experiments to retrieve the borders of the partitions.

References

1. Y. Bertrand, G. Damiand, and C. Fiorio. Topological map: Minimal encoding of 3d segmented images. In J. M. Jolion, W. Kropatsch, and M. Vento, editors, 3rd *Workshop on Graph-based Representations in Pattern Recognition*, pages 64–73, Ischia(Italy), May 2001. IAPR-TC15, CUEN.

2. L. Brun. *Traitement d'images couleur et pyramides combinatoires*. Habilitation à diriger des recherches, Université de Reims, 2002.

3. L. Brun and W. Kropatsch. The construction of pyramids with combinatorial maps. Technical Report 63, Institute of Computer Aided Design, Vienna University of Technology, lstr. 3/1832,A-1040 Vienna AUSTRIA, June 2000.

4. L. Brun and W. Kropatsch. Contraction kernels and combinatorial maps. In J. M. Jolion, W. Kropatsch, and M. Vento, editors, 3rd *IAPR-TC15 Workshop on Graph-based Representations in Pattern Recognition*, pages 12–21, Ischia Italy, May 2001. IAPR-TC15, CUEN.

5. R. Diestel. *Graph theory*. Springer-Verlag, New York, 2 edition, 2000.

6. C. Fiorio and J. Gustedt. Two linear time union-find strategies for image processing. Technical Report 375/1994, Technische Universität Berlin, 1994.

7. A. J. Gareth and D. Singerman. Theory of maps on orientable surfaces. *Proceedings of the London Mathematical Society*, 3(37):273–307, 1978.

8. Y. Haxhimusa, R. Glantz, M. Saib, G. Langs, and W. G. Kropatsch. Logarithmic Tapering Graph Pyramid. In L. Van Gool, editor, *Pattern Recognition, 24th DAGM Symposium*, volume 2449 of *Lecture Notes in Computer Science*, pages 117–124, Zurich, Switzerland, September 2002. Springer, Berlin Heidelberg.

9. J.-M. Jolion. Data driven decimation of graphs. In J.-M. Jolion, W. Kropatsch, and M. Vento, editors, *Proceedings of 3rd IAPR-TC15 Workshop on Graph based Representation in Pattern Recognition*, pages 105–114, Ischia-Italy, May 2001.

10. J. M. Jolion and A. Montanvert. The adaptive pyramid: A framework for 2d image analysis. *Computer Vision, Graphics, and Image Processing*, 55(3):339–348, May 1992.

11. P. Lienhardt. Topological models for boundary representations: a comparison with n-dimensional generalized maps. *Computer-Aided Design*, 23(1):59–82, 1991.

12. P. Meer. Stochastic image pyramids. *Computer Vision Graphics Image Processing*, 45:269–294, 1989.

13. A. Montanvert, P. Meer, and A. Rosenfeld. Hierarchical image analysis using irregular tessellations. *IEEE Transactions on Pattern Analysis and Machine Intelligence*, 13(4):307–316, APRIL 1991.

14. R. E. Tarjan. Efficiency of a good but non linear set union algorithm. *J. Assoc. Comput. System Mach.*, 22:215–225, 1975.

15. D. Willersinn and W. G. Kropatsch. Dual graph contraction for irregular pyramids. In *International Conference on Pattern Recogntion D: Parallel Computing*, pages 251–256, Jerusalem, Israel, 1994. International Association for Pattern Recognition.

16. S. Wong, S. Wan, and P. Prusinkiewicz. Monochrome image quantization. In *Canadian conference on electrical and computer engineering*, pages 17–20, September 1989.

On Graphs with Unique Node Labels

Peter J. Dickinson[1], Horst Bunke[2], Arek Dadej[3], and Miro Kraetzl[1]

[1] ISR Division, DSTO, PO Box 1500, Edinburgh SA 5111, Australia
{peter.dickinson,miro.kraetzl}@dsto.defence.gov.au
[2] Institut für Informatik und angewandte Mathematik, Universität Bern,
Neubrückstrasse 10, CH-3012 Bern, Switzerland
bunke@iam.unibe.ch
[3] ITR, University of South Australia,
Mawson Lakes SA 5095, Australia
arek.dadej@unisa.edu.au

Abstract. A special class of graphs is introduced in this paper. The graphs belonging to this class are characterised by the existence of unique node labels. A number of matching algorithms for graphs with unique node labels are developed. It is shown that problems such as graph isomorphism, subgraph isomorphism, maximum common subgraph and others have a computational complexity that is only quadratic in the number of nodes. We also discuss some potential applications of the considered class of graphs.

1 Introduction

Graph matching has become a very active field of research [1,2]. In its most general form, graph matching refers to the problem of finding a mapping f from the nodes of one given graph g_1 to the nodes of another given graph g_2, that satisfies some constraints or optimality criteria. For example, in graph isomorphism detection [3], mapping f is a bijection that preserves all edges and labels. In subgraph isomorphism detection [4], mapping f is requested to be injective such that all edges of g_1 are included in g_2 and all labels are preserved. Other graph matching problems that require the constructions of a mapping f with particular properties are maximum common subgraph detection [5,6] and graph edit distance computation [7,8].

The main problem with graph matching is its high computational complexity, which arises from the fact that it is usually very costly to find mapping f for a pair of given graphs. As a matter of fact, the detection of a subgraph isomorphism or a maximum common subgraph, as well as the computation of graph edit distance are known to be NP-complete. If the graphs in the application under consideration are small, optimal algorithms can be used. These algorithms are usually based on an exhaustive enumeration of all possible mappings f between the two graphs under consideration. For the matching of large graphs one needs to resort to suboptimal matching strategies. Methods of this type are characterised by a (often low-order) polynomial time complexity, but

E. Hancock and M. Vento (Eds.): IAPR Workshop GbRPR 2003, LNCS 2726, pp. 13–23, 2003.

they are no longer guaranteed to find the optimal solution to a given problem [11,12].

Another possibility to overcome the problem arising from the exponential complexity of graph matching is to focus on classes of graphs with an inherently lower computational complexity of the matching task. Some examples are given in [13,14]. Most recently, in the field of pattern recognition and computer vision, the class of trees has received considerable attention [15].

In the present paper another special class of graphs will be introduced. The graphs belonging to this class are characterised by the existence of unique node labels, which means that each node in a graph possesses a node label that is different from all other node labels in that graph. This condition implies that, whenever two graphs are being matched with each other, each node has at most one candidate for possible assignment under function f in the other graph. This candidate is uniquely defined through its node label. Consequently, the most costly step in graph matching, which is the exploration of all possible mappings between the nodes of the two graphs under consideration, is no longer needed. The present paper introduces matching algorithms for this special class of graphs and provides an analysis of their computational complexity. Particular attention will be directed to the computation of graph isomorphism, subgraph isomorphism, maximum common subgraph, and graph edit distance.

If constraints are imposed on a class of graphs, we usually lose some representational power. The class of graphs considered in this paper is restricted by the requirement of each node label being unique. Despite this restriction, there exist some interesting applications for this class of graphs. From the general point of view, graphs with unique node labels seem to be appropriate whenever the objects from the problem domain, which are modelled through nodes, possess properties that can be used to uniquely identify them. In the present paper we review one particular application of this class of graphs from the domain of computer network monitoring. Another application where the constraint is met is web document analysis [16].

The remainder of this paper is organised as follows. In Section 2, we introduce our basic concepts and terminology. Graphs with unique node labels and related matching strategies are presented in Section 3. Potential applications of this class of graphs are discussed in Section 4. In Section 5, we present the results of an experimental study where the run time of some of the proposed algorithms was measured. Finally, conclusions from this work are drawn in Section 6.

2   Basic Concepts and Notation

In this section the basic concepts and terminology used throughout the paper are introduced. The objects under consideration are directed graphs with labelled nodes and edges. Let L_V and L_E denote sets of node and edge labels, respectively. A *graph* $g = (V, E, \alpha, \beta)$ is a 4-tuple where V is the finite set of *vertices*, $E \subseteq V \times V$ is the set of *edges*, $\alpha : V \longrightarrow L_V$ is a function assigning labels to the nodes, and $\beta : E \longrightarrow L_E$ is a function assigning labels to edges. Edge $(x, y) \in E$

originates at node $x \in V$ and terminates at node $y \in V$. An undirected graph is obtained as a special case if there exists an edge $(y, x) \in E$ for every edge $(x, y) \in E$ with $\beta(x, y) = \beta(y, x)$.

Let $g = (V, E, \alpha, \beta)$ and $g' = (V', E', \alpha', \beta')$ be graphs; g' is a *subgraph* of g, $g' \subseteq g$, if $V' \subseteq V, E' \subseteq E, \alpha(x) = \alpha'(x)$ for all $x \in V'$, and $\beta(x, y) = \beta'(x, y)$ for all $(x, y) \in E'$. Let $g \subseteq g'$ and $g \subseteq g''$. Then g is called a *common subgraph* of g' and g''. Furthermore, g is called a *maximum common subgraph* (notation: *mcs*) of g' and g'' if there exists no other common subgraph of g' and g'' that has more nodes and, for a given number of nodes, more edges than g.

For graphs g and g', a *graph isomorphism* is any bijection $f : V \longrightarrow V'$ such that:

(i) $\alpha(x) = \alpha'(x)$ for all $x \in V$; and
(ii) for any edge $(x, y) \in E$ there exists $(f(x), f(y)) \in E'$ with $\beta(x, y) = \beta'(f(x), f(y))$ and for any edge $(x, y) \in E'$ there exists an edge $(f^{-1}(x'), f^{-1}(y)) \in E$ with $\beta'(x', y') = \beta(f^{-1}(x'), f^{-1}(y))$.

If $f : V \longrightarrow V'$ is a graph isomorphism between graphs g and g', and g' is a subgraph of another graph g'', i.e. $g' \subseteq g''$, then f is called a *subgraph isomorphism* from g to g''.

Finally we introduce the concept of graph edit distance, which is based on graph edit operations. We consider six types of edit operations: substitution of a node label, substitution of an edge label, insertion of a node, insertion of an edge, deletion of a node, and deletion of an edge. A cost (i.e. a non-negative real number) is assigned to each edit operation. Let e be an edit operation and $c(e)$ its cost. The cost of a sequence of edit operations, $s = e_1 \ldots e_n$, is given by the sum of all its individual costs, i.e. $c(s) = \sum_{i=1}^{n} c(e_i)$. Now the *edit distance* of two graphs g_1 and g_2 is equal to:

$$d(g_1, g_2) = \min\{c(s) | s \text{ is a sequence of edit operations transforming } g_1 \text{ to } g_2\}.$$

Henceforth, $d(g_1, g_2)$ is the *minimum cost*, taken over all sequences of edit operations, that transform g_1 into g_2.

3   Graphs with Unique Node Labels

In this section we introduce a special class of graphs that are characterised by the existence of unique node labels. Formally, we require that for any graph g and any pair $x, y \in V$ the condition $\alpha(x) \neq \alpha(y)$ holds if $x \neq y$. Furthermore, we assume that the underlying alphabet of node labels is an ordered set, for example, the integers, i.e. $L_V = \{1, 2, 3, \ldots\}$. Throughout the rest of this paper we consider graphs from this class only, unless otherwise mentioned.

Definition 1. *Let $g = (V, E, \alpha, \beta)$ be a graph. The* label representation *of g, $\rho(g)$, is given by $\rho(g) = (L, C, \lambda)$, where:*

(i) $L = \{\alpha(x)|x \in V\}$,
(ii) $C = \{(\alpha(x), \alpha(y))|(x, y) \in E\}$, *and*
(iii) $\lambda : C \longrightarrow L_E$ *with* $\lambda(\alpha(x), \alpha(y)) = \beta(x, y)$ *for all* $(x, y) \in E$.

According to this definition the label representation of a graph g is obtained by representing each node of g by its (unique) label and dropping set V. From the formal point of view, $\rho(g)$ defines the equivalence class of all graphs that are isomorphic to g. The individual members of this class can be obtained by assigning an arbitrary node, or more precisely an arbitrary node name, to each unique node label, i.e. to each element from L.

Example 1. Let $L_V = \{1, 2, 3, 4, 5\}$ and $g = (V, E, \alpha, \beta)$ where $V = \{a, b, c, d, e\}$, $E = \{(a, b), (b, e), (e, d), (d, a), (a, c), (b, c), (d, c), (e, c), (a, e), (b, d)\}$, $\alpha : a \mapsto 1, b \mapsto 2, c \mapsto 5, d \mapsto 4, e \mapsto 3$, $\beta : (x, y) \mapsto 1$ for all $(x, y) \in E$. A graphical illustration of g is shown in Fig. 1(a), where the node names (i.e. the elements of V) appear inside the nodes and the corresponding labels outside. Because all edge labels are identical, they have been omitted. The label representation $\rho(g)$ of g is then given by the following quantities: $L = \{1, 2, 3, 4, 5\}$, $C = \{(1, 3), (2, 3), (3, 4), (4, 1), (1, 5), (2, 5), (4, 5), (3, 5), (1, 3), (2, 4)\}$, $\lambda : (i, j) \mapsto 1$ for all $(i, j) \in C$.

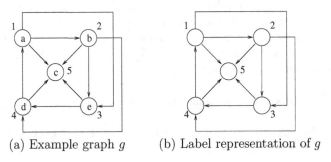

(a) Example graph g (b) Label representation of g

Fig. 1. Example graph g and its label representation

Intuitively, we can interpret the label representation $\rho(g)$ of any graph g, as a graph identical to g up to the fact that all node names are left unspecified. Hence $\rho(g)$ can be conveniently graphically represented in the same way as g is represented. For example, a graphical representation of $\rho(g)$, where g is shown in Fig. 1(a), is given in Fig. 1(b).

Lemma 1. Let $g_1 = (V_1, E_1, \alpha_1, \beta_1)$, $g_2 = (V_2, E_2, \alpha_2, \beta_2)$ be two graphs and $\rho(g_1) = (L_1, C_1, \lambda_1)$, $\rho(g_2) = (L_2, C_2, \lambda_2)$ their label representations. There exists a graph isomorphism between g_1 and g_2 if and only if $\rho(g_1) = \rho(g_2)$ (i.e. $L_1 = L_2$, $C_1 = C_2$, and $\lambda_1 = \lambda_2$).

Proof. Assume that there exists a graph isomorphism $f : V_1 \longrightarrow V_2$. Then $\alpha_1(x) = \alpha_2(f(x))$ for all $x \in V_1$. As f is bijective, it follows that $L_1 = L_2$. Furthermore, because of the conditions on the edges that are imposed by graph isomorphism f we conclude $C_1 = C_2$ and $\lambda_1 = \lambda_2$. Conversely, assume $\rho(g_1) = \rho(g_2)$. Construct now the mapping $f : V_1 \longrightarrow V_2$ such that $f(x) = y$ if and only if $\alpha_1(x) = \alpha_2(y)$. Because $L_1 = L_2$ and the node labels in both g_1 and g_2 are unique, this mapping is a bijection that satisfies the conditions of graph isomorphism imposed on the edges and edge labels in g_1 and g_2.

Based on this lemma, we can examine two graphs for isomorphism by simply generating their label representations and checking the conditions $L_1 = L_2$, $C_1 = C_2$, and $\lambda_1 = \lambda_2$. Assume $n = \max\{|V_1|, |V_2|\}$. Then $|L_1| = |L_2| = O(n)$, $|E_1| = |C_1| = O(n^2)$ and $|E_2| = |C_2| = O(n^2)$. Testing two ordered sets for equality is an operation that is linear in the number of elements. Hence the computational complexity of testing two graphs with unique node labels for isomorphism amounts to $O(n^2)$.

Lemma 2. *Let g_1, g_2, $\rho(g_1)$ and $\rho(g_2)$ be the same as in Lemma 1. There exists a subgraph isomorphism from g_1 to g_2 if and only if $L_1 \subseteq L_2$, $C_1 \subseteq C_2$, $\lambda_1(i,j) = \lambda_2(i,j)$ for all $(i,j) \in C_1$.*

Proof. Firstly, we assume that there exists a subgraph isomorphism $f : V_1 \longrightarrow V_2$. Then $\alpha_1(x) = \alpha_2(f(x))$ for all $x \in V_1$. As f is injective, it follows that $L_1 \subseteq L_2$. Similarly to the proof of Lemma 1, we conclude $C_1 \subseteq C_2$ and $\lambda_1(i,j) = \lambda_2(i,j)$ for all $(i,j) \in C_1$. Conversely, assume $L_1 \subseteq L_2$, $C_1 \subseteq C_2$ and $\lambda_1(i,j) = \lambda_2(i,j)$ for all $(i,j) \in C_1$. Then we can construct an injective mapping $f : V_1 \longrightarrow V_2$ such that $f(x) = y$ if and only if $\alpha_1(x) = \alpha_2(y)$. Similarly to the proof of Lemma 1, it follows that this mapping is a subgraph isomorphism from g_1 to g_2.

Using Lemma 2, testing two graphs for subgraph isomorphism reduces to examining the corresponding label representations for the three conditions $L_1 \subseteq L_2$, $C_1 \subseteq C_2$, and $\lambda_1(i,j) = \lambda_2(i,j)$ for all $(i,j) \in C_1$. The third condition can be checked in $O(|C_1|) = O(n^2)$ time. Checking whether an ordered set is a subset of another ordered set is linear in the size of the larger of the two sets. Hence the computational complexity of subgraph isomorphism of graphs with unique node labels amounts to $O(n^2)$.

Lemma 3. *Let g_1, g_2, $\rho(g_1)$ and $\rho(g_2)$ be the same as in Lemma 1. Let g be a graph with $\rho(g) = (L, C, \lambda)$ such that $L = L_1 \cap L_2$, $C = \{(i,j)|(i,j) \in C_1 \cap C_2$ and $\lambda_1(i,j) = \lambda_2(i,j)\}$, $\lambda(i,j) = \lambda_1(i,j)$ for all $(i,j) \in C$. Then g is an mcs of g_1 and g_2.*

Proof. First we note that $L \subseteq L_1$ and $L \subseteq L_2$. Hence $V \subseteq V_1$ and $V \subseteq V_2$ for any graph g with label representation $\rho(g) = (L, C, \lambda)$. Similarly, because C includes a pair (i,j) if and only if a corresponding edge with identical label exists in both g_1 and g_2, we observe $E \subseteq E_1$ and $E \subseteq E_2$ for any such graph g. Thirdly the labels of edges (x,y) occurring in both g_1 and g_2 are preserved under λ. Hence g is a subgraph of both g_1 and g_2. Now assume that g is not an *mcs*. In this case there must exist another subgraph g', of both g_1 and g_2 with either more nodes than g, or the same number of nodes, but more edges. The first case contradicts the way set L is constructed; if g' has more nodes than g, then $L \neq L_1 \cap L_2$. The second case is in conflict with the construction of C and λ, i.e. if g' has the same number of nodes as g, but more edges, then C and λ must be different from their values stated in Lemma 3. Hence g must be indeed an *mcs* of g_1 and g_2.

Possible computational procedures implied by Lemma 3 are again based on the intersection of two ordered sets. Hence the complexity of computing the *mcs* of two graphs with unique node labels is $O(n^2)$.

In [17] a detailed analysis was provided showing how graph edit distance depends on the costs associated with the individual edit operations. A set of edit operations together with their cost is also called a cost function. In this paper we focus our attention on the following cost function: $c_{nd}(x) = c_{ni}(x) = 1$, $c_{ns}(x) = \infty$, $c_{ed}(x, y) = c_{ei}(x, y) = c_{es}(x, y) = 1$, where $c_{nd}(x)$, $c_{ni}(x)$ and $c_{ns}(x)$ denote the cost associated with the deletion, insertion, and substitution of node x, respectively. Similarly, $c_{ed}(x, y)$, $c_{ei}(x, y)$ and $c_{es}(x, y)$ denote the cost associated with the deletion, insertion, and substitution of edge (x, y), respectively. This cost function is simple in the sense that each edit operation has a cost equal to one, except for node substitutions, which have infinite cost. As there will always exist a sequence of edit operations to transform g_1 to g_2 with a finite total cost, node substitution can be excluded. The exclusion of node substitutions for graphs with unique node labels makes sense because node label substitutions may generate graphs with non-unique node labels, i.e. graphs that do not belong to the class of graphs under consideration.

Lemma 4. *Let g_1, g_2, $\rho(g_1)$, and $\rho(g_2)$ be the same as in Lemma 1. Furthermore, let $C_0 = \{(i,j)|(i,j) \in C_1 \cap C_2 \text{ and } \lambda_1(i,j) = \lambda_2(i,j)\}$ and $C_0' = \{(i,j)|(i,j) \in C_1 \cap C_2 \text{ and } \lambda_1(i,j) \neq \lambda_2(i,j)\}$. Then $d(g_1, g_2) = |L_1| + |L_2| - 2|L_1 \cap L_2| + |C_1| + |C_2| - 2|C_0| + |C_0'|$.*

The proof is based on arguments similar to those used for Lemmas 1 to 3. Possible procedures for graph edit distance computation implied by Lemma 4 are again based on the intersection of two ordered sets. Hence, similarly to all other graph matching procedures considered before, the complexity of edit distance computation of graphs with unique node labels is $O(n^2)$.

The following theorem summarises all results of this section.

Theorem 1. *For the class of graphs with unique node labels there exist computational procedures that solve the following problems in quadratic time with respect to the number of nodes in the underlying graph: graph isomorphism, subgraph isomorphism, mcs, graph edit distance under the cost function introduced earlier in this section.*

In this section, we have assumed that there are $O(n^2)$ edges in a graph with n nodes. There exist, however, applications where the graphs are of bounded degree, i.e. the maximum number of edges incident to a node is bounded by a constant κ. In this case all expressions $O(n^2)$ reduce to $O(n)$.

4   Application to Computer Network Monitoring

The performance management of computer networks is becoming increasingly more important as computer networks grow in size and complexity. In [18], graph similarity measures for network monitoring and abnormal change detection were proposed. The basic idea is to represent a computer network by a graph where the nodes represent clients or servers, and the edges represent physical connections between clients or servers. A time series of graphs $g_1, g_2, \ldots, g_t, g_{t+1}, \ldots$ is obtained by measuring the state of a network at regular time intervals and representing each state by a graph. Graph matching techniques, as proposed in Section 3, can be applied to pairs of consecutive graphs to measure network change.

The class of graphs considered in this paper have the requirement that each node label must be unique. This constraint does not pose a problem when dealing with graphs constructed from data collected from computer networks. It is common in these networks that each node, such as a client, server or router, be uniquely identified. Take for example an intranet employing ethernet technology on one of its Local Area Network (LAN) segments. In this instance either the Media Access Control (MAC), or the Internet Protocol (IP) address, could be used to uniquely identify nodes on the local segment. As a consequence, such networks belong to the class of graphs considered in this paper and efficient graph matching algorithms described in Theorem 1, can be implemented.

Using Def. 1, the time series of graphs has label representation $\rho(g_1)$, $\rho(g_2)$, ..., $\rho(g_t)$, $\rho(g_{t+1})$, ... Measures of network difference that use graph matching algorithms, such as mcs and graph edit distance, were introduced in [18]. These measures can be applied between consecutive points g_t and g_{t+1}, in the time series of graphs. In order to gain a significant saving in computation time, we use the label representations $\rho(g_t)$, $\rho(g_{t+1})$, and apply the matching algorithms defined in Lemmas 3 and 4. The resulting distance $d(g_t, g_{t+1})$ can be compared to a user defined threshold; if the threshold is exceeded, it can be concluded that a significant network change has occurred at time $t+1$. A network administrator can then use other network management tools to determine whether the change represents an abnormal event.

The application of graph matching to computer network monitoring, whereby node labels are uniquely defined, is not a classical pattern recognition problem. It is, however, a classification problem since we want to discriminate between two types of events (i.e., normal and abnormal).

5   Experimental Results

The aim of the experiments described in this section is to verify the low computational complexity for the theoretical results proposed in Section 3. This is achieved by measuring the time taken to compute results for isomorphism, subgraph isomorphism, mcs and graph edit distance for graphs ranging in size from hundreds of nodes to tens of thousands of nodes. Experimental results are derived for both randomly generated graphs and real data derived from a computer network. Experiments on real data sets are used to confirm that computation times measured for randomly generated data sets can also be achieved for real data networks. The hardware specifications used for the experiments are irrelevant as we are only interested in relative time measurements.

5.1 Randomly Generated Graph

For the experiments described in this section, we first obtain a series S of randomly generated graphs, comprising 100, 1000, 5000, 10000 and 20000 nodes, respectively. The resulting graphs have directed edges with edge weights of one, and comprising, on average 2% of all possible edges. The value of 2% was selected to enable the larger graphs to fit into computing memory.

To measure the compute time for graph isomorphism we select the first graph g_1 from S, comprising one hundred unique nodes, and make an exact copy g_2. The fact that $g_2 = g_1$ guarantees that the graphs tested are in fact isomorphic to each other. The measurement of time does not include the time taken to derive the label representations $\rho(g_1)$ and $\rho(g_2)$, for graphs g_1 and g_2. This is the case for compute times measured for all matching algorithms. For the measurement of compute time for the subgraph isomorphism, we use the same graph g_1, together with graph g_3, obtained by removing 20% of the edges from g_1. The graph g_3 is obviously a subgraph of g_1. The measurements of time to compute both mcs and graph edit distance required another series S' of randomly generated graphs. To measure the time taken to execute these algorithms we again use g_1, and select the equivalent size graph from S'. The procedures outlined above were repeated for each of the graphs of size 1000, 5000, 10000 and 20000 nodes.

Table 1. Computation times for real and randomly generated data

	Compute Times for Experimental Data (seconds)					
	100	1000	5000	10000	20000	Real Data
Isomorphism	0.009	0.600	22.860	96.717	419.700	0.834
Subgraph ISO	0.006	0.288	12.827	56.508	245.369	0.552
MCS	0.006	0.385	14.560	65.332	280.294	0.498
ged	0.007	0.610	14.275	66.166	279.980	0.577

The results of all time measurements are shown in Table 1. As predicted, the measured computation complexity of all matching algorithms is $O(n^2)$. Fig. 2 illustrates this observation; the x-axis corresponds to the number of nodes in a graph and the y-axis represents the time, in seconds, to compute each graph matching algorithm. Compute times of graph isomorphisms were the longest whilst those for subgraph isomorphism were the shortest. The computation times for both *mcs* and graph edit distance, as observed in Fig. 2, are almost indistinguishable. This is not surprising since the computation steps, proposed in Lemmas 3 and 4, are almost identical.

5.2   Real Data from Computer Networks

The experimental data used in this section was acquired from a large enterprise data network using network performance monitoring tools. The network employs static IP addresses, hence its suitability for representation by the class of graphs defined in this paper. A time series of graphs has been produced, with each graph in the time series representing a single days traffic on the computer network. Graph vertices represent IP addresses, edges represent logical links, and edge weights represent the number of bytes communicated over a link.

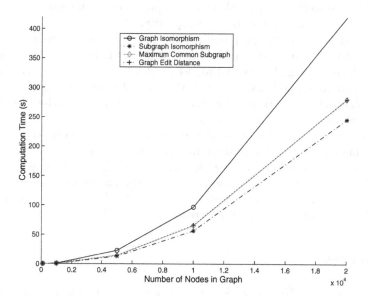

Fig. 2. Computation times for graph matching algorithms

Two consecutive graphs were arbitrarily selected from this data set for use in the computation time measurements. The label representation was obtained for both graphs. The graphs each comprised of approximately 9000 unique nodes and 35000 unique edges. This equates to 0.04% of all possible edges present in the graphs or, approximately one hundred times fewer edges than those in the randomly generated data set. Measurements on real data shown in Table 1, are closest to times for randomly generated graphs with 1000 nodes. This is largely due to the near equivalent number of edges in each. Since the number of edges in a graph has a dominant effect on the computation of the matching algorithms this result was expected. Hence results for real data sets are consistent with synthetic data measurements.

6   Summary and Conclusions

Graph matching is finding many applications in the fields of science and engineering. In this paper we considered a special class of graphs where all nodes in a graph are uniquely labelled. A number of matching algorithms for graphs with unique node labels were proposed and their computational complexity analysed. A label representation of a graph was defined such that each node is represented by its label, but vertex names are left unspecified. It was shown that the label representation of graphs with n nodes in this class enables the computation of graph isomorphism, subgraph isomorphism, maximum common subgraph, and graph edit distance in $O(n^2)$ time. This provides significant computational savings compared to the generalised class of graphs where such matching algorithms have an exponential computational complexity.

The application of the proposed matching algorithms to computer networks was considered. The constraint of unique node labels is satisfied for these networks. The matching algorithms proposed can be used to measure changes that occur in a computer network over time. Measures of network change provide good indicators of when abnormal network events have occurred. Such techniques can greatly enhance network monitoring.

Experiments conducted on randomly generated graphs verified that computation time for the matching algorithms was quadratic in the number of nodes. Further experimentation on network data, derived from a computer network, verified that the computation times for real data was in accordance with that of the simulated data sets. Future work will extend the number of matching algorithms developed for the considered class of graphs.

References

1. *IEEE Trans. PAMI*, Special Section on Graph Algorithms and Computer Vision, Vol. 23(10), 2001, 1049–1151
2. *Pattern Recognition Letters*, Special Issue on Graph Based Representation, 2003, to appear
3. McKay, B.D.: Practical Graph Isomorphism, *Congressus Numerantium*, Vol. 30, 1981, 45–87
4. Ullman, J.R.: An algorihm for subgraph isomorphism, *Journal of the ACM*, Vol. 23(1), 1976, 31–42
5. McGregor, J.: Backtrack search algorithms and the maximal common subgraph problem, *Software-Practice and Experience*, Vol. 12, 1982, 23–13
6. Levi, G.: A note on the derivation of maximal common subgraphs of two directed or undirected graphs, *Calcolo 9*, 1972, 341–354
7. Sanfeliu, A., Fu, K.S.: A Distance Measure Between Attributed Relational Graphs for Pattern Recognition, *IEEE Trans. SMC*, Vol. 13(3), 1983, 353–363

8. Messmer, B.T., Bunke, H.: A New Algorithm for Error-Tolerant Subgraph Isomorphism Detection, *IEEE Trans. PAMI*, Vol. 20(5), 1998, 493–507

9. Cordella, L.P., Foggia, P., Sansome, C., Vento, M.: An Improved Algorithm for Matching Large Graphs, *Proceedings of the 3rd IAPR-TC15 Workshop on Graph-based Representations in Pattern Recognition*, 2001, 149–159

10. Larrosa, J., Valiente, G.: Constraint satisfaction algorithms for graph pattern matching, *Math. Struct. in Computer Science*, Vol. 12, 2002, 403–422

11. Christmas, W.J., Kittler, J., Petrou, M.: Structural matching in computer vision using probabilistic relaxation, *IEEE Trans. PAMI*, Vol. 17(8), 1995, 749–764

12. Wang, I., Fan, K-C., Horng, J-T.: Genetic-based search for error-correcting graph isomorphism, *IEEE Trans. SMC*, Vol. 27(4), 1997, 588–597

13. Hopcroft, J. E., Wong, J.: Linear time algorithm for isomorphism of planar graphs, *Proceedings of the 6th Annual ACM Symposium on Theory of Computing*, 1974, 172–184

14. Luks, E.M.: Isomorphism of Graphs of Bounded Valence Can Be Tested In Polynomial Time, *Journal of Computer and Systems Sciences*, Vol. 25, 1982, 42–65

15. Pelillo, M.: Matching Free Trees, Maximal Cliques and Monotone Game Dynamics, *IEEE Trans. PAMI*, Vol. 24(11), 2002, 1535–1541

16. Schenker, A., Last, M., Bunke, H., Kandel, A.: Clustering of web documents using a graph model, In *Antonacopoulos, A., Hu, Jianying (eds), Web Document Analysis: Challenges and Opportunities*, World Scientific, 2003, to appear

17. Bunke, H.: Error correcting graph matching: on the influence of the underlying cost function, *IEEE Trans. PAMI*, Vol. 21, 1999, 917–922

18. Shoubridge, P.J., Kraetzl, M., Wallis, W.D., Bunke, H.: Abnormal Change Detection of Abnormal Change in a Time Series of Graphs, *Journal of Interconnection Networks*, Vol. 3, 2002, 85–101

Constructing Stochastic Pyramids by MIDES – Maximal Independent Directed Edge Set[*]

Yll Haxhimusa[1], Roland Glantz[2], and Walter G. Kropatsch[1]

[1] Pattern Recognition and Image Processing Group 183/2
Institute for Computer Aided Automation
Vienna University of Technology
Favoritenstrasse 9, A-1040 Vienna, Austria
{yll,krw}@prip.tuwien.ac.at
[2] Dipartimento di Informatica
Università di Ca' Foscari di Venezia
Via Torino 155, 30172 Mestre (VE), Italy
glantz@dsi.unive.it

Abstract. We present a new method (MIDES) to determine contraction kernels for the construction of graph pyramids. Experimentally the new method has a reduction factor higher than 2.0. Thus, the new method yields a higher reduction factor than the stochastic decimation algorithm (MIS) and maximal independent edge set (MIES), in all tests. This means the number of vertices in the subgraph induced by any set of contractible edges is reduced to half or less by a single parallel contraction. The lower bound of the reduction factor becomes crucial with large images.

Keywords: irregular graph pyramids, maximal independent set, maximal independent directed edge set, topology preserving contraction.

1 Introduction

In a regular image pyramid the number of pixels at any level k, is λ times higher than the number of pixels at the next reduced level $k+1$. The so called reduction factor λ is larger than 1 and it is the same for all levels k. If P denotes the number of pixels in an image I, the number of new levels on top of I amounts to $log_\lambda(P)$ (Figure 1(a)). Thus, the regular image pyramid may be an efficient structure to access image objects in a top-down process. For more in depth on the subject see [20].

However, regular image pyramids are confined to globally defined sampling grids and lack shift invariance [1]. In [19,10] it was shown how these drawbacks can be avoided by irregular image pyramids, the so called adaptive pyramids, where the hierarchical structure (vertical network) of the pyramid was not a priori known but recursively built based on the data. Moreover in [18,4,16,2] it

---

[*] This paper has been supported by the Austrian Science Fund under grants P14445-MAT and P14662-INF

E. Hancock and M. Vento (Eds.): IAPR Workshop GbRPR 2003, LNCS 2726, pp. 24–34, 2003.

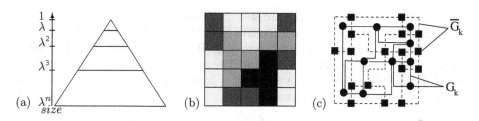

Fig. 1. (a) Pyramid concept. (b) Partition of pixel set into cells. (c) Representation of the cells and their neighborhood relations by a dual pair $(G, \overline{G})$ of plane graphs.

was shown that the irregular pyramid can be used for segmentation and feature detection.

Each level represents a partition of the pixel set into cells, i.e. 4-connected subsets of pixels. The construction of an irregular image pyramid is iteratively local [17,9]:

- the cells have no information about their global position.
- the cells are connected only to (direct) neighbors.
- the cells cannot distinguish the spatial positions of the neighbors.

On the base level (level 0) of an irregular image pyramid the cells represent single pixels and the neighborhood of the cells is defined by the 4-connectivity of the pixels. A cell on level $k+1$ (parent) is the union of some neighboring cells on level k (children). This union is controlled by so called contraction kernels (decimation parameters [13]). For more in depth on the subject see the book of Jolion [11]. Neighborhoods on level $k+1$, are derived from neighborhoods on level k. Two cells c_1 and c_2 on level $k+1$ are neighbors if there exist children p_1 of c_1 and p_2 of c_2 such that p_1 and p_2 are neighbors in level k, Figure 1(b). We assume that on each level $l+1$ ($l \geq 0$) there exists at least one cell not contained in level l. In particular, there exists a highest level h. Furthermore, we restrict ourselves to irregular pyramids with an apex, i.e. level h contains one cell.

In this paper we represent the levels as dual pairs $(G_k, \overline{G_k})$ of plane graphs G_k and $\overline{G_k}$, Figure 1(c). The vertices of G_k represent the cells and the edges of G_k represent the neighborhood relations of the cells on level k, depicted with circle vertices and solid edges in Figure 1(c). The edges of $\overline{G_k}$ represent the borders of the cells on level k, depicted with dashed lines in Figure 1(c), possibly including so called pseudo edges needed to represent the neighborhood relation to a cell completely surrounded by another cell. Finally, the vertices of $\overline{G_k}$, the squares in Figure 1(c), represent points where at least three edges from $\overline{G_k}$ meet. The sequence $(G_k, \overline{G_k})$, $0 \leq k \leq h$ is called (dual) graph pyramid.

The homogeneous region does not offer additional information to be used to contract this region. Thus the stochastic selection principle is used to contract a homogeneous region. The aim of this paper is to combine the advantage of regular pyramids (logarithmic tapering) with the advantages of irregular graph pyramids (their purely local construction and shift invariance). The aim is reached by exchanging the stochastic selection method (MIS) for contraction kernels proposed in [17] by another iteratively local method (MIDES) that has a reduction factor higher than 2.0 in all out experiments. The other goal is not to be limited by the direction of contraction, which is a drawback of MIES [7]. Experiments with both selection methods show that:

- the MIS method does not lead necessarily to logarithmic tapering graph pyramids, i.e. the reduction factors of graph pyramids built by the MIS can get arbitrarily close to 1.0 [19].
- the sizes of the receptive fields from the new method (MIDES) are much more uniform.

Not only stochastic decimation [17], but also connected component analysis [15] gains from the new method.

The plan of the paper is as follows. In Section 2 we recall the main idea of the stochastic pyramid algorithm and in Section 2.3 we see that graph pyramids from maximal independent vertex sets (MIS) may have a very small reduction factor. Moreover, experiments show that small reduction factors are likely, especially when the images are large. We propose a new method (MIDES), in Section 3, based on directed graphs and show in Section 3.2 that this method has a reduction factor larger than 2.0.

2 Maximal Independent Vertex Set

In the following the iterated stochastic construction of the irregular image pyramid in [17,4,16] is described in the language of graph pyramids. The main idea is to first calculate a so called *maximal independent vertex set*[1] [5]. Let V_k and E_k denote the vertex set and the edge set of G_k, respectively and let $\iota(\cdot)$ be the mapping from an edge to its set of end vertices. The neighborhood $\Gamma_k(v)$ of a vertex $v \in V_k$ is defined by

$$\Gamma_k(v) = \{v\} \cup \{w \in V_k \mid \exists e \in E_k \text{ such that } v, w \in \iota(e)\}.$$

A subset W_k of V_k is called maximal independent vertex set if:

1. $w_1 \notin \Gamma_k(w_2)$ for all $w_1 \neq w_2 \in W_k$,
2. for all $v \in V_k \setminus W_k$ there exists $w \in W_k$ such that $v \in \Gamma_k(w)$.

Put in words, two members (survivors) of maximal independent vertex set cannot be neighbors (condition 1) and every non-member (non-survivor) is in the neighborhood of at least one member (condition 2). An example of a maximal independent vertex set is shown with black vertices in Figure 2(a), the arrows indicate a corresponding collection of contraction kernels.

---

[1] also called maximal stable set; we distinguish maximal from maximum independent set, whose construction is NP-complete. See [5] for algorithmic complexity.

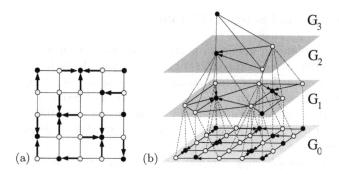

Fig. 2. (a) Maximal independent vertex set. (b) A graph pyramid from maximal independent vertex set.

2.1 Maximal Independent Vertex Set Algorithm (MIS)

The maximal independent vertex set (MIS) problem was solved using parallel, probabilistic symmetry breaking algorithm [17]. The number of iterations to complete maximal independent set converges in most of the cases very fast, only few iterations for correction are needed [17] .

MIS may be generated as follows.

Algorithm 1 – MIS Algorithm

1: Mark every vertex v of V_l as *candidate*.
2: **while** there are candidates **do**
3:    Assign random numbers to the candidates of V_l.
4:    Determine the candidates whose random numbers are larger than the random numbers of all neighboring candidates and mark them as *member* (of the maximal independent set) and as *non-candidate*. Also mark every neighbor of every new member as *non-candidate* and as *non-member*.
5: **end while**
6: In each neighborhood of a vertex that is not a member there will now be a member. Let each non-member choose its neighboring member, say the one with the maximal random number (we assume that no two random numbers are equal).

The assignment of the non-members to their members determines a collection of *contraction kernels*: each non-member is contracted toward its member and all contractions can be done in a single parallel step. In Figure 2(a) the contractions are indicated by arrows. A stochastic graph pyramid with MIS can be seen in Figure 2(b), where $G_0, G_1, \ldots$ etc. represent graphs on different levels of the pyramid. Note that we remove parallel edges and self-loops that emerge from the contractions, if they are not needed to encode inclusion of regions by other regions (in the example of Figure 2(b) we do not need loops nor parallel edges). This can be done by the dual graph contraction algorithm [13].

2.2   Experimental Setup and Evaluation

Uniformly distributed random numbers are assigned to vertices or edges in the base grid graphs. We generated 1000 graphs, on top of which we build stochastic graph pyramids. In our experiments we use graphs of sizes 10000 and 40000 nodes, which correspond to image sizes 100×100 and 200×200 pixels, respectively. Figure 3 summarizes the result of the first 100 of 1000 tests. Data in Table 1 and Table 2 were derived using graphs of size 200×200 nodes with 1000 experiments.

We extract the following parameters: the height of the pyramid (*height*); the maximum and the mean of the degree of vertices [2] (*max-degree* and *mean-degree*); and the number of iterations for correction (*correction*) to complete maximal independent set for any graph in the contraction process (Table 1). In Table 2 are shown reduction factors for vertices ($|V_k|/|V_{k+1}|$) and edges ($|E_k|/|E_{k+1}|$). We average these values on the whole data set (μ-mean and σ-standard deviation). The degree of the vertex is of importance because it is directly related to the memory costs for the graph's representation [9]. We compare the quality of selection methods in Section 2.3 and 3.2. A full documentation of the experiments can be found in the technical report [8].

2.3   Experiments with MIS

The number of levels needed to reduce the graph at the base level (level 0) to an apex (top of the pyramid) are given in Figure 3(a),(b). The vertical axis indicates the number of nodes on the levels indicated by the horizontal axis. The slopes of the lines correspond to the reduction factors. From Figure 3(a),(b) we see that the **height** of the pyramid **cannot be guaranteed to be logarithmic**, except for some good cases.

In the worst case the pyramid had 22 levels for the 100×100, respectively 41 levels for the 200×200 graphs. See [8] for numerical details. In these cases we have a very poor reduction factor. A **poor reduction factor** is likely, as can be seen in Figure 3(a),(b), especially when the **images are large**. This is due to the evolution of larger and larger variations between the vertex degrees

Table 1. Comparison of MIS, MIES and MIDES.

Process	height μ	σ	max-degree μ	σ	mean-degree μ	σ	correction μ	σ
MIS	20.78	5.13	70.69	23.88	4.84	0.23	2.95	0.81
MIES	14.01	0.10	11.74	0.71	4.78	0.07	4.06	1.17
MIDES	12.07	0.46	13.29	1.06	4.68	0.14	2.82	1.07

---

[2] the number of edges incident to a vertex, i.e the number of non survivor contracted into the survivor.

in the contracted graphs (Table 1 *max-degree* and *mean-degree* columns). The absolute maximum vertex degree was 148. The *a priori* probability of a vertex being the local maximum dependents on its neighborhood. The larger the neighborhood the smaller is the *a priori* probability that a vertex will survive. The number of iterations necessary for correction are the same as reported by [17] (Table 1 *correction* columns).

To summarize, a **constant reduction factor cannot be guaranteed**.

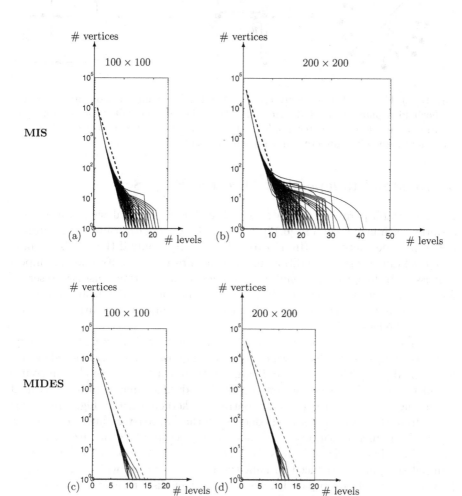

Fig. 3. Comparing MIS and MIDES. Number of vertices in levels of MIS and MIDES pyramids. The base levels are rectangular grid graphs containing 100×100 and 200×200 vertices. Dashed lines represent the theoretical reduction factor of 2.0 (MIES [7]).

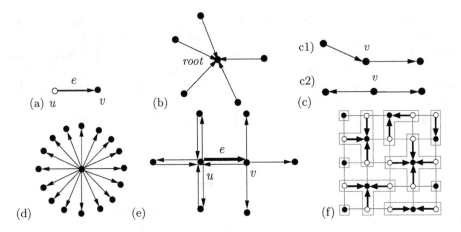

Fig. 4. (a) The direction of contraction. (b) A legal configuration of directed edges. (c) Forbidden pairs of directed edges. (d) The reduction factor of a star with n edges pointing away from the center is $n/(n-1)$. (e) The directed neighborhood $N(e)$. (f) Maximal independent directed edge set with respect to $N(e)$.

3   Maximal Independent Directed Edge Set

In many graph pyramid applications such as line image analysis [3,14], the description of image structure [6,12], a directed edge e with source u and target $v \neq u$ must be contracted (from u to v, Figure 4(a)), only if the attributes of e, u, and v fulfill a certain condition, making the direction of contraction an important issue. In line drawings end point of lines or intersection must be preserved for geometric accuracy reasons. In particular, the condition depends on u being the source and v being the target. The edges that fulfill the condition are called *preselected* edges.

From now on the plane graphs in the pyramid have (bi)directed edges. Typically, the edges in the base level of the pyramid form pairs of reverse edges, i.e. for each edge e with source u and target v there exists an edge e' with source v and target u. However, the set of preselected edges may contain e without containing e'. The goal is to build contraction kernels with a "high" reduction factor from the set of preselected edges, such that a predetermined target v survives. The reduction will always be determined according to the directed graph induced by the preselected edges. For example, if the number of vertices in the induced subgraph is reduced to half, the reduction factor will be 2.0. To perform the contractions in parallel, we need a vertex disjoint union of contraction kernels. The plan is to define such a union in terms of independent directed edges, where "independent" means that no pair of directed edges belongs to the same neighborhood $N(e)$ (see Definition 3). Then, dealing with edges instead of vertices, we may find the contraction kernels as in MIS.

Definition 1. *A contraction kernel is a vertex disjoint rooted tree of depth one or zero (single vertex), each edge of which is directed toward the root (survivor).*

Note that the direction of edges uniquely determines which vertex survives on the next level of the pyramid, i.e determines the contraction kernel (decimation parameter [13]). Figure 4(b) shows a configuration of directed edges, which yields a contraction kernel.

Definition 2. *Let v be a vertex of bi-directed graph G. We define the out-degree as*

$$s(v) := \#\{e \in G \,|v \text{ is source of } e\},$$

and in-degree as

$$t(v) := \#\{e \in G \,|v \text{ is target of } e\}.$$

In Figure 4(b) the number of edges with target in *root* is $t(root) = 5$; and for the center vertex with n edges pointing away $s(center) = n$, Figure 4(d).

Proposition 1. *Let D denote a set of directed edges from a bi-directed graph G and G_D denote the subgraph induced by D. Then the following statements are equivalent:*

(a) $s(v) < 2 \,\wedge\, s(v) \cdot t(v) = 0, \quad \forall v \in G_D.$
(b) G_D *is a vertex disjoint union of contraction kernels.*

Proof. **i)** (a) $\Rightarrow$ (b): Let $R := \{r|\ r \text{ is target of some } e \in D \wedge s(r) = 0\}$ be the set of roots. Furthermore, set $E_r := \{e \in D|\ r \text{ is target of } e\}$, $r \in R$. Then $E = \bigcup_{r \in R} E_r$ and E_r induces a contraction kernel C_r for any $r \in R$. It remains to show that the C_r are vertex disjoint. Assume the opposite, i.e. there exists a vertex u contained in C_v and in C_w for some $v \neq w \in R$. From (a) it follows that $u \notin \{v, w\}$. Hence, there exist edges $(u, v), (u, w) \in D$, a contradiction to $s(u) < 2$.

ii) (b) $\Rightarrow$ (a): Let T be the set of roots of the vertex disjoint contraction kernels and let C_t denote the unique contraction kernel with root t, $t \in T$. Furthermore, let $v \in G_D$. Since the C_t are vertex disjoint, exactly one of the following holds:

1. $v \in T$ and $s(v) = 0$.
2. $v \notin T$ and $s(v) = 1, t(v) = 0$.

Examples of kernels which do not fulfill the Definition 1 are shown in Figure 4(c1) where $s(v) = 1$ and $t(v) = 1$; and (c2) where $s(v) = 0$ and $t(v) = 2$. From the example in Figure 4(d) it is clear that only one edge can be contracted (otherwise one ends with forbidden contraction kernels), which means in general, vertex reduction factor can get arbitrarily close to 1.0.

Note that, in contrast to MIS, the roots of two contraction kernels may be neighbors. Condition (a) in Proposition 1 is fulfilled, if and only if no pair of directed edges from D belongs to the same $N(e)$, $e \in D$, where $N(e)$ is defined in Definition 3. Hence, a maximal vertex disjoint union of contraction kernels may be found via a maximal set of directed edges that are independent with respect to $N(e)$. A parallel algorithm to find a maximal independent set with respect to $N(e)$ is specified in the next section.

Definition 3. *Let* $e = (u, v)$ *be a directed edge of* G. *Then the directed neighborhood* $N(e)$ *is given by all directed edges* $e' = (u', v')$ *such that*

$$u \in \{u'\} \cup \{v'\} \vee u' = v.$$

Neighborhood $N(e)$ of e is given by edges which point toward the source of e, edges with the same source u as e and the edges the source of which is the target of e. Note that edges pointing towards the target of e are not part of the directed neighborhood. Figure 4(e) depicts $N(e)$ in case of u and v both having 4 neighbors.

3.1 Maximal Independent Directed Edge Set Algorithm (MIDES)

To find a maximal (independent) set of directed edges (MIDES) forming vertex disjoint rooted trees of depth zero or one, we proceed analogously to the generation of maximal independent vertex sets (MIS), as explained in the Section 2. Let E_k denote the set of bi-directed edges in the graph G_k of the graph pyramid. We proceed as follows.

Algorithm 2 – MIDES Algorithm

1: Mark every directed edge e of E_k as *candidate*.
2: **while** there are candidates **do**
3:    Assign random numbers to the candidates of E_k.
4:    Determine the candidates e whose random numbers are larger than the random numbers of all candidates in its directed neighborhood $N(e) \setminus \{e\}$ and mark them as *member* (of a contraction kernel) and as *non-candidate*. Also mark every $e' \in N(e)$ of every new member e as *non-candidate*.
5: **end while**

Since the direction of edges uniquely determines the roots of the contraction kernels (the survivors), all the vertices which are the sources of directed edges are marked as non-survivor. An example of a set of contraction kernels C found by MIDES is given in Figure 4(f) (the survivors are depicted with black and non-survivors with white).

3.2 Experiments with Maximal Independent Directed Edge Sets

The same set of 1000 graphs (Section 2.2) was used to test MIDES. The numbers of levels needed to reduce the graph on the base level to an apex of the pyramid are shown in Figure 3 (c),(d). Again the vertical axis indicates the number of vertices in the levels indicated by the horizontal axis. The experiments show that the **reduction factor** of MIDES is indeed **higher** than 2.0 (indicated by the dashed line in Figure 3(c),(d), even in the worst case (see [8] for details on numerical results). Also the *in-degrees* of the vertices is much smaller than for MIS (*max-degree* column in Table 1). For the case of the graph with size

Table 2. Reduction factors of vertices and edges.

Process	$\frac{\|V_k\|}{\|V_{k+1}\|}$		$\frac{\|E_k\|}{\|E_{k+1}\|}$	
	μ	σ	μ	σ
MIS	1.94	1.49	1.78	0.69
MIES	2.27	0.21	2.57	1.21
MIDES	2.62	0.36	3.09	1.41

200×200 vertices, MIDES needs less levels than MIS and MIES and the number of iterations needed to complete the maximal independent set was comparable with the one of MIS (Table 1, *correction* column). In Table 2, statistics of reduction factors for vertices ($|V_k|/|V_{k+1}|$) and edges ($|E_k|/|E_{k+1}|$) are given. From this table one can read that **MIDES** algorithm shows a **better reduction factor** than MIES [7] and MIS.

To summarize, in our experiments the reduction factor is always **higher** than 2.0 (outperforming MIS and MIES), even though the theoretical lower bound is 1.0.

4   Conclusion and Outlook

Experiments with (stochastic) irregular image pyramids using maximal independent vertex sets (MIS) showed that the reduction factor can get arbitrarily close to 1.0 for large images. After an initial phase of strong reduction, the reduction decreases dramatically. This is due to the evolution of larger and larger variations between the vertex degrees in the contracted graphs. To overcome this problem we proposed a new method (MIDES). Experimentally, this method has a reduction factor larger than 2.0. Moreover, MIDES, in contrast to MIES, allows us to take into account constraints on the directions of the contractions.

References

1. M. Bister, J. Cornelis, and A. Rosenfeld. A critical view of pyramid segmentation algorithms. *Pattern Recognition Letters*, Vol. 11(No. 9):pp. 605–617, September 1990.
2. M. Borowy and J. Jolion. A pyramidal framework for fast feature detection. In *Proc. of 4th Int. Workshop on Parellel Image Analysis*, pages 193–202, 1995.
3. M. Burge and W. G. Kropatsch. Run Graphs and MLPP Graphs in Line Image Encoding. In W. G. Kropatsch and J.-M. Jolion, editors, *2nd IAPR-TC-15 Workshop on Graph-based Representation*, pages 11–19. OCG-Schriftenreihe, Österreichische Computer Gesellschaft, 1999. Band 126.
4. K. Cho and P. Meer. Image Segmentation from Consensus Information. *CVGIP: Image Understanding*, 68 (1):72–89, 1997.
5. N. Christofides. *Graph Theory - An Algorithmic Approach*. Academic Press, New York, London, San Francisco, 1975.

6. R. Glantz, R. Englert, and W. G. Kropatsch. Representation of Image Structure by a Pair of Dual Graphs. In W. G. Kropatsch and J.-M. Jolion, editors, *2nd IAPR-TC-15 Workshop on Graph-based Representation*, pages 155–163. OCG-Schriftenreihe, Österreichische Computer Gesellschaft, 1999. Band 126.
7. Y. Haxhimusa, R. Glantz, M. Saib, Langs, and W. G. Kropatsch. Logarithmic Tapering Graph Pyramid. In L. van Gool, editor, *Proceedings of 24th DAGM Symposium*, pages 117–124, Swiss, 2002. Springer Verlag LNCS 2449.
8. Y. Haxhimusa and W. G. Kropatsch. Experimental Results of MIS, MIES, MIDES and D3P. Technical Report No.78, PRIP, Vienna University of Technology. http://www.prip.tuwien.ac.at/ftp/pub/publications/trs, 2003.
9. J.-M. Jolion. Stochastic pyramid revisited. *Pattern Recognition Letters*, 24(8):pp. 1035–1042, 2003.
10. J.-M. Jolion and A. Montanvert. The adaptive pyramid, a framework for 2D image analysis. *Computer Vision, Graphics, and Image Processing: Image Understanding*, 55(3):pp.339–348, May 1992.
11. J.-M. Jolion and A. Rosenfeld. *A Pyramid Framework for Early Vision*. Kluwer Academic Publishers, 1994.
12. P. Kammerer and R. Glantz. Using Graphs for Segmenting Crosshatched Brush Strokes. In J.-M. Jolion, W. G. Kropatsch, and M. Vento, editors, *3nd IAPR-TC-15 Workshop on Graph-based Representation*, pages 74–83. CUEN, 2001.
13. W. G. Kropatsch. Building Irregular Pyramids by Dual Graph Contraction. *IEE-Proc. Vision, Image and Signal Processing*, Vol. 142(No. 6):pp. 366–374, December 1995.
14. W. G. Kropatsch and M. Burge. Minimizing the Topological Structure of Line Images. In A. Amin, D. Dori, P. Pudil, and H. Freeman, editors, *Advances in Pattern Recognition, Joint IAPR International Workshops SSPR'98 and SPR'98*, volume Vol. 1451 of *Lecture Notes in Computer Science*, pages 149–158, Sydney, Australia, August 1998. Springer, Berlin Heidelberg, New York.
15. W. G. Kropatsch and H. Macho. Finding the structure of connected components using dual irregular pyramids. In *Cinquième Colloque DGCI*, pages 147–158. LLAIC1, Université d'Auvergne, ISBN 2-87663-040-0, September 1995.
16. C. Mathieu, I. E. Magnin, and C. Baldy-Porcher. Optimal stochastic pyramid: segmentation of MRI data. *Proc. Med. Imaging VI: Image Processing*, SPIE Vol.1652:pp.14–22, Feb. 1992.
17. P. Meer. Stochastic image pyramids. *Computer Vision, Graphics, and Image Processing*, Vol. 45(No. 3):pp.269–294, March 1989.
18. P. Meer, D. Mintz, A. Montanvert, and A. Rosenfeld. Consensus vision. In *AAAI-90 Workshop on Qualitative Vision*, Boston, Massachusetts, USA, July 29 1990.
19. A. Montanvert, P. Meer, and A. Rosenfeld. Hierarchical image analysis using irregular tesselations. *IEEE Transactions on Pattern Analysis and Machine Intelligence*, PAMI-13(No.4):pp.307–316, April 1991.
20. A. Rosenfeld, editor. *Multiresolution Image Processing and Analysis*. Springer, Berlin, 1984.

Functional Modeling of Structured Images*

Jocelyn Marchadier and Walter G. Kropatsch

Pattern Recognition and Image Processing Group(PRIP)
Favoritenstraße 9/1832, A-1040 Wien, AUSTRIA,
jm@prip.tuwien.ac.at

Abstract. Functional Graphical Models (FGM) describe functional
dependence between variables by means of implicit equations. They
offer a convenient way to represent, code, and analyze many problems
in computer vision. By explicitly modeling functional dependences by a
hypergraph, we obtain a structure well-adapted to information retrieval
and processing. Thanks to the functional dependences, we show how
all the variables involved in a functional graphical model can be stored
efficiently. We derive from that result a description length of general
FGMs which can be used to achieve model selection for example. We
demonstrate their relevance for capturing regularities in data by giving
examples of functional models coding 1D signals and 2D images.

Keywords: functional model, hypergraph, graph representation, code,
model selection

1 Introduction

A primitive is an object approximating a certain aspect of reality. In general,
basic primitives can be seen as points forming a subset of $\mathbb{R}^n$ [6]. A structure
to encode and to analyze primitives is often required in many complex image
analysis tasks, such as in image registration and matching, in document analysis,
in camera calibration, and in model reconstruction.

Sampled data can be encoded by some function with known parameters,
which describe a "deterministic" behavior of the data and can be viewed as
primitives, and some stochastic model which takes into account variations of the
data around the approximation. This kind of representation has already been
used in [3][7][12] to represent different image objects, as well as complicated 3D
models. The parameters of the approximate model may be encoded using the
same principle, reducing even more the description of the model. This is the
basic idea behind functional model based encoding.

In order to avoid pitfalls of continuous models which fail to take into account
discontinuities, a piecewise continuous function can be used to approximate the
samples. In such a case, a model is required to encode the discontinuities of the

---

* This paper has been supported by the Austrian Science Fund (FWF) under grants
P14445-MAT and P14662-INF

E. Hancock and M. Vento (Eds.): IAPR Workshop GbRPR 2003, LNCS 2726, pp. 35–46, 2003.

function. This model is topological in the sense that it has to be invariant under geometric bijective transformations (which preserve discontinuities).

In this contribution, we propose to represent functional dependences between primitives in a coherent framework by means of functional graphical models (FGM, section 2). We first introduce some needed notions (section 2.1) and derive definitions related to FGMs (section 2.2). We then show how functional based encoding of real valued variables can be achieved efficiently (section 2.3). We add a stochastic part adapted to real modeling problems, and propose a cost function based on the minimum description length (MDL) principle which enable a quantitative evaluation of the adaptation of a model to given data (section 2.4). Finally, we show how to encode a 1D time varying signal (section 2.5). This application demonstrates the need for a multi-part FGM driven by topology.

Section 3 generalizes the previous results to 2D images. In section 3.1, we show how to encode topological information of an image. In section 3.2, we propose several functional models that can be put together in order to efficiently encode an image. Those combined models bridge the gap between a continuous and a discrete representation of an image. In section 3.3, we extend the preceding functional models to encode global information which leads to even smaller codes. In section 3.4, we deduce the description length of a functional graphical model based image code.

2   Functional Graphical Models (FGM)

A functional model is an implicit equation system relating measurements and parameters. Functional graphical models [13][14] (FGM for short) may be used to design and analyze functional models via functional dependences between the involved variables. "Graphical" is related to the graph structure underlying a functional model (see section 2.2).

2.1   Variables and Implicit Equations

A *(real-valued) variable* V_i is a symbol representing a point in a subspace $D_i \subset \mathbb{R}^{d_i}$. D is the *domain* of V_i and represents the admissible values of V_i. For convenience, we use the vectorial notation $V_i = (c_1 \ ... \ c_{d_i})^t$. $c_1, ..., c_{d_i}$ are referred in the following text as *components* of the variable V_i. The *dimension* of V_i is the number of components of V_i, noted $dim(V_i) = d_i$. An *instance* of a variable V_i is a constant vector $v_i \in D_i$. A variable describes a set of admissible parameters associated to a primitive without specifying them. A variable may be viewed as a means to identify a specific primitive. In the following text, we do not distinguish between the terms primitive and variable.

A real-valued function f_j in $\mathbb{R}^{m_j}$ and of variables $V_{j_1}, ..., V_{j_{n_j}}$, with $j_1, ..., j_{n_j}$ a permutation of n_j integers, the function defined as :

$$f_j : D_{j_1} \times ... \times D_{j_{n_j}} \to \mathbb{R}^{m_j}$$
$$(v_{j_1}, ..., v_{j_{n_j}}) \mapsto f_j(v_{j_1}, ..., v_{j_{n_j}})$$

An *functional constraint* of function f and of variables $V_{j_1}, ..., V_{j_{n_j}}$ is an n_j-ary relation characterized by the $(n_j + 1)$-tuple $F_j = (f_j, V_{j_1}, ..., V_{j_{n_j}})$. We say that F_j is *satisfied* for $(v_{j_1}, ..., v_{j_{n_j}}) \in D_{j_1} \times ... \times D_{j_{n_j}}$ iff $f_j(v_{j_1}, ..., v_{j_{n_j}}) = 0$. The *domain set* of the functional constraint $F_j = (f_j, V_{j_1}, ..., V_{j_{n_j}})$ is the set of variables $Dom(F_j) = \{V_{j_1}, ..., V_{j_{n_j}}\}$. For convenience, a functional constraint $F_j = (f_j, V_{j_1}, ..., V_{j_{n_j}})$ is denoted $F_j = [f_j(V_{j_1}, ..., V_{j_{n_j}}) = 0]$. We will note $dim(F_j) = m_j$, the dimension of function f_j, and we will note abusively $|F_j| = n_j$, the cardinal of the domain set of F_j.

We will need, for a thin analysis, that the constraints depend effectively on all the components of all the variables belonging to their domains. Without giving further analytical details, we say that for a *well-formed* constraint F_j, the validity of F_j depends on all the components of the variables of $Dom(F_j)$. From an analytical point of view which is only outlined here, F_j is well-formed if for every instances satisfying it, we can apply the implicit function theorem [6] with all the components of the variables and all the monodimensional functions used to define F_j.

2.2   Functional Graphical Model: Definitions

A *functional graphical model* is a couple $M = (\mathcal{V}, \mathcal{F})$ with : $\mathcal{V}$ a set of primitives $V_i \in \mathcal{V}$, $\mathcal{F}$ a set of functional constraints $F_j \in \mathcal{F}$ of the form $F_j = [f_j(V_{j_1}, ..., V_{j_{n_j}}) = 0]$, where $V_{j_i} \in \mathcal{V}$ and f_j is a function in $\mathbb{R}_j^d$. M is said to be *well-formed* iff all its functional constraints are well-formed.

An FGM can be describe by a hypergraph [2], with possibly isolated vertices and several connected components. This hypergraph can be used in turn in structural pattern recognition techniques. FGMs capture some knowledge which can be stated as functional dependences between variables. They can be used to define and analyze properties of implicit systems [13].

A simple FGM enables us to proceed to the analysis component by component. A *simple* FGM $M = (\mathcal{V}, \mathcal{F})$ is an FGM such that each variable has only one component and each constraint corresponds to only one equation. A simple FGM M' can be constructed from any FGM M by considering all the components of the variables of M and all its constraints separately.

We derive the notion of partial model (which is related to partial hypergraph [13]). The *partial FGM* of an FGM $M = (\mathcal{V}, \mathcal{F})$ generated by a family of functional constraints $\mathcal{E} \subset \mathcal{F}$ is an FGM $(\mathcal{V}_\mathcal{E}, \mathcal{E})$ where $\mathcal{V}_\mathcal{E} = \cup_{F_i \in \mathcal{E}} Dom(F_i)$, A partial model is constituted by a subset of constraints of the original model.

The union of two FGMs $M_1 = (\mathcal{V}_1, \mathcal{F}_1)$ and $M_2 = (\mathcal{V}_2, \mathcal{F}_2)$ is the FGM $M = (\mathcal{V}_1 \cup \mathcal{V}_2, \mathcal{F}_1 \cup \mathcal{F}_2)$. Two FGMs $M_1 = (\mathcal{V}_1, \mathcal{F}_1)$ and $M_2 = (\mathcal{V}_2, \mathcal{F}_2)$ are *disjoint* iff $\mathcal{F}_1 \cap \mathcal{F}_2 = \emptyset$. Two FGMs are *independent* iff $\mathcal{V}_1 \cap \mathcal{V}_2 = \emptyset$. Remark that two independent FGMs are also disjoint.

An FGM only describes the structure of functional constraints between unknown variables. When confronted with reality, variables need to be instantiated, i.e. values for the variables are given. An *instance* $(M|\mathcal{V} = v)$ of an FGM $M = (\mathcal{V}, \mathcal{F})$ is the attributed hypergraph $(v, \mathcal{G}, r)$ obtained by identifying the primitives of M to the real valued vectors of the set $v = \{v_1, ..., v_{|V|}\}$. The relations from $\mathcal{G}$ are subsets corresponding to the domain-sets of the constraints from $\mathcal{F}$ and are attributed by the real valued vectors $r(G_j) = -f_j(v_{j_1}, ..., v_{j_{n_j}})$ (the relations' residuals). In [1], there is a similar distinction between Function-Described graphs which represent an ensemble of attributed graphs (which could be considered in our formalism as instances of the previous models).

2.3 Encoding Primitives with FGMs

Certain variable's instances of an FGM can be calculated knowing the values of other variable's instances and the functional constraints of the FGM. This is done by putting a functional constraint under one of its explicit form (when the conditions of the implicit function theorem are fulfilled), i.e. a relation $f(x, y, z) = 0$ (f,x,y,z being of dimension 1) can be used to calculate either x, y, or z using respectively the equivalent forms $x = g_1(y, z)$, $y = g_2(x, z)$ or $z = g_3(x, y)$.

Let us consider a simple well-formed FGM $M = (\mathcal{V}, \mathcal{F})$, and a realization $(M|\mathcal{V} = v)$ satisfying each equation of the model. Each relation $F_i \in \mathcal{F}$ can be put, considering that the hypotheses of the implicit equation theorem hold, under the form $V_i = g_j(V_{i-j_1}, ..., V_{i-j_{n_j}-1})$.

With these explicit forms, an ordered sequence of variables of M can be constructed. Thus, by giving the instances $v_1, ..., v_k$ of the first k variables $V_1, ..., V_k$ of the sequence, and by using the given explicit forms of some relations of the model, we can compute the instances of the remaining variables $V_{k+1}, ..., V_{|V|}$ of the sequence. Let $M = (\mathcal{V}, \mathcal{F})$ be a simple well-formed FGM. The set $\mathcal{G} \subset \mathcal{V}$ is a *generating set* of M iff there exists an order $V_1, ..., V_{|\mathcal{G}|}, V_{|\mathcal{G}|+1}..., V_{|\mathcal{V}|}$ of all the variables of $\mathcal{V}$ such that : $\mathcal{G} = \{V_1, ..., V_{|\mathcal{G}|}\}$, $\forall k > |\mathcal{G}|$, $\exists F_j \in \mathcal{F}$ such that $V_k \in Dom(F_j)$ and $Dom(F_j) \cap \{V_{k+1}, ..., V_{|\mathcal{V}|}\} = \emptyset$. $\mathcal{G}$ is said to be *minimal* if there exists no generating set smaller than $\mathcal{G}$.

Considering a simple well-formed FGM M, and instances of one of its minimum generating set variables satisfying the constraints of M, we can in theory compute entirely the instance $(M|\mathcal{V} = v)$ such that the constrains of M are satisfied. Thus, when an FGM is known both by a coder and a decoder, the code of an instance is given by the code of the corresponding instance of one of the FGM's minimum generating set.

2.4 FGMs and Description Length

In practice, variables' instances do not satisfy exactly the model's constraints due to quantification and measurement noise. Practical applications need to model the errors of variables and relations, especially when a model is confronted with real world measurements. The uncertainty of variables can be taken into account explicitly in the model by associating probability distributions to vertices and hyperedges of the functional model.

We can then define different costs that evaluate instances of FGMs, such as a model selection criterion which evaluates the appropriateness of an instance of an FGM to a data set, and which can be used to compare several FGMs that do not have necessarily the same structure. A multipurpose cost may be, for example, the MDL which is widely used in computer vision [9][11].

The description length of a two-part code of an instance of an FGM $M = (\mathcal{V}, \mathcal{F})$ such that its simple derived FGM M' is well-formed and $\mathcal{G}$ is a minimum generating set of M' is approximated by :

$$k(M|\mathcal{V} = v) = -\sum_{F_j \in \mathcal{F}} log_2\left(p\left(F_j\right)\right) + g\ sizeof(Real)$$

where $sizeof$ is the size (in bits) of the given type, $p\left(F_j\right) = p\left(f_j(V_{j_1}, ..., V_{j_{n_j}}) = 0\right)$ is the likelihood of F_j given the instances from v of the variables, and $g = |\mathcal{G}|$ is the size of a minimum generating set of M. Note that this is also the MDL criterion of a set $\mathcal{D}$ of variables knowing a FGM M when its minimum partial model encoding $\mathcal{D}$ is M itself.

We can compute the MDL criterion for any instance of an FGM $M = \cup_{i=1}^{m} M_i$, where M_i are independent FGMs. This cost is given by the sum of all the costs of the instances of the models M_i. This is due to the following result.

Theorem 1. *A minimum generating set of the model $M = \cup_{i=1}^{m} M_i$, where M_i are independent FGMs, is given by the union of minimum generating sets of the models M_i.*

Proof. The proof is straightforward, as we can remark that if the models are independent, they cannot share variables.

2.5 Multi-part FGMs: Encoding a Digital 1D Signal

We want to encode a digital 1D-signal. Measurements associated to each sample of the digital signal are vectors of the form $m_i = (a_i\ t_i)^t, i = 1, ..., n$ where a_i is the amplitude of the signal at time t_i. Functions can be used to model regularities of the samples.

An FGM describing a signal is shown in Fig. 1.

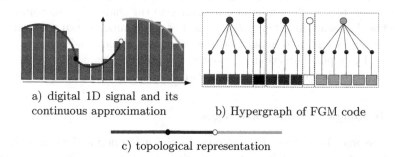

a) digital 1D signal and its
continuous approximation

b) Hypergraph of FGM code

c) topological representation

Fig. 1. FGM of a signal

Fig. 1.a describes a time varying digital 1D-signal and a possible continuous description of that signal. The lower part of the drawing represents the hypergraph structure of an FGM coding the relations between the groups of parameters used to draw the continuous signal, and the measurements associated with the digital signal. The dashed boxes define different independent partial models.

The relations have the form $F_1 = [f(m, p_1) = 0]$, $F_2 = [f(m, p_2) = 0]$ and $F_3 = [f(m, p_3) = 0]$ where m is one of the measurements explained by each curve, p_1, p_2 and p_3 are the parameters of the continuous curves representing each part of the digital signal, and f is a quadratic polynomial of t.

The given continuous primitives do not permit the direct retrieval of the time and the amplitude from the functional model, as the partial model associated with each continuous primitive is valid only on a time interval. In order to avoid to store explicitly the time component of each sample, we associate with the FGM a topological model (Fig. 1.c), which is used to retrieved the set of samples on which apply a specific continuous model. The vector of parameters associated with the points of discontinuities of the previous continuous model have the form $p' = (a\ t)^t$, and the simple equation that permits one to encode the digital measurement associated with that primitive has the form $g(m, p') = m - p' = 0$.

Thanks to both models, the overall model (including the measurements) may be encoded in a compact representation consisting in the set of instances of the generating set of each stored independent model excluding the variables already encoded implicitly by the topological model, the set of relation confidence, and the labeled topological model which indicates in which order the elementary models have to be applied. An extra parameter which indicate the sampling frequency of the signal can be stored in order to retrieve exactly the original signal. The compact representation can be used to recover the entire FGM.

3 FGM Based Encoding of 2D Images

An image may be seen as a piecewise continuous function that has been digitized. The discontinuities of the original function are of primary importance for image understanding. Completely encoding an image by a piecewise continuous model can be achieved by describing domains where some encoded measurements vary continuously (e.g. radiometric information), and the shape of those domains (i.e. geometrical information of an image). Most of the common continuous variations observed on a structured image (such as an image of man-made objects) can be captured by implicit functions. However, a topological model is required when encoding both the geometry and other information with implicit functions.

3.1 Topological Model of Images

An image may be defined by different specialized functional models that can be put together to explain simultaneously interesting measures (geometry of discontinuities and radiometry, for example). A topological model, such as the topological map described below, is then required in order to explain on which domain each functional model applies.

A topological map [5] is defined as the decomposition of an orientable surface into (Fig. 2) : a finite set S of *points*, a finite set A of Jordan *curves* whose

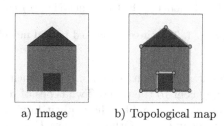

a) Image   b) Topological map

Fig. 2. Continuous topological map encoding implicit functions

extremities are elements of S, and a finite set F of simply connected domains or *faces*, whose boundaries are unions of elements of S and A. Points, curves and faces are called topological primitives of the topological model. Each of those primitive captures coherent continuous variations of measurements. For example, curves, which capture the (geometric) boundaries shapes, are subdivided into curve segments described by smooth functions, and the variation of color on a face is captured by a single implicit equation.

A topological map can be encoded by a combinatorial map [5,4], with a structure of size $k_T = (6a)sizeof(reference)$, where a is the number of arcs, and $sizeof(reference)$ is the size of a reference.

3.2 Local Models and Composition

An image I may be defined as a function $I : \mathbb{Z}^2 \to E$ where E is a set of measurements. For illustrating image encoding, we consider a graylevel image associated with measurements that are vectors of the form $I(i,j) = (x\ y\ n\ dx\ dy)^t$, where x and y are (subpixel) coordinates associated with the digital point (i,j), (dx, dy) is the computed gradient at the considered pixel, and n is the gray value of the original image at the point (x, y).

A functional based encoding of images describes relationships between a piecewise continuous representation and a finite set of measurements. The relation between a continuous piecewise function and the measurement sets of its digital embedding can be described by three different kinds of partial FGMs which involve primitives describing points, curves, and faces. Fig. 3 shows the hypergraph structure of those three partial models. Each big disk represents a primitive of the depicted model, and the functional constraints are depicted as small disks, connected to the primitives involved in the relation by lines. The dashed arrows represent topological constraints that are discussed below.

A *model of a point* (Fig. 3.a) is composed of a primitive $I(i,j)$ which is the measurement associated with a pixel, a primitive S that codes the information

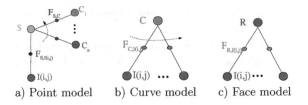

a) Point model   b) Curve model   c) Face model

Fig. 3. The three connected partial models

associated with a point and primitives C_1, ..., C_n which are the parameters of the curve segments adjacent to the point. Two types of relations are involved in the model. A relation $F_{S,I(i,j)}$ enables the explicit linking of a primitive S and its associated measurement $I(i,j)$. Other relations F_{S,C_i} link a junction S to the adjacent curves C_i, and do not need to be stored explicitly (they are redundant with the relations of the topological model). For example, we can use a point primitive $S = (x_S\ y_S)$, and the following relations.

$$F_{S,I(i,j)} = \left[\begin{pmatrix} x_S - x \\ y_S - y \end{pmatrix} = 0 \right]$$
$$F_{S,C_i} = [f(x_S, y_S, C_i) = 0]$$

where C_i is the primitive associated with a curve adjacent to the vertex, and f is a monodimensional implicit function associated with the curve C_i (see below). The second relation enables the reduction of the number of parameters associated with a vertex primitive, as one of its coordinates can be deduced by giving one relation, and the two coordinates are deduced by two relations.

A *model of curve* (Fig. 3.b) is composed of a primitive C which encodes the parameters of the continuous curve, and a set of digital points. The digitization of a continuous curve results in a well-composed digital curve [8]. In order to have a coherent digital embedding, we should impose that the set of measurements associated to a curve primitive forms a well-composed digital curve. The following relation can be used :

$$F_{C,I(i,j)} = \left[\begin{pmatrix} f(x,y,C) \\ \sin\left(\theta_L - \arctan(\frac{\delta f}{\delta y}(x,y,C)/\frac{\delta f}{\delta x}(x,y,C))\right) \end{pmatrix} = 0 \right]$$

where f is an implicit function associated with C. For example, a line segment may be represented by a primitive $C = (\theta_C\ d_C)$, where d_C is the distance of the origin to the line, and the function f would be $f(x,y,C) = x\cos\theta + y\sin\theta - d$. The number of parameters needed to store a curve segment is the number of components of C, as vertices of the curve segment are already stored by the model.

The two preceding models encode the geometrical information of the image, and do not encode the radiometrical part of the associated measurements. They can be encoded by other relations involving the face primitives described below.

A *model of face* (Fig. 3.c) links each measurement of a region to a primitive R that are parameters of the smooth function encoding the variations of the measurements on R. Each relation $F_{R,I(i,j)}$ relating R to a measurement $I(i,j)$ inside the region can then be described by :

$$F_{R,I(i,j)} = \left[\begin{pmatrix} g(x,y,R) - n \\ \frac{\delta g}{\delta x}(x,y,R) - dx \\ \frac{\delta g}{\delta y}(x,y,R) - dy \end{pmatrix} = 0 \right]$$

where g is a smooth function, whose parameters are encoded by R.

As demonstrated in Fig. 4, the union of those models forms an FGM encoding the measurements of an image. We can derive from that model a compact representation independent of the digital embedding. The compact representation involves only continuous primitives and their domain of validity. This compact representation can then be used in order to generate the complete model for a specific embedding, when for example a model instance is given.

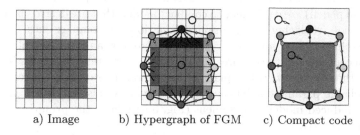

a) Image   b) Hypergraph of FGM   c) Compact code

Fig. 4. An image FGM example

3.3   Global Models

The preceding section was dedicated to local partial models. Some other models may be defined in order to code global information, such as alignments of line segments, pencil of lines, circles formed by circle arcs, concentric circles, etc. In this section, we are interested in the alignment and pencil models which form an FGM that has a tree structure (Fig. 5.a).

The *model of alignment* is composed of constraints relating the parameters of the line associated with each segment C, and the parameter of the line $L = (\theta_L \; d_L)$ describing the alignment. Those constraints can simply be stated as :

$$F_{C,L} = [C - L = 0]$$

This generalization can be extended to any implicit curve, or to regions. It takes naturally into account effects of occlusion, and can be viewed as "perceptual organization" of segments [10].

The *model of pencil* involves line primitives and the pencil primitive of the form $P = (x_P \; y_P)^t$, where x_P and y_P are the coordinates of the point common to each line. This form cannot represent parallel lines. The relation involved in the pencil model is :

$$F_{P,L} = [x_P \; \cos\theta_L + y_P \; \sin\theta_L - d_L = 0]$$

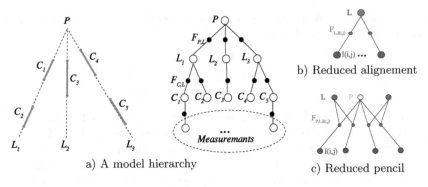

a) A model hierarchy

b) Reduced alignement

c) Reduced pencil

Fig. 5. Model of a pencil of aligned line segments

When the relations are treated as constraints, the preceding models (alignment and pencil) are equivalent to compact models that are derived by parameter substitution, replacing simple implicit equations by complex ones. This is done by an explicit form of some other equation related by a model variable [13], which has as a consequence the suppression of variables and relations. Substituted models can then be optimized in place of the overall hierarchical model, leading to more reliable parameters. Their generating sets can also be determined.

The reduced model of alignment (Fig. 5.b) relates the primitive L and each measurement $I(i, j)$:

$$F_{L, I(i,j)} = \left[\left(\begin{array}{c} x \cos \theta_L + y \sin \theta_L - d_L \\ \sin \left(\theta_L - \arctan(dy/dx) \right) \end{array} \right) = 0 \right]$$

The reduced model of pencil (Fig. 5.c) is composed of ternary relations : :

$$F_{P, L, I(i,j)} = \left[\left(\begin{array}{c} \cos \theta_L (x - x_P) + \sin \theta_L (y - y_P) \\ \sin \left(\theta_L - \arctan(dy/dx) \right) \end{array} \right) = 0 \right]$$

3.4   Description Length of FGM of an Image

The code of an instance of an FGM M encoding an image contains the codes of all the instances of the partial models involved, and the code of the topological model. Its description length, for the previously defined partial model set takes the following form :

$$k_T(M) = \sum_{M_V} k(M_V) + \sum_{M_L} k(M_L) + \sum_{M_P} k(M_P) + \sum_{M_R} k(M_R) + k_t$$

where $k(.) = -\sum_{F_j \in \mathcal{F}} log_2 \left(p\left(F_j \right) \right) + g \; sizeof(Real)$ is the description length of each partial model (c.f. section 2.4). However, each partial model is not independent, as for example vertex and curve models. The preceding approximation is computable with :

- $k(M_V)$ is the length of the shortest description of the point model M_V. The size of a point model generating set is $g = 1$ when the encoded point is adjacent to one curve, and $g = 0$ when it is adjacent to more than one curve.

- $k(M_L)$ is the length of the shortest description of the curve model M_L, such that M_L is not a partial model of a pencil model. Two real valued parameters define a line ($g = 2$).
- $k(M_P)$ is the length of the shortest description of the pencil model M_P, and $g = 2 + n$, where n is the number of lines involved in the pencil model.
- $k(M_R)$ is the length of the shortest description of the face model M_R. The size of the generating set is then the dimension of the vector R encoding the parameters of the smooth function used for modeling the measurement variation on the region.
- k_T is the size of the topological model (c.f. section 3.1).

The given cost function can be used to estimate the parameters of a model given measurements on the original image, and also to choose the best model of a given image. Model construction or recognition turn out to be in that context a combinatorial optimization problem. Fig. 6 illustrates the encoding of an outline image with the FGM instance depicted on the right. The number of parameters has been reduced by a factor of approximately 10 by encoding the outlines by a model derived from the exposed approach.

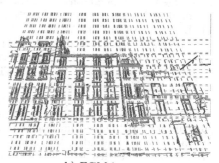

a) Image with superimposed outlines   b) FGM instance

Fig. 6. FGM instance (left) encoding an outline (right)

4   Perspectives

In this contribution, we introduced functional graphical models. They offer a convenient way to represent functional relations between variables. The variables involved in such a description can be encoded efficiently, as only a subset of variable instances of a specific model can be theoretically used to retrieve all the instances of the variables involved in the model. Thus, this code enables a significant reduction of the the space needed to store data with regularities. A description length has been derived. It can be used as a cost function for problems such as parameter estimation and model selection.

When considering a topological model and a functional model, one can describe regularities of images efficiently. The FGM-based models can encode simple and local regularities as well as complex ones, leading to a generic mathematical framework describing coherently geometric and radiometric information. The description length of such a model can be in turn used to define primitive extraction and shape recognition in a combinatorial optimization paradigm. The extraction of all the regularities of the data can then achieved by optimizing a single cost function. The presented framework can be extended in several ways. We could extend the framework for modeling images of higher dimensions, such as 3D images. We can also design models of stereoscopic system for space orientation and 3D reconstruction. Image sequences can also be encoded by FGMs, for compression applications for example. Although we demonstrate here how to efficiently encode images with FGMs, their hypergraph structure is also relevant for many other structural pattern recognition techniques, ranging from graph matching to high level information retrieval.

References

1. Alquézar R, Serratosa F., Sanfeliu A., "Distance between Attributed Graphs and Function-Described Graphs Relaxing 2nd Order Restrictions", SSPR2000&SPR 2000, FJ. Ferri & al. eds, LNCS vol. 1876 (2000), 277–286
2. Berge C., "Hypergraphs, Combinatorics of Finitite Sets", North-Holland Mathematical Library, vol. 45 (1989) 255 pages
3. Baker S., Nayar S. K.,Murase H., "Parametric Feature Detection", Int. Journal of Computer Vision vol. 27(1) (1998) 27–50
4. Braquelaire J.-P., Brun L. "Image Segmentation with Topological Maps and Interpixel Representation", Journal of Visual Communication and Image Representation, vol. 9(1) (1998) 62–79
5. Cori R., "Un code pour les graphes planaires et ses applications2, Astérisque, vol. 27, Société de Mathématiques de France, Paris, 1975
6. Faugeras O., "Three-Dimensional Computer Vision, A Geometric Viewpoint", The MIT Press, Cambridge, Massachusetts, (1999), 663 pages
7. Förstner, W., "Reliability Analysis of Parameter Estimation in Linear Models with Application to the Mensuration Problems in Computer Vision", Computer Vision, Graphics and Image Processing, vol. 40 (1987) 273–310
8. Latecki L., "Discrete Representation of Spatial Objects in Computer Vision", Computational Imaging and Vision vol. 11, Kluwer (1999) 216 pages
9. Leclerc, Y. G., "Constructing Simple Stable Descriptions for Image Partitioning", International Journal of Computer Vision, vol. 3 (1989) 73–102
10. Lee, J.-W., Kweon, I.S., "Extraction of Line Features in a Noisy Image", Pattern Recognition, vol. 30(10) (1997) 1651–1660
11. Lindeberg, T., Li, M.-X., "Segmentation and Classification of Edges Using Minimum Description Length Approximation and Complementary Junction Cues", Tech. Rep. ISRN KTH/NA/P–96/01–SE, Jan. 1996.
12. Lowe, D. G, "Fitting Parametrized Three-Dimensional Models to Images", IEEE Trans. on Pattern Analysis and Machine Intelligence, vol. 13(5) (1991) 441–450
13. Jocelyn Marchadier, "Functional Graphical Models", Technical Report 76, Vienna University of Technology (2002), 24 pages
14. Jocelyn Marchadier, "Modélisation fonctionnelle et topologique pour la vision par ordinateur : Application au relèvement de clichés urbains", PhD dissertation, Université de Marne la Vallée, IGM 2002-09, 183 pages

Building of Symbolic Hierarchical Graphs for Feature Extraction

Mickaël Melki and Jean-Michel Jolion

Lyon Research Center for Images and Information Systems (LIRIS),
Bât. J. Verne, INSA Lyon, 69621 Villeurbanne Cedex, France
{melki,jolion}@rfv.insa-lyon.fr

Abstract. This paper presents a new approach for feature extraction based on symbolic hierarchical graphs. We will focus on the construction model, which uses simple rules considering both the graph's structure and the symbols stored in each node. We will also consider some trends to dynamically modify those rules in order to implement a learning process or to be able to adapt the rules to one given problem or to the input image.

1 Introduction

As the computing capability of today's computers grows higher, more and more complex tools can now be used in image analysis and pattern recognition. Among those tools are graphs, which can be used in many areas, like image segmentation (hierarchical graphs), pattern recognition (graph matching), clustering (graph manipulation), structural analysis (conceptual graphs)...

Many previous works showed how useful graphs were in domains like segmentation, and more generally in image analysis. We are interested in this work in hierarchical graphs, which have already been used for this kind of applications [5, 9,7,1]. Our work here aims at describing a new model based upon graphs, more precisely upon graph pyramids, which will be used to built symbolic descriptions of an image featuring high adaption and learning capabilities.

Hierarchical graphs, also called graph pyramids, are ordered sets of graphs called *levels* of the pyramid, such that each level contains fewer nodes than the next one in the pyramid. Each node of a given level is considered as being originated from one node in the level below, and is called *surviving node*. Each node also has a set of *children*, which contains the surviving nodes from which this node is originated. Different kind of approaches has been proposed for the design of graph pyramids : the decimation strategy, like the one proposed by P.Meer [9], or the contraction strategy, like the one proposed by W.Kropatsch [7]. Our model will refer to the last one.

Our aim was to define a process for producing the contraction kernels required by the dual-graph contraction algorithm from the data in the image. We have chosen to use a symbolic representation of this data, since it grants us the ability to keep much more information than a classical numeric representation. The

E. Hancock and M. Vento (Eds.): IAPR Workshop GbRPR 2003, LNCS 2726, pp. 47–58, 2003.

potential of such representation has already been studied in previous works like the "n-cell description language" that was used for representing curves [8], or the 3×3 masks introduced by Canning *et al.* [3] that can be used for detecting angles and concavities [4] or texture elements [2].

The process we have defined uses a set of reduction rules for finding the appropriate contraction kernels. Those rules are part of the knowledge used by the algorithm, as well as the symbols themselves. All the knowledge elements are stored in an appropriate flexible structure, that can easily be dynamically improved by learning or adaption algorithms.

In the first part of this paper, we explain how this structure is defined, and how it is used to produce the contraction-kernels and build the pyramid. We then propose a simple example to see how the whole algorithm is working. Finally, we focus on the knowledge and describe possible implementations using learning and adaption algorithms.

2   The Pyramid Framework

In order to describe the pyramid framework, we need to explain first the *knowledge base* that contains all the knowledge used by our algorithm. After that, the process for building a pyramid will be explained.

2.1   The Knowledge Base

We need two kind of entities : the *symbols*, and the *rules*. Both will define the knowledge base.

The Symbols. Symbols are spawned into many *abstraction levels*, according to their complexity. The *Low-level symbols* are used to build the first level of the pyramid from the picture. They directly express some basic properties of the pixels, like their color, the shape of their immediate neighborhood, and so on... They are like what Tsotsos called the "tokens" in [10]. They also have some analogy to B. Julesz's "textons" used for texture characterization [6]. The *High-level* symbols represent the more complex notions we want to get at the top of the pyramid. The problem to solve is defined as a property on a graph whose nodes are valuated by such symbols. They can be related to Tsotsos' "visual prototypes". The *In-between level* symbols represent notions of intermediate complexity, and are used to go from low level symbols to high level ones by using more and mode complex symbols. They are themselves spawned into many levels.

Some symbols can be valuated. A node which is associated to a valuated symbol should also get another attribute used to store the value required by the symbol (it can be coordinates, colors, size, etc)...

For each level of abstraction, we build a graph, whose nodes represent symbols and whose edges define some kind of neighborhood between symbols. This

neighborhood express the similarity between symbols, and the edges are valuated by values that express how likely one symbol can be misunderstood for his
neighbor.

   This structure is then extended to hypergraph by adding vertical edges from
nodes from one level to nodes from one upper level if there is at least one possible
reduction from the lower level symbol to the higher level one. The edge is then
labelled by the set of *rules* that can perform this reduction. We can also allow
self loops labelled by rules that just contract the graph, without producing more
abstraction.

The Rules. Rules are used to build the *reduction hypotheses* that will be used
to build the next level, using the data in some nodes of the graph. For that, a rule
needs to generate a the resulting symbol, a contraction kernel, and an estimation
of the "confidence" of the reduction. An hypothesis is just the two first elements,
plus a priority value, computed from the "confidence" of the reduction and from
the importance of the rule in regards to the problem, and the set of the nodes
that were used by the rule to produce this hypothesis.

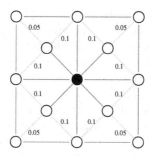

Fig. 1. A simple configuration : the filled node is the main node of the configuration,
the non-valuated edges define the skeleton, and the valuated edges are the extra-edges.

   To define which nodes are needed to produce this information, we defined
the *configuration*, composed of a connected graph called the *skeleton* and some
valuated extra-edges. One of the node of the skeleton will be the *main node* of
the configuration. A simple configuration is shown on figure 1.

   A rule holds for a node n of the graph if the skeleton can be matched to
the image graph in such a way that the main node is matched to n. The extra-
edges do not need to be matched for the rule to be applicable, but the value of
each missing extra-edge will be added to the *structural error*. But the rule can
really be applied only if the symbol in the nodes that have been matched are
in the domain of the rule. In that case, a *content error* is computed from those
symbols : for example, a rule that holds for an ellipse may also accept a circle
in its domain, but in that case the structural error will be higher. Then, the
outputs of the rule are produced : the output symbol, the contraction kernel,

and the confidence value (which is the product of the structural and the content errors). Note that one rule can be applied many time to the same node, as there can be many matching subgraphs, like in figure 2.

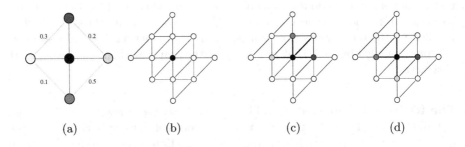

(a)   (b)   (c)   (d)

Fig. 2. Application of a rule : the left part shows both the configuration (a), and the graph we are working on (b). The right part presents two possible applications : on (c) the error is 0.2, and on (d) it is 0.8.

Practically, we will limit us to rather simple rules at first : we will only consider rules whose skeleton is a star graph of radius 2 or less in order to make the matching process simpler, and the domain of the rule will always be a singleton, which means that the rule will except only one possible symbol for each node of the matched subgraph in order to simplify the definition of a rule and to make it easier to be learned. It also means that the content error and the output symbol of a rule will remain constant. That also means that a rule will be defined only by : a configuration, an input symbol for each node of the skeleton, an output symbol, a contraction kernel, and a priority value. We will consider that a rule reduce the input symbol of the main node into the output symbol, and so it is stored as an attribute of the edge linking those two symbols in the knowledge base.

2.2   Construction of a Level

The construction of a level of the pyramid is composed of three phases : hypotheses are produced during the *exploration phase*, then some are selected during the *selection phase*, and finally applied during the *reduction phase*.

The Exploration Phase. The aim of this phase is to produce all the possible hypotheses. That means that for each node of the graph, we try to apply every rule to every possible subgraph matching the configuration of the rule and that each time the rule holds, an hypothesis is added to the list of the hypotheses related to this node. One node can have any number of hypotheses, each related to a particular matched subgraph or used rule, like in figure 3.

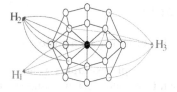

Fig. 3. Exploration phase : here, we have three rules (or maybe three times the same rule) that produce three hypotheses for the central node. The arrows here show the nodes that has been used by the rule that produced each hypothesis (that is, the nodes of the matched subgraph).

The Selection Phase. Since some hypotheses cannot both be used at the same time, we need to find a subset of hypotheses such that any hypothesis is compatible[1] with every other one. From all such subsets, we want to keep the best[2] one. Unfortunately, this is an NP problem, so we had to use an heuristic that usually produce a good set but not always the best : we first select the hypothesis with maximal priority, then we delete any hypothesis it conflicts with, and finally we repeat the process on the remaining hypothesis, until all the hypotheses have been either selected or deleted.

The Reduction Phase. Finally, we can apply the contractions which kernels are given by the selected hypotheses. We thus end up with a graph that will be the next level of the pyramid. The symbols of the nodes of that graph are given by the hypotheses (nodes that belong to no hypothesis, keep their symbol unchanged). Lastly, each node is linked to its sons, that is the set of node used in the reduction, given by the hypothesis (nodes that belong to no hypothesis are linked to their counterpart in the previous level).

2.3   Building the Pyramid

We have seen how to build the level $n+1$ from the level n of a pyramid. To build the whole pyramid, we first define a level 0 using the data from the input image, and then we iteratively build each level using the process described previously. We stop when no more hypothesis is generated, or when the selected hypothesis does not make any change to the graph.

Also, each time a level is built, we add edges between the current level and new built one, in order to link each node of the new level to the nodes of the previous level that where in the subgraph that was used by the rule that produced the corresponding hypothesis (if a node is not originated from any hypothesis, it is linked to the corresponding surviving node in the previous level).

---

[1] Two hypotheses are incompatible if their associated contraction kernels are intersecting

[2] The goodness of a set can be defined by means of the sum of all the individual priorities of the hypotheses belonging to the set

3   A Simple Example

3.1   Presentation of the Chosen Example

In this section, we will use a simple example to show how the reduction process is working. We have chosen to present how we can use our process to detect filled polygons whose edges are only vertical or horizontal (other orientations would have need to add more rules and the example would be much more complex). We will also suppose that polygons are always at least two pixels wide.

Even though the problem itself is not difficult to solve using traditional methods, it will allow us to see how our method is working and to present some of its limits (its advantages will be shown in the next section).

3.2   The Symbols Used in Our Example

For solving our problem, we will need symbols spawned in four levels, representing the different notions we need :

- The Low-Level Symbols : they represent the shape of the local neighborhood of the pixel or the set of pixels represented by the node (a pixel whose 3×3 neighborhood matches no symbol gets one of the two last ones, depending on the color of the pixel) :

- The Polyline-Level Symbols : they represent polygonal lines, with an attribute for storing their first and last corners.
- The Polygon-Level Symbols : they represent particular polygons which number of edges is preset (quadrilaterals, pentagons, etc)...
- The High-Level symbol : it represents a general polygon, that is a polygon with any number of edges (this number being stored as an attribute).

Symbols representing corners are valuated by the coordinates of the corner, and those representing polygonal lines are valuated by the coordinates of the corners at their end-points and by the number of corners contained in the polygonal line.

3.3   The Rules Used in Our Example

The rules we will use are the following, ordered by decreasing priority value :

- A corner can merge with some other corners and a filled area to produce a particular polygon.
- A corner or a polygonal line can absorb one or two incident edges.
- An edge can absorb an adjacent edge.
- A corner can merge with another corner to produce a polygonal line.
- A polygonal line can merge with a closing corner to produce a polygon.
- A polygonal line can absorb a corner that does not close the line.

- The two previous rules but with a second polygonal line instead of a corner.
- A black or white area can absorb another area of the same kind.
- If we just want to detect polygons without being interested in their number of edges, we can add a rule that transforms a polygon with fixed number of edges into a general polygon.

We can see that there are three kind of rules : those which absorb some neighbor nodes without producing a new symbol, those which merge one node with some other nodes to produce a new symbol, and those that transforms the symbol of a node into another symbol, without absorbing any other node.

3.4   An Example Showing the Building of a Pyramid

Here, we present how the building process works if we start from a 5 × 5 image composed of only one filled 3 × 3 square in the center, outlined by white pixels.

The rules that are applicable here are the ones that merge the outside areas, and the rules that make a corner absorb one or two adjacent lines (the latter having an higher priority). From all the possible hypotheses produced by those rules, a subset of compatible hypotheses is selected. In the following figure, we can see both the contraction kernels of the selected hypotheses (left), and the contracted graph with the symbols updated (right) :

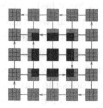 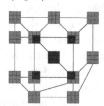

The rules that can now apply are those that merge the outside areas, and those that merge two angles to produce a polygonal line (those rules have higher priority). The contraction kernels related to one possible selection and the resulting graph is given by the next two figures (note the emergence of a symbol from the polyline level, representing in that case a polygonal line with two corners) :

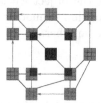 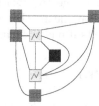

Now, the rules that still apply are the ones that merge outside areas, and the one that will make the merging of the two polygonal lines to produce a quadrilateral (this one has a priority higher than the first ones). Again, the following figure represents the contraction kernels of a possible selected set of hypotheses, and the result of the contraction (and this time a polygon-level symbol, "quadrilateral", appears) :

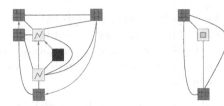

Finally, the two rules that can apply are the one which make the merging of the two outside areas that are still here, and the one that transforms the quadrilateral into a general polygon without doing any contraction. So, the final graph at the top of the pyramid will have only two nodes : one with the symbol "white area" that represents the background, and one with the symbol "general polygon" representing the square.

3.5   Limits of the Method

We can see with this example how complex this algorithm is, not only in terms of memory usage, since many hypotheses has to be stored for each node, but also in terms of computing power, as for each node many rules need to be applied many times. Even though we had chosen to select a very simple example, the number of rules and symbols is not so low, and we can easily predict that for more complex problems, this number will grow very fast. For that reason, it will not be possible to manually define each rule when we want to solve complex problem, which means that we will have to define a learning process that would be able to automatically learn the rules. Since the number of symbols grows fast too, we would also need to have the same kind of learning process for the symbols. However, even if we will present some possible ways to learn the rules in the next section, we still do not have defined an efficient process for the learning of symbols.

One other limit of the way we build the pyramid is due to one of the property that our algorithm has : it preserves topology. Even though this is usually an advantage, there is some cases where it may become a limitation. For example, if we have an object "split" into two parts due to an occluding second object, we have no way to merge the two parts of the object without absorbing the occluding object. That means that our algorithm cannot detect both parts of two important objects if one is occluding one another.

4   Dynamic Processes

The knowledge base has been defined such that it is a relatively flexible structure which contains all the needed information to build a pyramid. For that reason, it can be easily modified automatically so that the building process becomes more and more efficient for the problem we have to solve. We have defined three kind of modification process : the learning of the rules, the targeting of the rules to a given problem, and the adaption of the rules in respect with a given input image.

4.1   Learning

In our model, a rule is defined by a configuration (that is, a skeleton and a set of extra-edges), a contraction kernel, its input symbols, its output symbol, and its priority. Among those parameters, the priority of the rule does not modify the behavior of the rule, but only the probability it has to be used. For that reason, this parameter will not have to be learned during the learning process, but should instead be adjusted during the targeting process described in the next section.

The learning of the structural part of the rules (configuration and contraction kernel) is a difficult operation, but can be done by starting from a simple minimal structure, that is a configuration and a kernel composed of a single node (the main node), and then by making it grow. For the configuration skeleton, we can add a neighbor to the main node, add a neighbor to one of the neighbor of the main node (since the skeleton can be a star of radius 2), delete a node at distance 2 from the main node, or delete a neighbor of the main node, if this node does not have another neighbor except the main node. For the contraction kernel, we can add any node from the current skeleton, or remove any node except the main node, provided that the kernel remains connected. Note that if a node is removed from the skeleton, it has to be removed from the skeleton too.

For the learning of the content part of the rules (that is their input and output symbols), we propose a genetic-based approach : for each rule, we define a gene consisting in the set of its input symbols, and its output symbol. Then, we try to make those genes evolve using a genetic algorithm. We thus need to define a goal function, which should measure the quality of a set of genes. When we want to define such a function, we had to make a choice between different approaches : we can take into account only the final reduction (the graph on the top of the pyramid), but we also can take into account all the in-between levels. Even though the first approach is simpler, not taking into account the reductions that occurred at the lower levels of the pyramid means that even a good set of rule can be considered bad if only one rule is disturbing the result at the top of the pyramid. Anyway, once we are able to implement the first approach, we can easily implement the second one, by computing the qualities of each level of the pyramid, and then by returning a weighted mean of them.

4.2   Targeting

This learning process allows us to learn the behavior of the rules, or the laws that govern the reduction. But some of those laws, even if they produce good results, may not be appropriate for solving a given problem. The targeting process aims at increasing the importance of the rules that are the more useful in order to get the information required to solve the problem, and to make the building of the pyramid faster.

The focusing can be done at two occasions. First, when we define the problem to solve, we can give higher priority to the rules that can produce symbols useful for solving our problem. We thus have to define a set of useful high-level symbols, and then we give a malus to each rule whose input symbol is a descendant of a useful symbol, but which outputs a symbol which is not a descendant of a useful symbol, and we give a bonus to each rule that outputs a symbol which is a descendant of a useful symbol (the higher the level of the symbol is, the higher the bonus will be).

The second focusing occasion is during the solving of the problem, when we build the pyramid : when the pyramid is build, we give some quality values to the nodes at the top of the pyramid expressing how useful this node is for solving the problem. Then, we just go down the pyramid to adjust the rules that were used to produce each node of the top-level according to its quality. This operation has to be done when we are solving the problem, so it has to be done after the adaption process described in the next section (since the problem is solved after the end of the adaption process).

4.3   Adaption

When the pyramid is being built, all the rules produce hypotheses but only those which seem to be the more plausible are applied. But this is still an approximation, since it is not always possible to select the best hypothesis during selection phase, as some higher level information can be needed to make the proper choice. That is why we defined an adaption process, that consists in going backward down the pyramid once it has been built in order to adjust the priority of the rules by using the high-level information. Also, this process can be used to correct some bad assertions that have been done when generating the graph from the input image, due to the noise in the image for instance.

This process starts from the detection of ambiguous nodes of the apex of the pyramid. Ambiguous nodes are nodes which symbol is at rather short distance to another symbol that would change the result to something more plausible. For instance, if a face has been defined as a set of nodes representing the mouth, the nose, the eyes, and the outline of the face, and after reducing an image we obtain exactly a graph representing a face, except that one of the two nodes that should be representing eyes is representing a circle, this node will be considered ambiguous. If so, we try to change the symbol to the expected symbol. For levels other than the base level of the pyramid, the adaption process is done as described in figure 4 : if some rule can be used to produce the requested symbol, its weight is increased, else we try to change the symbol of the son, by repeating recursively the algorithm. If we have to adapt a node from the base level, we have to look back at the picture. If the image part related to the node can be reinterpreted as the requested symbol, the symbol of the node is changed to the requested one. Note that in all cases, we use a tolerance value which is decreased each time an operation like the increase of the priority of a rule, or the change of a symbol is performed. This tolerance has to remain positive for the adaption to be valid.

Once the adaption process has ended, the pyramid can be built again from the lowest level that was involved during the adaption process, by using the new priority values for the rules, and the new symbols if the symbols have been

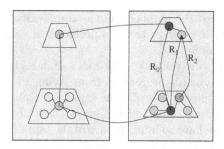

Fig. 4. Adaption for a node from an in-between level of the pyramid : we try to replace the symbol of node at the top by the other symbol represented at the same level of the knowledge base. To do that, we look at the son of the node : if a rule like R_1 exists and is applicable, its priority is increased so that it becomes higher than the one of R_0. If that is not the case, we try recursively to adapt the symbol of the node's son to some symbol such that there is an applicable rule like R_2.

changed (if the adaption has gone down to the base level). This step is needed, as the change in the rules may produce conflicts with hypotheses that were once selected and hence some global changes can happen, which could induce lots of changes in the top level. Finally, if no more ambiguous node remains, we can solve the problem, else the process is repeated. Though, because cycles are possible, the process will be actually repeated only a limited number of times : once this number is reached, the more frequent answer to the problem will be used.

5   Conclusion

In this paper, we have presented a possible framework to define a process to build a graph pyramid using rules working on symbolic data stored in the nodes, and taking into account both those data and the local structure of the graph. When defining this process, we stressed in grouping all the knowledge that was needed to perform the building into one single flexible structure that was the more independent as possible from the data, so that dynamic evolutions can be easily performed. We thus defined the knowledge base, an hypergraph which nodes represent the symbols and which edges represent the rules. Its hypergraph structure makes it easy to be adapted.

We proposed some possible ways to learn the rules or to adapt the rules to a given problem. We also explained how the rules can be adapted in order to take into account some high-level information when the pyramid has been built once. This process can significantly improve the results, as it can limit dependence to noise on the input image, but also because the high-level information can help discriminating which of two equally plausible reductions should be taken into account. But even though this is the process we are focusing on, there are no way we can make useful experiments until the learning process has been perfectly defined. Indeed the number and the complexity of the rules that have to be defined is large, if we want to solve problems of complexity which makes the adaption process useful.

The complexity of our process is its main drawback : not only the large number of rules and symbols prevent us from trying to manually define them when the problem is complex, but also the exploration phase of the construction needs to perform many matching between graphs, and produces lots of hypotheses. This means that we have to limit the size of the input images because of both high memory usage and high computing power required. It is also a problem for the learning process, as it requires to build the pyramid hundreds of times.

Finally, if we have proposed some possible way to learn the rules, we still do not have defined any useful method to learn new symbols, even though the structure of the knowledge base wouldn't make it difficult to add new symbols, remove old symbols or even change relations between symbols. But if it is still acceptable to let the user define the low-level symbols that are used to describe the input image and the high-level symbols that are used to solve the problem, the definition of in-between level symbols may become very fastidious, so that some work concerning the learning of those symbols must be done for our model to become more useful.

References

1. E. Duchesnay. *Agents situés dans l'image organisés en pyramide irrégulière, contribution à la segmentation par une approche d'agrégation coopérative et adaptative.* PhD thesis, Laboratoire de Traitement du Signal et de l'Image, University of Rennes I, France, 2001.
2. C. Duperthuy and J.M. Jolion. Toward a generalized primal sketch. In *Advances in computing*, pages 109–118. Springer Verlag, 1997.
3. N.S. Netanyahu J. Canning, J.J. Kim and A. Rosenfeld. Symbolic pixel labeling for curvilinear feature detection. *Pattern Recognition Letters*, 8:299–308, 1998.
4. JM. Jolion. Concavity detection using mask-based approach. In P. Pudil A.Amin, D.Dori and H.Freeman, editors, *Lecture Notes in Computer Science*, pages 302–311. Springer, 1998.
5. JM. Jolion and A. Montanvert. The adaptive pyramid : a framework for 2d image analysis. *Computer Vision, Graphics, and Image Processing : Image Understanding*, 55(3):339–248, 1992.
6. B. Julesz. Textons, the elements of texture perception and their interaction. *Nature*, 290:91–97, 1981.
7. W.G. Kropatsch. Building irregular pyramids by dual graph contraction. *IEEE Proc. Vision, Image and Signal Processing*, 142(6):366–374, 1995.
8. W.G. Kropatsch. Property preserving hierarchical graph transformations. In L.P. Cordella C. Arcelli and G. Sanniti di Baja, editors, *Advances in Visual Form Analysis*, pages 340–349. World Scientific Publishing Company, 1997.
9. P. Meer. Stochastic image pyramids. *Computer Vision, Graphics, and Image Processing*, 45(3):269–294, 1989.
10. J.K. Tsotsos. A "complexity level" analysis of immediate vision. *Int. Journal of Computer Vision*, 1(4):303–320, 1988.

Comparison and Convergence of Two Topological Models for 3D Image Segmentation

Achille Braquelaire[1], Guillaume Damiand[2], Jean-Philippe Domenger[1], and Frédéric Vidil[2]

[1] LaBRI, Bordeaux 1 University, 351 cours de la Libération, F-33405 Talence
[2] IRCOM-SIC, Poitiers University, Bt SP2MI, BP30179, F-86962 Futuroscope Cedex
{braquelaire,domenger}@labri.fr, {damiand,vidil}@sic.univ-poitiers.fr

Abstract. In this paper we compare two topological models of 3D segmented images representation. These models are based on a collaboration between a topological representation and a geometrical representation of the regions of the segmented image. In both models the description of the topology lays on topological maps with a combinatorial representation. A geometrical embedding of maps is used to describe the geometry of regions. Both models differ in the way of defining this embedding. The aim of this paper is to compare these two models from the point of view of their interest for image segmentation, and to explore the possibility and the interest of making both these models converge.

1 Introduction

A 3D image is a 3-dimensional set of elements called voxels. A voxel can be seen either as a valuated point or as a valuated unit cube. The second approach is more convenient when considering the decomposition of an image into volumetric objects. In this approach, the space is recursively decomposed into 3D, 2D, 1D and 0D unit elements called respectively *3-cells*, *2-cells*, *1-cells*, and *0-cells*. A voxel is then a valuated *3-cell*. An object of the 3D segmented image is defined as a set V of 6-connected voxels (2 voxels are 6-connected if they share a common *2-cell*), and the boundary of V is the set $B(V)$ of faces of voxels defined by $B(V) = \{b = v_1 \cap v_2, \ with \ v_1 \in V \ and \ v_2 \notin V\}$. Such voxel faces are called *s-cells* (for *surface cells*). This definition of boundary is consistent from a topological point of view and can be easily extended to the decomposition of an image into several objects, and to images of higher dimensions.

The problem of image segmentation is to define and implement some process for decomposing images into several objects (or regions) according to some homogeneity criteria. This problem has been widely explored in dimension 2, and it has been shown that topological maps provide a powerful framework for split and merge segmentation [2,4]. On the other hand, the segmentation of 3D image is most of the time restricted to the extraction of a foreground component from a background. In order to extend 2D segmentation methods to 3-dimensional images it is necessary to provide the set of 3D images with a structure allowing:

E. Hancock and M. Vento (Eds.): IAPR Workshop GbRPR 2003, LNCS 2726, pp. 59–70, 2003.
© Springer-Verlag Berlin Heidelberg 2003

1. The topological and geometrical representation of the segmented image: definition of boundary, neighborhood, surface shape, etc.
2. The extraction of features involved in the segmentation process: boundary and volume reconstruction, determination of adjacency (such as adjacent or included volume, common boundary), etc.
3. The updates of the representation required by the segmentation process, mainly when splitting a volume into sub-volume or when merging adjacent volumes.

Two models have been proposed for solve this problem which are both based on a topological representation of the decomposition, associated with a geometry embedding. In both models the topology is described by 3D topological maps and based on a combinatorial representation, also called combinatorial maps. These models differ in the way of defining the geometrical embedding of the topological map and the correspondence between topological and geometrical levels. One of them uses a global geometrical embedding [3,7] and the other one a local and hierarchical geometrical embedding [1,6]. In this paper we call the first one *topological map with global embedding* (in short GE topological map) and the second one *topological map with hierarchical local embedding* (in short HLE topological map). The aim of this paper is to compare both models and to investigate the possibility and the interest of making these models converge in the context of 3D image segmentation.

2   Combinatorial Representation of the Topology

In both models the topological representation is based on 3D combinatorial maps or 3-maps. Combinatorial maps [9,5] are a mathematical model allowing the topological representation of a space subdivision in any dimension. They were first introduced in 2D as a planar graph representation model [8,11], then extended in dimension n to represent orientable or not-orientable quasi-manifold [10]. They encode the space subdivision and the incidence relations between subdivision elements. A combinatorial map is a set of atomic elements, called *darts*, that represent the space subdivision, and functions between these elements that represent the incidence relations. Intuitively, a dart corresponds to a minimal oriented element of the boundary of the decomposition.

Definition 1. *A 3-map, is an 4-tuple $M = (\boldsymbol{D}, \delta_1, \delta_2, \delta_3)$ where:*

1. *$\boldsymbol{D}$ is a finite set of darts;*
2. *δ_1 is a permutation on $\boldsymbol{D}$;*
3. *δ_2, δ_3, and $\delta_1\delta_3$ are involutions on $\boldsymbol{D}$;*

The incidence relations can be represented in different ways with the permutation and the two involutions. The 3D combinatorial map used in the GE model is defined by $M_{GE} = (\boldsymbol{D}, \gamma, \sigma, \alpha)$. The permutation γ links darts that belong to adjacent faces around a same edge. The involution σ links two darts incident to the same vertex and that belong to a same face. Finally the involution α links two opposite darts of the same edge and that belong to a same face.

The combinatorial map used in the HLE model is defined by $M_{HLE} = (\boldsymbol{D}, \beta_1, \beta_2, \beta_3)$. The permutation β_1 links darts that belong to consecutive edges around a same face. The involution β_2 links two darts incident to the same edge and that belong to a same volume. Finally the involution β_3 is the same that the involution α and links two darts incident to the same face and that belong to the same edge.

These two representations are equivalent. Indeed, both set of darts $\boldsymbol{D}$ are equal and all permutations of each model can be obtained by a combination of permutations of the other model. The conversion from a M_{GE} map into the corresponding M_{HLE} map, is given by: $\gamma = \beta_3\beta_2$, $\sigma = \beta_3\beta_1$, $\alpha = \beta_3$ and the inverse conversion is given by: $\beta_1 = \sigma\alpha$, $\beta_2 = \gamma^{-1}\alpha$, $\beta_3 = \alpha$.

In Fig. 1, we present an example of a 3D object and the corresponding combinatorial maps used in both models.

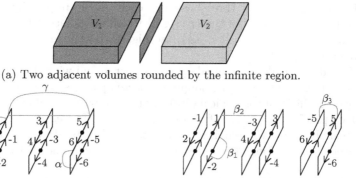

(a) Two adjacent volumes rounded by the infinite region.

(b) Combinatorial map in the GE model.   (c) Combinatorial map in the HLE model.

Fig. 1. The topological elements of the 3D subdivision are: 3 volumes, 6 oriented faces, 2 edges, and 2 vertices. The relations between darts are partially represented.

Topological cells (vertex, edges, faces, volumes) are implicitly represented in 3-maps. Each cell corresponds to the set of darts that are incident to the cell. These sets of darts are the connected components of particular 2-maps that represent the partitions of the space into vertices, edges, faces and volumes:

	Vertices	Edges	Faces	Volumes
GE model	$(\boldsymbol{D}, \gamma, \sigma)$	$(\boldsymbol{D}, \gamma, \alpha)$	$(\boldsymbol{D}, \sigma, \alpha)$	$(\boldsymbol{D}, \gamma^{-1}\sigma, \gamma^{-1}\alpha)$
HLE model	$(\boldsymbol{D}, \beta_2\beta_1, \beta_3\beta_1)$	$(\boldsymbol{D}, \beta_2, \beta_3)$	$(\boldsymbol{D}, \beta_1, \beta_3)$	$(\boldsymbol{D}, \beta_1, \beta_2)$

When a volume is totally included into another one V, the boundary of V is composed of more than one oriented closed surface. Each closed surface is associated with a connected component of the 3-maps. One of its closed surface has a positive orientation (the exterior boundary of V), the others have negative orientation (the interior boundaries). We need to keep the relative positions of these surfaces that is not represented in the combinatorial map. A classical solution consists in using an *inclusion tree* that gives associations between a volume and its cavities and that allows to preserve the inclusion relation.

3   The Model of GE Topological Maps

The model of topological maps with global geometrical embedding is based on the global encoding of the geometry of the boundary of the segmented image and on the definition of the geometrical analogous of topological dart. This definition is used to build the correspondence between the topological representation of the segmented image and the global geometric description of regions and region boundaries.

Let I be a 3D segmented image I of size $\Delta_x \times \Delta_y \times \Delta_z$. The *s-cells* of I are encoded by a 3D array B_I of size $(\Delta_x + 1) \times (\Delta_y + 1) \times (\Delta_z + 1)$. The entry $B_I(i, j, k)$ is a tuple of three booleans, each one corresponding to one of the three faces of the voxel of center (i, j, k) which are adjacent to the voxel vertex of coordinate $(i - \frac{1}{2}, j - \frac{1}{2}, k - \frac{1}{2})$ (see Fig. 2(a)). The array $B(I)$ is called the *boundary image* of I.

In order to make the boundary image correspond with the topological maps we have to decompose it into simply connected patches of surface (sets of *s-cells* homeomorphic to a topological disc). By this way it is possible to associate each surface patch with a face of a topological map. To decompose the boundary into simply connected patches, we cut the surface by drawing closed paths of adjacent *1-cells* along the boundary. Each cutting is encoded by marking the related *1-cell*.

The decomposition of the boundary into patches requires two different cuttings. The first one is the *inter-region cutting*. It consists in cutting the boundary along the border of any maximal part of surface shared by two volumes. The *1-cells* to be marked by inter-object cutting are easy to find because they are the *1-cells* adjacent to more than two *s-cells*. The second cutting is the *intra-object* cutting. It is used to decompose the boundary components resulting from the inter-object cutting. A boundary element has to be cut once if it is homeomorphic to a sphere, twice if it is homeomorphic to a one hole torus, and so on. This cutting is based on topological region growing. The algorithm used to perform this region growing requires conditions on the segmented images that lead us to restrict this model to weakly well composed segmented images.

The result of the cuttings is a set of *1-cells* of the boundaries of the objects of segmented image. These *1-cells* are called *l-cells*. Each connected component of these *l-cells* forms a graph. We call *p-cells* the nodes of this graph.

The topological map is built from the boundary image by traversing the boundary and associating each surface patch with a face of the surface map. Since the topological encoding lays on topological dart it is necessary to define the geometrical analogous of a dart. A *geometrical dart* is a tuple $\prec p, l, s \succ$

such that p is a *p-cell*, l is *l-cell* adjacent to p and s is a *s-cell* adjacent to l (see Fig. 2(b)).

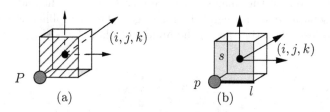

Fig. 2. (a) Each entry of the boundary image encodes the boundary of the segmented image at the neighborhood of the voxel vertex of coordinates resulting from a translation of the voxel center by $(-\frac{1}{2}, -\frac{1}{2}, -\frac{1}{2})$. (b) The tuple $\prec p, l, s \succ$ is a geometrical dart.

For this construction we consider the $\alpha\sigma\gamma$ representation. The construction of the map is done by following each chain of *l-cells* linking two *p-cells* with respect to the orientation of the surface. By this way it is possible to associate the two geometrical darts at the ends of such a chain. If one of these geometrical darts (say g) is associated with the topological dart d the other one (say g') is associated with $\alpha(d)$. Then we consider the sequence of geometrical darts encountered when turning around the *l-cell* of g and the sequence of geometrical darts encountered when turning in the opposite sense around the *l-cell* of g'. We can verify that both sequences have the same length. Let $g'_1 \dots g'_k$ and $g''_1 \dots g''_k$ be these sequences. We associate each g'_i with a new dart d_i and g''_i with $\alpha(d_i)$. Each sequence $d_1 \dots d_k$ is a cycle of γ. The permutation σ can be initialized from the neighborhood of each geometrical dart.

This construction associates each geometrical dart with a topological one, and conversely associates each topological dart which its geometrical embedding. Since any topological element, whatever the dimension, is represented by a topological dart, it is possible to retrieve the geometrical embedding of any topological element from the associated geometrical dart.

4   The Model of HLE Topological Maps

The main principle of the HLE topological maps definition consists in a progressive simplification of a combinatorial map that represents all the voxels of the image. For this simplification, we mainly use the *merging operation* which takes two adjacent *i-cells* and merge them in a unique *i-cell*.

The first combinatorial map, which is the starting point of the simplification process, just represents all the voxels of the image (called *level 0 map*). This map has an implicit canonical embedding (the usual geometry of the voxel grid). We are going to simplify progressively this map by using this implicit geometry. In order to simplify, we use the term of *coplanar faces* to design faces that have coplanar embedding (and the same for *collinear edges*). First we merge all adjacent volumes of level 0 map that belong to the same region of the image (Fig. 3(b)). This removes all the interior faces and we so obtain the boundaries of all the regions of the image. Then we merge all adjacent and coplanar faces that are separated by a degree 1 or 2 edge[1]. This simplification allows us to remove useless edges that are in the middle of plane faces (Fig. 3(c)).

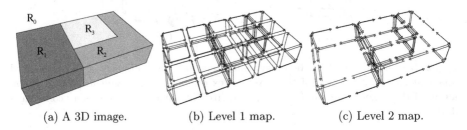

(a) A 3D image.   (b) Level 1 map.   (c) Level 2 map.

Fig. 3. A 3D region segmented image and the corresponding levels 1 and 2 maps.

Then we merge all adjacent and collinear edges that are separated by a degree 2 vertex[2] (Fig. 4(a)). In order to obtain the minimal representation of a 3D image that does not depend on the geometry but only on the topology, we need to simplify the non-coplanar faces and the non-collinear edges by similar merging operations. So, starting from the level 3 map, we merge all adjacent and non-coplanar faces that are separated by a degree 1 or 2 edge (Fig. 4(b)), and at last we merge all adjacent and non-collinear edges that are separated by a degree 2 vertex (Fig. 4(c)).

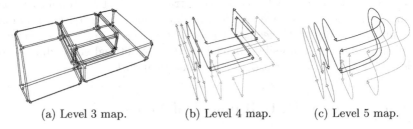

(a) Level 3 map.   (b) Level 4 map.   (c) Level 5 map.

Fig. 4. Levels 3, 4 and 5 maps.

The face merging can lead to the disconnection of the map into several connected components, for faces that have more than one border After this disconnection, it is not possible to place the different connected components relating

---

[1] The degree of an edge is the number of incident faces.

[2] The degree of a vertex is the number of incident edges.

to each others: we have lost topological information. In order to solve this problem, a solution consists in adding particular edges that keep the map connected. These edges are called *fictive edges* because they do not represent the border of a face unlike the other edges.

The last map obtained is the minimal map that represents the topology of the 3D image. Indeed, we can prove that all the different simplifications do not lead to a modification of any topological characteristic. Moreover, it is not possible to remove anything in this map without modifying its topological characteristics. This is why this map is called *topological map*.

There are several ways to add geometrical information to the topological map and the choice of one of this way depends on the application and the needed operations. We use here a local and hierarchical embedding which is interesting when the 3D images are composed with many large plane surfaces. To achieve this embedding, we associate each face of the topological map with an embedded 2-map (which has a straightforward linear embedding) that represents the geometry of the face (we can see Fig. 5(b) an example of such an embedding).

To obtain this embedding, we begin by linking to each face of the level 0 map a 2-map that represents the geometry of the corresponding voxel face. Then, we just need to update the embedding during each merging operations: when we merge two faces, we merge the corresponding embeddings when they are coplanar (merge operation to obtain the level 2 map) and we just sew the embedding when they are not (to obtain the level 4 map). This is the same principle for the edge merging, where we need to perform similar simplifications in the embedding maps. For this reason, the updating of the embedding can be achieved easily with a very efficient algorithm in complexity time. Moreover, this principle can be extended in upper dimension where each i-dimensional cell could be embedded with a i-map.

5   Comparison of Both Models

Both models can be compared from several points of view: power of expression, cost, reconstruction, extraction of features, point localization, and modifications of the subdivision. First we are going to present how some useful operations can be performed on both models, then we summarize the advantages of each one.

5.1   Presentation of Basic Operations in Both Models

Both models structure images decomposed into 6-connected objects. The HLE model makes possible to analyze any segmented image, whereas with the GE model there is a restriction to weekly well composed image. This restriction is not imposed by the model but by the cutting algorithm which fails on some pathological cases. These cases correspond to some configurations of edge adjacent volumes for which the topology cannot be reconstructed by a local analysis.

With the GE model, the size of the data structure is approximatively twice the one of a 256 gray level image. In the HLE model the size of data structure depends of several image properties (mainly the planarity of faces). Conversely the cost in time of the map construction is lower with the HLE model where each voxel is processed once. In the GE model some preprocessings involved in the construction (for instance to get the weekly well formed condition) or in updates may require several processings for each voxel. Both complexities are linear but on large images the difference of run times may be noticeable.

There are two important reconstruction operations: the reconstruction of the boundary and the reconstruction of the set of voxels of a volume or of a union of volumes. Each boundary patch shared by two adjacent volumes is described by a set of topological darts. In the GE model each of these topological darts is associated with a geometrical one, each geometrical dart corresponding to a *s-cell* in the boundary image. Thus the boundary element can be reconstructed by traversing the boundary image from any of these *s-cell*. In the HLE model each topological dart is associated with an embedded 2-map. The traversal of this 2-map gives planar patches of surfaces that are decomposed in rectangles and then reconstructed.

With the GE model the reconstruction of a volume can be achieved directly in the image boundary by traversing it from a starting voxel given by a geometrical dart. With the HLE model this reconstruction needs first to reconstruct the boundary and then to perform a similar processing.

The extraction of topological features is based on the analysis of the 3-maps and thus is equivalent in both models. Topological characteristics, like the genus or the number of cavities, come directly from the maps. The topological maps also provides extraction of features like volumes adjacent to a given one, included volumes, faces shared by two adjacent volumes. These features can then be used to initiate boundary or volume reconstruction.

The point localization consists in retrieving a volume from a given voxel of the image. In the GE model it is done by traversing the boundary image until reaching a *s-cell s*, then by founding an *l-cell* of the surface element containing s and following adjacent *l-cells* until reaching a geometrical dart g. The volume is given by the topological dart associated with g. This localization needs a more heavy algorithm that consists in computing intersections between a half-line and every 2-face of the HLE.

The main modifications involved in the segmentation process are the splitting of a volume into two or more sub-volumes and the merging of two or more adjacent volumes. Both these operations require topological and geometrical updates. The topological updates are equivalent in both models. In the GE model the geometrical update of the merge operation requires to traverse the *s-cells* of the removed boundary elements in order to erase them in the boundary image. In the HLE model, it is enough to traverse the *l-cells* of the border of these boundary elements in order to merge related embedded 2-maps. The geometrical update of the splitting of a volume into two adjacent sub-volumes consists in adding a boundary element separating these sub-volumes. This operation is the inverse of the previous one. The splitting of a volume into more than two sub-volumes is more complex and consists in applying the initial construction described in Sec. 3 and 4, but on the split volume instead of on the whole image.

5.2 Advantages of Each Model

Since the GE and HLE models mainly differ in the way of coding the geometrical embedding, pure topological operations are totally equivalent in both of them. But for most of operations that have to access to the geometry, we can establish which is the most efficient model. So advantages and inconveniences of each one are strongly linked to the use and the kind of operations from the point of view of segmentation.

The GE model is more interesting for volume traverse because it explicitly describes each border cell and locates it in the global embedding. So while moving inside the image, it is easy to detect when a volume boundary is crossed. Therefore, point localization is quite simple both on algorithmic and time complexity aspects, unlike with the HLE model in which a localization requires heavy processings. The GE model is also more efficient for global geometrical characteristics determination like the size of a volume. Boundary and volume reconstructions are also very simple. All of these features can be computed using a simple 2D or 3D floodfill algorithm. In the other model, volumetric computation and volume reconstruction are quite difficult operations because the geometrical description is more concise and does not encode explicitly all the *s-cells*. Last, the split of an image by a discrete plane (useful operation in medical imaging) is also more efficient with a global embedding because intersections can be found locally by exploring the plane surface, whereas with the HLE there is no other way but computing the intersection of each 2-face with the plane.

On the other hand the HLE model has no restriction on segmented images, the simplicity of the construction makes its implementation easier, and the multilevel representation allows us to define multi-level operations. Actually, each operation can be decomposed in several levels, and each level is made of only a few number of basic processes. By the fact that most constituent elements of boundaries are given implicitly, local operations are lighter than with the GE model. This is particulary obvious for the volume merging operation, in which we need to remove every surfel on the deleted face. This process is not needed in the HLE model. For the same reason, the inverse operation which consists in splitting a volume by inserting a face along consecutive edges located on its boundary is more efficient with the HLE model than with the GE one.

The interest of the integration of both models in a unique framework is of course to get the advantages of both models. But the problem is to avoid getting back inconveniences of each one. It seems difficult to achieve, but could be done by allowing to switch on or off the update of several parts of the geometry to prevent some operations from getting a bad efficiency. The goal is to obtain a more generic model, more adaptable to different applications and also to gather different connected works in order to test and combine several operations and algorithms and define new ones.

6   Integration of Both Models

It follows from the previous sections that both models presents their own advantages and drawbacks. It is thus natural to try to make both models converge in order to get the advantages of each one. A first solution consists in making both models collaborate, and maintaining in parallel both data structures. A second solution, more ambitious, is to integrate both approaches in a unique model.

Conversion between both models. The collaboration of both models has been investigated by defining the operation of conversions between each representation. These conversions have been partially implemented.

The conversion operations are shared out among topological operations and geometrical operations. For both models, the topology is described by a combinatorial map and an inclusion tree. We have seen in Sec. 2 how to switch between both kind of map. So the topological conversions are obvious. But we have to notice that the topological representation of an image is not unique, because of the edges that result from the intra-object cutting (for GE) and fictive edges (for HLE) can generally be variously placed. But except for the position of these particular edges, the conversion of the representation obtained using a model gives the representation of the other one.

The conversion for the geometrical part requires particular algorithms. We study both possible directions of conversion, basing on the example in Fig. 5.

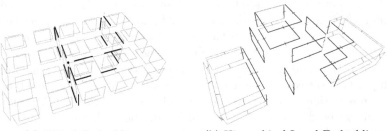

(a) Global Embedding.   (b) Hierarchical Local Embedding.

Fig. 5. GE and HLE correspondence

- Global Embedding → Hierarchical Local Embedding
  This conversion consists in traversing the s-cells of each topological face, and building a geometry equivalent to a level 3 map by merging aligned cells, first coplanar faces and then collinear edges (see Fig. 6). Mergings that would cause disconnections should not be applied. To obtain a minimal description of the geometry, it can be necessary to shift fictive edges.
- Hierarchical Local Embedding → Global Embedding
  This conversion consists in processing one by one the faces of the embedding 2-map (which we know that they are plane) and to cut them in convex

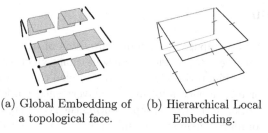

(a) Global Embedding of   (b) Hierarchical Local
a topological face.   Embedding.

Fig. 6. Extraction of HLE.

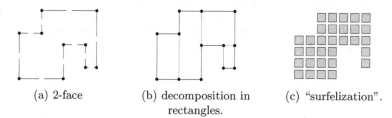

(a) 2-face   (b) decomposition in   (c) "surfelization".
rectangles.

Fig. 7. Extraction of GE.

polygons, or better in rectangles, from which we can enumerate all s-cells composing the 2-face (see Fig. 7); l-cells are given by traversing the 2-map borders (see Fig. 5), and each vertex of the topological map correspond to a p-cell .

Toward a unification of both models. It follows from the previous sections that the global approach is more efficient from the point of view of geometrical implementation and reconstructions and that the local approach is more simple and more general from the point of view of the topological construction. The interest of making both models converge and to unify data structures is thus to make all the features available in a unique environment and to optimize constructions and updates.

We think that it is interesting to maintain the global embedding which provides more efficient reconstructions and which size complexity is bounded. But the construction is not bounded in time and imposes restrictions on the set of segmented images. We conjecture that these drawbacks could be drawn by using the construction by levels developed in the HLE model. By this way we hope that it will be possible to maintain the global embedding on any segmented image with a construction bounded in time. This is the future direction of this unifying work. On the other hand it will be interesting to investigate other kinds of embeddings and to compare the efficiency of each solution from the point of view of specific applications.

7 Conclusion

In this paper we have compared two representations of 3D segmented images. Both models use equivalent topological representations which are 3D topological maps described by combinatorial maps. We have given the relation between the models and the correspondence allowing to convert each model into the other one. These models differ in the way of representing the geometry of 3D segmented images. We have compared geometrical embeddings in use in each model and we have defined correspondence between themselves. Finally we have proposed to unify these models in order to get the advantages of each of them. A promising direction is to adapt the map construction used in the HLE model to improve and generalize the construction of the boundary image. Another aspect that has not been considered in this paper but that must be developed is the contribution to the unification of both models to their extension to 4D images.

References

[1] Y. Bertrand, G. Damiand, and C. Fiorio. Topological encoding of 3d segmented images. In *Discrete Geometry for Computer Imagery*, number 1953 in Lecture Notes in Computer Science, pages 311–324, Uppsala, Sweden, december 2000.

[2] J.P. Braquelaire and L. Brun. Image segmentation with topological maps and inter-pixel representation. *Journal on Visual Communication and Image Representation*, 9(1):62–79, 1998.

[3] J.P. Braquelaire, P. Desbarats, J.P. Domenger, and C.A. Wüthrich. A topological structuring for aggregates of 3D discrete objects. In *Proc. of the 2nd IAPR-TC-15 Workshop on Graph Based Representation*, pages 193–202. Österreichische Computer Gesellschaft, 1999. ISBN 3-85804-126-2.

[4] L. Brun, J.P. Domenger, and J.P. Braquelaire. Discrete maps : a framework for region segmentation algorithms. In J.-M. Jolion and W.G. Kropatsch, editors, *Graph Based Representation in Pattern Recognition, GBR'97*, volume 12 of *Computing Suplement*, pages 83–92. SpingerWienNewYork, 1998.

[5] R. Cori. Un code pour les graphes planaires et ses applications. In *Astérisque*, volume 27. Soc. Math. de France, Paris, France, 1975.

[6] G. Damiand. *Définition et étude d'un modèle topologique minimal de représentation d'images 2D et 3D*. Thèse de doctorat, Université Montpellier II, décembre 2001.

[7] P. Desbarats. *Strucuration d'images segmentées 3D discrètes*. Thèse de doctorat, Université Bordeaux 1, décembre 2001.

[8] J. Edmonds. A combinatorial representation for polyhedral surfaces. *Notices of the American Mathematical Society*, 7, 1960.

[9] A. Jacques. Constellations et graphes topologiques. In *Combinatorial Theory and Applications*, volume 2, pages 657–673, 1970.

[10] P. Lienhardt. Topological models for boundary representation: a comparison with n-dimensional generalized maps. *Commputer Aided Design*, 23(1):59–82, 1991.

[11] W.T. Tutte. A census of planar maps. *Canad. J. Math.*, 15:249–271, 1963.

Tree Edit Distance from Information Theory

Andrea Torsello and Edwin R. Hancock

Dept. of Computer Science, University of York
Heslington, York, YO10 5DD, UK
atorsell@cs.york.ac.uk

Abstract. This paper presents a method for estimating the cost of tree edit operations. The approach poses the problem as that of estimating a generative model for a set of tree samples. The generative model uses the tree-union as the structural archetype for every tree in the distribution and assigns to each node in the archetype the probability that the node is present in a sample. A minimum descriptor length formulation is then used to estimate the structure and parameters of this tree model as well as the node-correspondences between trees in the sample-set and the tree model.

1 Introduction

The task of measuring the similarity or distance between pairs of graphs has attracted considerable interest over the past three decades [1,2,5,17,3]. It is an important problem since the availability of reliable distance information allows a number of practical problems involving graphs to be addressed. These include graph-matching, graph retrieval and graph clustering.

Some of the earliest work on the problem was presented by Shapiro and Haralick [19] who demonstrated how counting consistent subgraphs could lead to a metric distance between graphs. Fu and his co-workers [5,17] devoted considerable effort to extending the concept of string edit distance to graphs. The contribution here was to show how the distance between graphs could be gauged using a series of costs for edge and node relabeling, and for edge and node insertion or deletion. However, the estimation of the necessary costs remains an open problem.

This early work, drew its inspiration from structural pattern recognition, which was at the time a field in its infancy. Subsequent work, aimed to take a more principled approach to the problem by drawing on information theory and probabilistic models. For instance, adopting an information theoretic framework, Wong and You [25] have shown how the cost of edit operations can be measured using changes in entropy. The probabilistic analysis of the graph-matching problem has been the focus of recent activity. For instance, Christmas, Kittler and Petrou [4] have developed a probabilistic relaxation scheme which measures the compatibility of matching using a distribution function for pairwise attributes residing on the edges of graphs. Wilson and Hancock [24], have developed a discrete relaxation algorithm in which the distribution of correspondence errors and structural errors are modeled probabilistically.

E. Hancock and M. Vento (Eds.): IAPR Workshop GbRPR 2003, LNCS 2726, pp. 71–82, 2003.

Recently there has been renewed interest in placing the concept of graph edit distance on a more rigorous footing, One of the criticisms that can be leveled at the early work is that it lacks the formality of the treatment which underpins the string edit distance. Bunke and Kandel [3] have demonstrated the relationship between the size of the maximum common subgraph and edit distance under the assumption of uniform edit costs. Myers, Wilson and Hancock [14] have shown how string edit distance can be used to model the distribution of structural error in graph neighborhoods. Robles-Kelly and Hancock [16] have shown how to use eigenvectors of the adjacency matrix to convert graphs to strings, so that their similarity may be assessed using string edit distance.

However, despite this recent work the question of how to estimate edit costs remains largely unanswered. This can be viewed as a problem of machine learning, where the costs associated with edit operations can be estimated from the frequency of occurrence of the corresponding edit operation in a set of training data. In this paper, we describe an information theoretic framework which can be used to estimate tree edit distances. The approach is as follows. We pose the problem of learning the edit operations as that of constructing a union-tree over a set of example trees. Associated with each node in the union structure is a probability which indicates the frequency of the node in the tree sample. The merging of trees to form the union structure is cast as the problem of locating a structure that minimizes description length. The recovery of the optimal union structure involves three update steps. First, correspondences between the sample trees and the current union structure must be located. Second, the node probabilities must be updated, Thirdly, and finally, the set of edit operations that result in the optimal union must be identified. These three steps may each be formulated in terms of the description length. Moreover, the cost associated with the edit operations is related to the description length advantage, and this in turn is given by the node probabilities. Hence, by estimating node probabilities we learn the edit costs.

2   Tree Edit-Distance

The idea behind edit distance is that it is possible to identify a set of basic edit operations on the nodes and edges of a structure, and to associate with these operations a cost. The edit-distance is found by searching for the sequence of edit operations that will make the two graphs isomorphic with one-another and which have minimum cost. In previous work we have shown that the optimal sequence can be found using only structure reducing operations [23]. This can be explained by the fact that we can transform node insertions in one tree into node removals in the other. This means that the edit distance between two trees is completely determined by the subset of residual nodes left after the optimal removal sequence, or, equivalently, by the nodes that are in correspondence.

The edit-distance between two trees t and t' can be defined in terms of the matching nodes:

$$D(t, t') = \sum_{i \notin \mathrm{Dom}(\mathcal{M})} r_i + \sum_{j \notin \mathrm{Im}(\mathcal{M})} r_j + \sum_{<i,j> \in \mathcal{M}} m_{ij}. \tag{1}$$

Here r_i and r_j are the costs of removing i and j respectively, $\mathcal{M}$ is the set of pairs of nodes from t and t' that match, $m_{i,j}$ is the cost of matching i to j, and $\mathrm{Dom}(\mathcal{M})$ and $\mathrm{Im}(\mathcal{M})$ are the domain and image of the relation $\mathcal{M}$. Letting $\mathcal{N}^t$ be the set of nodes of tree t, the distance can be rewritten as:

$$D(t, t') = \sum_{u \in \mathcal{N}^t} r_u + \sum_{v \in \mathcal{N}^{t'}} r_v + \sum_{(u,v) \in \mathcal{M}} (m_{uv} - r_u - r_v).$$

Hence the distance is minimized by the set of correspondences that maximizes the utility

$$\mathcal{U}(\mathcal{M}) = \sum_{(u,v) \in \mathcal{M}} (r_u + r_v - m_{uv}). \tag{2}$$

3   Generative Tree Model

The aim of this paper is to learn the cost of the tree edit operations described above. We do this by fitting a generative model of tree structures to a set of sample-trees. This allows us to globally estimate the node correspondences solving the identification problem, and determine the modes of structural variation in the distribution of tree structures. To this end, we assume that there is an underlying "structure model", which determines a distribution of tree structures, and that each tree is a training sample drawn from that distribution. In this way edit operations are linked to sampling error, and their cost to the error probability. We, then, need a way to estimate the underlying structural model.

Consider the set or sample of trees $\mathcal{D} = \{t_1, t_2, \ldots, t_n\}$. Our aim is to learn a generative model for the distribution of trees in a pattern space. The tree model must be capable of capturing the structural variations for the sample trees using a probability distribution. The information on the structural variation can then be used to determine the cost of the edit operations. The resulting approach assigns low cost to operations associated to common variations, and higher cost to more unlikely variations.

In prior work, we have described how tree unions can be used as structural models for samples of trees [22]. However, the union is constructed so as to minimize tree-edit distance. Here we intend to use the union structure as a class model, and use the model to derive the distance. This approach extends the idea in two important ways. First, we pose the recovery of the union tree in an information theoretic setting. Second, we aim to characterize uncertainties in the structure by assigning probabilities to nodes.

A tree model $\mathcal{T}$ is a structural archetype derived from the tree-union over the set of trees constituting a class. Associated with the archetype is a probability distribution which captures the variations in tree structure within the class. Hence, the learning process involves estimating the union structure and the parameters of the associated probability distribution for the class model $\mathcal{T}$. As a prerequisite, we require the set of node correspondences $\mathcal{C}$ between sample trees and the union tree.

The learning process is cast into an information theoretic setting and the estimation of the required class models is effected using optimization methods. The quantity to be optimized is the descriptor length for the sample-data set $\mathcal{D}$. The parameters to be optimized include the structural archetype of the model $\mathcal{T}$ as well as the node correspondences $\mathcal{C}$ between samples and the archetype. The inter-sample node correspondences are not assumed to be known a priori. Since the correspondences are unknown, we must solve two interdependent optimization problems. These are the optimization of the union structure given a set of correspondences, and the optimization of the correspondences given the tree structure. These dual optimization steps are approximated by greedily merging similar tree-models.

The basic ingredients of our structural learning approach are:
1. A structural model of tree variation.
2. A probability distribution on the said model.
3. A structural optimization algorithm that allows us to merge two structural models in a way that minimizes the description length.

Hence, the structural model is provided by the tree-union of the set of samples, while the frequencies with which nodes from the sample set are mapped to nodes in the model provide the probability distribution. By adopting this information theoretic approach we demonstrate that the tree-edit distance, and hence the costs for the edit operations used to merge trees, are related to the entropies associated with the node probabilities. As a result, we provide a framework in which tree edit distances are learned. This has been a longstanding problem since Fu and his co-workers introduced the idea of edit distance in the early 1980's [17,5].

The basis of the proposed structural learning approach is a generative model of trees which allows us to assign a probability distribution to a sample of hierarchical trees. A hierarchical tree t is defined by a set of nodes $\mathcal{N}^t$ and a tree-order relation $\mathcal{O}^t \subset \mathcal{N}^t \times \mathcal{N}^t$ between the nodes. A tree-order relation $\mathcal{O}^t$ is an order relation with the added constraint that if $(x, y) \in \mathcal{O}^t$ and $(z, y) \in \mathcal{O}^t$, then either $(x, z) \in \mathcal{O}^t$ or $(z, x) \in \mathcal{O}^t$. A node b is said to be a *descendent* of a, or $a \rightsquigarrow b$, if $(a, b) \in \mathcal{O}^t$, furthermore, b descendent of a is also a *child* of a if there is no node x such that $a \rightsquigarrow x$ and $x \rightsquigarrow b$, that is there is non node between a and b in the tree-order.

Given this definition, we can construct a generative model for a subset of the samples $\mathcal{D}_c \subset \mathcal{D}$. This model $\mathcal{T}$ is an instance of a set of nodes $\mathcal{N}$. Associated with the set of nodes is a tree order relation $\mathcal{O} \subset \mathcal{N} \times \mathcal{N}$ and a set $\Theta = \{\theta^i, i \in \mathcal{N}\}$ of sampling probabilities θ^i for each node $i \in \mathcal{N}$.

$$\mathcal{T} = (\mathcal{N}, \mathcal{O}, \Theta) \tag{3}$$

A sample from this model is a hierarchical tree $t = (\mathcal{N}^t, \mathcal{O}^t)$ with node set $\mathcal{N}^t \subset \mathcal{N}$ and a node hierarchy $\mathcal{O}^t$ that is the restriction to $\mathcal{N}^t$ of $\mathcal{O}$. The probability of observing the sample tree t given the model tree $\mathcal{T}$ is:

$$P\{t|\mathcal{T}\} = \prod_{i\in\mathcal{N}^t} \theta^i \prod_{j\in(\mathcal{N}\backslash\mathcal{N}^t)} (1-\theta^j) \tag{4}$$

The model underpinning this probability distribution is as follows. First, we assume that the set of nodes $\mathcal{N}$ for the union structure $\mathcal{T}$ spans all the nodes that might be encountered in the set of sample trees. Second, we assume that the sampling error acts only on nodes, while the hierarchical relations are always sampled correctly. That is, if nodes i and j satisfy the relation $i\mathcal{O}j$, node i will be an ancestor of node j in each tree-sample that has both nodes. This assumption implies that two nodes will always satisfy the same hierarchical relation whenever they are both present in a sample tree. A consequences of this assumptions is that the structure of a sample tree is completely determined by restricting the order relation of the model $\mathcal{O}$ to the nodes observed in the sample tree. Hence, the links in the sampled tree can be seen as the minimal representation of the order relation between the nodes. The sampling process is equivalent to the application of a set of node removal operations to the archetypical structure $\mathcal{T} = (\mathcal{N}, \mathcal{O}, \Theta)$, which makes the archetype a union of the set of all possible tree samples.

The definition of the structural distribution assumes that we know the correspondences between the nodes in the sample tree t and the nodes in the class-model $\mathcal{T}$. When extracting a sample from the model this assumption obviously holds. However, given a tree t, the probability that this tree is a sample from the class model $\mathcal{T}$ depends on the tree, the model, but also on the way we map the nodes of the tree to the corresponding nodes of the model. To capture this correspondence problem, we define a map $\mathcal{C} : \mathcal{N}^t \to \mathcal{N}$ from the set $\mathcal{N}^t$ of the nodes of t, to the nodes of the model.

The mapping induces a sample-correspondence for each node $i \in \mathcal{N}$. The correspondence probability for the node i is

$$\phi(i|t,\mathcal{T},\mathcal{C}) = \begin{cases} \theta^i & \text{if } \exists j \in \mathcal{N}^t | \mathcal{C}(j) = i \\ 1-\theta^i & \text{otherwise.} \end{cases}$$

while the probability of sampling the tree t from the model $\mathcal{T}$ given the set of correspondences $\mathcal{C}$ is

$$\Phi(t|\mathcal{T},\mathcal{C}) = \begin{cases} \prod_{i\in\mathcal{N}} \phi(i|t,\mathcal{T},\mathcal{C}) & \text{if } \forall v,w \in \mathcal{N}^t, v \rightsquigarrow w \iff \mathcal{C}(v) \rightsquigarrow \mathcal{C}(w) \\ 0 & \text{otherwise.} \end{cases}$$

4   Minimizing the Descriptor Length

The generative tree model provides us with a probability distribution for tree structure. Given a set $\mathcal{D} = \{t_1, t_2, \ldots, t_n\}$ of sample trees, we would like to estimate the tree model $\mathcal{T}$ that generated the samples, and the correspondence $\mathcal{C}$ from the nodes of the sample trees to the nodes of the tree model.

Here we use a minimum descriptor length approach. The MDL principle [15] asserts that the model that best describes a set of data is that which minimizes the combined cost of encoding the model and the error between the model and the data.

By virtue of Shannon theorem, the cost of describing data $\mathcal{D}$ using model $\mathcal{T} = \{\mathcal{N}, \mathcal{O}, \Theta\}$ is $-\sum_{t \in \mathcal{D}} \log [\Phi(t|\mathcal{T}, \mathcal{C})] = -\mathcal{L}(\mathcal{D}|\mathcal{T}, \mathcal{C})$, while the cost incurred describing or encoding the model is $-\log [P(\mathcal{T})]$. Here we assume that the tree probability decays exponentially with the number of node. That is $P(\mathcal{T}) = \exp(-ld)$, where d is the number of nodes in $\mathcal{T}$. Using this prior, the cost incurred in describing the structure is proportional to the number of nodes and the per-node structural description cost is l.

Our model is described by the union structure $\mathcal{T} = \{\mathcal{N}, \mathcal{O}, \Theta\}$. The model consists of a set of nodes $\mathcal{N}$, an order relation $\mathcal{O}$ and a set of node probabilities $\Theta = \{\theta^i, i \in \mathcal{N}\}$, where θ^i is the probability for the node n in the union-tree indexed c. To describe or encode the fit of the model to the data, for each node in the model, we need to describe or encode whether or not the node was present in the sample.

Hence, given a model $\mathcal{T}$ consisting of d nodes, the descriptor length for the model, conditional on the set of correspondences is $\mathcal{C}$ is:

$$\mathrm{LL}(\mathcal{D}|\mathcal{T}, \mathcal{C}) = \sum_{i=1}^{d} [-\mathcal{L}(\mathcal{D}|\mathcal{T}, \mathcal{C}) + l] . \tag{5}$$

Our aim is to optimize the descriptor length with respect to two variables: the correspondence map $\mathcal{C}$ and the tree union model $\mathcal{T}$. These variables, though, are not independent. The reason for this is that they both depend on the node-set $\mathcal{N}$. A variation in the actual identity does not change the log-likelihood, hence the dependency to the node-set can be lifted by simply assuming that the node set is $Im(\mathcal{C})$, the image of the correspondence map. This can be done because, as we will see, nodes that are not mapped to do not affect the optimization process. With this simplification, the remaining variables are: the order relation $\mathcal{O}$, the set of sampling probabilities Θ, and the map $\mathcal{C}$.

Given $\mathcal{C}$, it is easy to minimize the descriptor length with respect to the remaining two sets of variables. The length of the description is minimized by any order relation $\mathcal{O}$ that is consistent with all the hierarchies for the sample trees (if any exists). Let $p_i(\mathcal{C})$ be the number of trees $t \in \mathcal{D}$ such that $\exists j | \mathcal{C}(j) = i$, that is there is a node that maps to i. Furthermore, let $m = \#\mathcal{D}$ be the number of trees in the data set, then the sampling probability θ^i for the node i that maximizes the likelihood function is

$$\theta^i = \frac{p_i(\mathcal{C})}{m}.$$

When the optimal sampling probabilities are substituted into the descriptor length, we have that

$$\hat{LL}(\mathcal{D}|\mathcal{C}) = m \sum_{i \in \mathcal{N}} \left[\frac{p_i(\mathcal{C})}{m} \log\left(\frac{p_i(\mathcal{C})}{m}\right) + \left(1 - \frac{p_i(\mathcal{C})}{m}\right) \log\left(1 - \frac{p_i(\mathcal{C})}{m}\right) \right] + nl =$$

$$- m \sum_{i \in \mathcal{N}} I(\theta^i) + nl, \quad (6)$$

where $I(\theta^i) = - \left[\theta^i \log(\theta^i) + (1 - \theta^i) \log(1 - \theta^i)\right]$ is the entropy of the sampling distribution for node i, and d is the number of nodes in $\mathcal{T}$. This equation holds assuming that there exists an order relation that is respected by every hierarchical tree in the sample set $\mathcal{D}$. If this is not the case then the descriptor length takes on the value ∞.

The structural component of the model is a tree union constructed from the trees in the sample $\mathcal{D}$ so as to maximize the likelihood function appearing in (6). In our previous work [22]. we have shown how the union tree may be constructed so that every tree in the sample set $\mathcal{D}$ may be obtained from it by using node removal operations alone. Hence every node in the tree sample is represented in the union structure. Moreover, the order-relations in the union structure are all preserved by pairs of nodes in the tree-samples in $\mathcal{D}$.

5   Computing the Model

Finding the global minimum of the descriptor length is an intractable combinatorial problem. Hence, we resort to a local search technique.

The main requirement of our description length minimization algorithm is that we can optimally merge two tree models. That is that we can find a structure from which it is possible to sample every tree previously assigned to the two models.

Interestingly, we can pose the model-merging problem as an instance of a particular minimum edit-distance problem.

Given two tree models $\mathcal{T}_1$ and $\mathcal{T}_2$, we wish to construct a union $\hat{\mathcal{T}}$ whose structure respects the hierarchical constraints present in both $\mathcal{T}_1$ and $\mathcal{T}_2$, and that also minimizes the quantity $LL(\hat{\mathcal{T}})$. Since the trees $\mathcal{T}_1$ and $\mathcal{T}_2$ already assign node correspondences $\mathcal{C}_1$ and $\mathcal{C}_2$ from the data samples to the model, we can simply find a map $\mathcal{M}$ from the nodes in $\mathcal{T}_1$ and $\mathcal{T}_2$ to $\hat{\mathcal{T}}$ and transitively extend the correspondences from the samples to the final model $\hat{\mathcal{T}}$ in such a way that $\hat{\mathcal{C}}(v) = \hat{\mathcal{C}}(w) \Leftrightarrow w = \mathcal{M}(v)$.

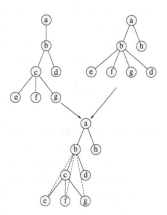

Fig. 1. The new model is obtained by merging matching nodes.

Reduced to the merge of two structures, the correspondence problem is reduced to finding the set of nodes in $\mathcal{T}_1$ and $\mathcal{T}_2$ that are in common. Starting with the two structures, we merge the sets of nodes that would reduce the descriptor length by the largest amount while still satisfying the hierarchical constraint.

That is we merge nodes v and w of $\mathcal{T}_1$ with node v' and w' of $\mathcal{T}_2$ respectively if and only if $v \rightsquigarrow w \Leftrightarrow v' \rightsquigarrow w'$, where $a \rightsquigarrow b$ indicates that a is an ancestor of b.

Let m_1 and m_2 be the number of tree samples from $\mathcal{D}$ that are respectively assigned to $\mathcal{T}_1$ and $\mathcal{T}_2$. Further let p_v and $p_{v'}$ be the number of times the nodes v and v' in $\mathcal{T}_1$ and $\mathcal{T}_2$ are respectively in correspondence with nodes of trees in the data sample $\mathcal{D}$. The sampling probabilities for the two nodes, if they are not merged, are $\theta v = \frac{p_v}{m_1+m_2}$ and $\theta v' = \frac{p'_v}{m_1+m_2}$ respectively, while the sampling probability of the merged node is $\theta vv' = \frac{p_v+p_{v'}}{m_1+m_2}$. Hence, the descriptor length advantage obtained by merging the nodes v and v' is:

$$\mathcal{A}(v,v') = (m_1 + m_2)\left[I(\theta v) + I(\theta v') - I(\theta vv')\right] + l. \tag{7}$$

This implies that the set of merges $\mathcal{M}$ that minimizes the descriptor length of the combined model maximizes the advantage function

$$\mathcal{A}(\mathcal{M}) = \sum_{(v,v')\in\mathcal{M}} \mathcal{A}(v,v') = \sum_{(v,v')\in\mathcal{M}} \left[(m_1 + m_2)\left[I(\theta v) + I(\theta v') - I(\theta vv')\right] + l\right]. \tag{8}$$

Assuming that the class archetypes $\mathcal{T}_1$ and $\mathcal{T}_2$ are trees, finding the set of nodes to be merged can be transformed into a tree-edit distance problem. That is, assigning particular costs to node removal and matching operations, the set of correspondences that minimize the edit distance between the archetypes of $\mathcal{T}_1$ and $\mathcal{T}_2$ also maximizes the utility of the merged model. The costs that allowed the problem to be posed as an edit distance problem are $r_v = (m_1+m_2)I(\theta v)+l$ for the removal of node v, and $m_{vv'} = (m_1 + m_2)I(\theta vv') + l$ for matching node v with node v'.

With these edit costs, the utility in (2) assumes the same value of the advantage in descriptor length:

$$\mathcal{U}(\mathcal{M}) = \sum_{(u,v)\in\mathcal{M}} \left[(m_1 + m_2)(I(\theta u) + I(\theta v) - I(\theta uv)) + l\right]. \tag{9}$$

Since the combinatorial problem underlying both edit-distance and model merge share the same hierarchical constraints and objective function, the solution to one problem can be derived from the solution to the other. In particular the set of common nodes obtained through the edit-distance approach is equal to the set of nodes to be merged to optimally merge the tree-models.

We initialize our algorithm by calculating the sum of the descriptor length of using a per tree-sample in $\mathcal{D}$. The descriptor length is given by $l\sum_{t\in\mathcal{D}} \#\mathcal{N}^t$, where $\#\mathcal{N}^t$ is the number of nodes in the tree-sample t. For each pair of initial union model we calculate the merged structure and the advantage in descriptor length arising from merging the models. From the set of potential merges, we can identify the one which reduces the descriptor cost by the greatest amount. At this point we calculate the union and the advantage in descriptor cost obtained by merging the newly obtained model with each of the remaining model, and we iterate the algorithm until we merge every tree model.

6   Computing the Distance

To find the set correspondences that minimizes the edit distance between two trees we make use of two results presented in [21]. We call $\Omega(t)$ the closure of tree t, $E_i(t)$ the edit operation that removes node i from t and $\mathcal{E}_i(\Omega(t))$ the equivalent edit operation that removes i from the closure. The first result is that edit and closure operations commute: $\mathcal{E}_i(\Omega(t)) = \Omega(E_i(t))$. For the second result we need some more definitions: We call a subtree s of $\Omega(t)$ *obtainable* if for each node i of s if there cannot be two children a and b so that (a, b) is in $\Omega(t)$. In other words, for s to be obtainable, there cannot be a path in t connecting two nodes that are siblings in s. We can, now, introduce the following:

Theorem 1. *A tree $\hat{t}$ can be generated from a tree t with a sequence of node removal operations if and only if $\hat{t}$ is an obtainable subtree of the directed acyclic graph $\Omega(t)$.*

By virtue of the theorem above, the node correspondences yielding the minimum edit distance between trees t and t' form an obtainable subtree of both $\Omega(t)$ and $\Omega(t')$, hence we reduce the problem to the search for a common substructure that maximizes the utility: the maximum common obtainable subtree (MCOS). That is, let O be the set of matches that satisfy the obtainability constraint, the node correspondence that minimized the edit distance is

$$\mathcal{M}^* = \operatorname*{argmax}_{\mathcal{M} \in O} \mathcal{U}(\mathcal{M}). \tag{10}$$

The solution to this problem is obtained by looking for the best matches at the leaves of the two trees, and by then propagating them upwards towards the roots. Let us assume that we know the utility of the best match rooted at every descendent of nodes i and j of t and t' respectively. To propagate the matches to i and j we need to find the set of siblings with greatest total utility. This problem can be transformed into a maximum weighted clique problem on a derived structure and then approximated using a heuristical algorithm. When the matches have been propagated to all the pairs of nodes drawn from t and t', the set of matches associated with the maximum utility give the solution to the maximum common obtainable subtree problem, and hence the edit-distance. We refer to [21] for a detailed explanation of the approach.

7   Experimental Results

The aim of this session is to illustrate the behavior of the tree merging algorithm and the effects of the estimated distances. To meet this goal we have randomly generated a prototype tree model and, from this tree, we generated some tree-samples. The procedure for generating the random tree prototype was as follows: we commence with an empty tree (i.e. one with no nodes) and we iteratively add the required number of nodes. At each iteration nodes are added as children of one of the existing nodes. The parents are randomly selected with uniform

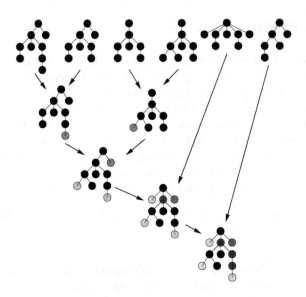

Fig. 2. Merging sample trees into a tree-model.

probability from among the existing nodes. The sampling probability of the nodes is assigned from a uniform distribution.

Figure 2 illustrates an example merge of 6 sample trees. The figure shows the structural archetype of the merged models after each stage. The color of the nodes in the tree models represents the sampling probability of that node: nodes sampled with higher probability are displayed in darker colors, while lighter colors are associated to nodes with low sampling probability.

In order to asses the quality of the extracted distance we compare clusters defined by performing pairwise clustering of the distance matrix [21,12]. We evaluate the approach on the problem of shock tree matching. We calculate the pairwise edit distance between shock-trees extracted from a database of 16 shapes.

c) Mixture of tree models

b) Pairwise clustering from edit-distance

Fig. 3. Comparison of clusters obtained from non-attributed edit-distance and mixture of trees.

Figure 3 shows the clusters using the two distance measures. The first column shows the clusters extracted fitting the generative model, while the second column displays the cluster extracted from the edit-distances with uniform cost.

While there is some merge and leakage, the cluster extracted by fitting a generative tree model clearly outperform those obtained using uniform edit-cost. The second to last cluster extracted using the tree-model approach deserves some explanation: the structure of the shock-trees of the tools in the cluster is identical. Hence the model, which uses only structural information, correctly clusters the shock-trees together.

8 Conclusions

This paper presented a method for estimating of the cost of tree edit operations. The approach poses the problem as that of estimating a generative model for a set of tree samples. The generative model uses the tree-union as the structural archetype of trees in the distribution and assigns to each node in the archetype the probability that the node is present in a sample. A minimum descriptor length formulation is then used to estimate the structure and parameters of this tree model as well as the node-correspondences between trees in the sample-set and the tree model. Finally, we use the extracted correspondences and sampling probability to set the removal cost for each node.

References

1. H. G. Barrow and R. M. Burstall, Subgraph isomorphism, matching relational structures and maximal cliques, *Inf. Proc. Letters*, Vol. 4, pp.83, 84, 1976.
2. H. Bunke and G. Allermann, Inexact graph matching for structural pattern recognition, *Pattern Recognition Letters*, Vol 1, pp. 245–253, 1983.
3. H. Bunke and A. Kandel, Mean and maximum common subgraph of two graphs, *Pattern Recognition Letters*, Vol. 21, pp. 163–168, 2000.
4. W. J. Christmas, J. V. Kittler, and M. Petrou, Structural matching in computer vision using probabilistic relaxation, *PAMI*, Vol. 17(8), pp. 749–764, 1995.
5. M. A. Eshera and K-S Fu, An image understanding system using attributed symbolic representation and inexact graph-matching, *PAMI*, Vol 8, pp. 604–618, 1986.
6. N. Friedman and D. Koller, Being Bayesian about Network Structure, *Machine Learning*, to appear, 2002
7. L. Getoor et al., Learning Probabilistic models of relational structure, in *8th Int. Conf. on Machine Learning*, 2001.
8. D. Heckerman, D. Geiger, and D. M. Chickering, Learning Bayesian networks: the combination of knowledge and statistical data, *Machine Learning*, Vol. 20(3), pp. 197–243, 1995.
9. S. Ioffe and D. A. Forsyth, Human Tracking with Mixtures of Trees, *ICCV*, Vol. I, pp. 690–695, 2001.
10. X. Jiang, A. Muenger, and H. Bunke, Computing the generalized mean of a set of graphs, in *Workshop on Graph-based Representations, GbR'99*, pp. 115–124, 2000.
11. T. Liu and D. Geiger, Approximate Tree Matching and Shape Similarity, *ICCV*, pp. 456–462, 1999.
12. B. Luo, et al., A probabilistic framework for graph clustering, in *CVPR*, pp. 912–919, 2001.
13. M. Meilă. *Learning with Mixtures of Trees*. PhD thesis, MIT, 1999.

14. R. Myers, R. C. Wilson, E. R. Hancock, Bayesian graph edit distance, *PAMI*, Vol. 22(6), p. 628–635, 2000.
15. J. Riassen, Stochastic complexity and modeling, *Annals of Statistics*, Vol. 14, pp. 1080–1100, 1986.
16. A. Robles-Kelly and E. R. Hancock, String edit distance, random walks and graph matching, in *Structural, Syntactic, and Statistical Pattern Recognition*, pp.104–112, LNCS 2396, 2002.
17. A. Sanfeliu and K. S. Fu. A distance measure between attributed relational graphs for pattern recognition. *IEEE Transactions on Systems, Man and Cybernetics*, 13:353–362, 1983.
18. S. Sclaroff and A. P. Pentland, Modal matching for correspondence and recognition, *PAMI*, Vol. 17, pp. 545–661, 1995.
19. L. G. Shapiro and R. M. Haralick, Structural descriptions and inexact matching, *PAMI*, Vol. 3(5), pp. 504–519, 1981.
20. A. Shokoufandeh, S. J. Dickinson, K. Siddiqi, and S. W. Zucker, Indexing using a spectral encoding of topological structure, in *CVPR*, 1999.
21. A. Torsello and E. R. Hancock, Efficiently computing weighted tree edit distance using relaxation labeling, in *EMMCVPR*, LNCS 2134, pp. 438–453, 2001.
22. A. Torsello and E. R. Hancock, Matching and embedding through edit-union of trees, in *ECCV*, LNCS 2352, pp. 822–836, 2002.
23. A. Torsello and E. R. Hancock, Computing approximate tree edit-distance using relaxation labeling, *Pattern REcognition Letters*, Vol. 24(8), pp. 1089–1097
24. R. C. Wilson and E. R. Hancock, Structural matching by discrete relaxation, *PAMI*, 1997.
25. A. K. C. Wong and M. You, Entropy and distance of random graphs with application to structural pattern recognition, *PAMI*, Vol. 7(5), pp. 599–609, 1985.

Self-Organizing Graph Edit Distance

Michel Neuhaus and Horst Bunke

Department of Computer Science, University of Bern,
Neubrückstrasse 10, CH-3012 Bern, Switzerland
{mneuhaus,bunke}@iam.unibe.ch

Abstract. This paper addresses the issue of learning graph edit distance cost functions for numerically labeled graphs from a corpus of sample graphs. We propose a system of self-organizing maps representing attribute distance spaces that encode edit operation costs. The self-organizing maps are iteratively adapted to minimize the edit distance of those graphs that are required to be similar. To demonstrate the learning effect, the distance model is applied to graphs representing line drawings and diatoms.

1 Introduction

Graphs are a powerful and universal concept to represent structured objects. Particularly in pattern recognition and related areas they have gained significant amounts of attention recently [1,2,3]. If both unknown objects and known prototypes of pattern classes are represented by graphs, then the problem of pattern classification turns into the problem of measuring the similarity (or, equivalently, the distance) of graphs. A number of graph similarity measures haven been proposed in the literature [4,5,6]. Among those measures, graph edit distance has become quite popular [7,8,9]. In graph edit distance, a set of graph edit operations are introduced. These edit operations are used to transform one of a pair of given graphs into the other. The edit distance of two graphs is defined as the length of the shortest sequence of edit operations required to transform one graph into the other. A more powerful graph distance model is obtained if a cost is assigned to each edit operation. The costs are to be defined in such a way that edit operations that are very likely to occur have a low cost, while the cost of infrequent edit operations is high. Using such a cost model, the edit distance now becomes the cost of the cheapest sequence of edit operations transforming one of the given graphs into the other. Graph edit distance is very powerful from the theoretical point of view. However, it has a severe practical limitation in that no automatic procedures are available to automatically infer edit operation costs based on a sample set of graphs. In all applications reported in the literature, the edit costs are derived manually, using knowledge of the problem domain and heuristics. Obviously, such a procedure is not feasible in complex tasks.

In the present paper we propose a procedure to infer the costs of a set of edit operations automatically from a sample set of graphs. We focus on a special class of graphs, which is characterized by the existence of node and/or edge labels, where each label is a vector from the n-dimensional space of real numbers.

E. Hancock and M. Vento (Eds.): IAPR Workshop GbRPR 2003, LNCS 2726, pp. 83–94, 2003.

This class of graphs is not universal, but suitable for a wide range of problems. In particular it is useful in applications where the objects under consideration are geometrical entities that are characterized by quantities such as location or spatial extent. Our approach to learning the cost of graph edit operations is based on the concept of self-organizing maps (SOM) [10]. The basic idea is to represent the distribution of the node and edge labels occurring in a population of graphs through a SOM, where distances of labels correspond to edit operation costs. The location of the labels are iteratively changed in a learning procedure such that the cost of frequent operations are reduced.

In the following section, we give a brief introduction to SOMs. Then the proposed procedure for learning the costs of graph edit operations is presented in Section 3. To show the feasibility of the model introduced in this paper, experimental results are reported in Section 4. Finally conclusions are drawn in Section 5.

2   Self-Organizing Maps

Self-organizing maps (SOMs) aim at modeling the structure of a sample set of patterns by means of an unsupervised learning process [10]. A SOM is a two-layer neural network with neurons representing map vertices. The first layer selectively forwards input network stimuli to neurons of the second layer. The second layer (the competitive layer) contains a finite collection of mutually connected neurons. These neurons are usually arranged in a homogeneous grid structure. A weight vector is assigned to each neuron specifying the neuron's position in the underlying pattern space.

Upon feeding a randomly chosen sample vector into the network, the second-layer neurons compete for the input element's position. The closest neuron and its neighbors are shifted towards the input vector to better represent the input element. The intensity of the adjustment and the radius of the area of influence decrease over time, so that the SOM eventually reaches a stable state. Thus the competitive neurons tend to migrate to areas with many input elements.

Two distance measures are defined in the competitive layer: the standard Euclidean distance d_v between neuron weight vectors, and the original distance d_n between neurons defined by means of neighborhood relations. Distance d_n is usually chosen to be the Euclidean distance in the initial grid structure. The winner neuron is determined with respect to distance d_v, whereas the neurons in the neighborhood are found using distance d_n.

Initial values of the intensity c_{init}, neighborhood radius r_{init}, and the decrease rates δc and δr have to be chosen. Furthermore, it is necessary to define a nonincreasing activation function $f_n : R^+ \to R^+$ such that $f_n(0) = 1$, $f_n(x) = 0$ for $x > 1$. In the edit distance model presented in Section 3 we choose the activation function given by

$$f_n(x) = \left(e^{-4x^2} - xe^{-4}\right) \cdot 1_{[0,1]}(x) \ ,$$

where $1_{[0,1]}(x) = 1$ for $x \in [0,1]$ and $1_{[0,1]}(x) = 0$ otherwise.

Algorithm 1 Self-organizing learning procedure

Input: Self-organizing map and the set of input vectors.
Output: Trained map.
Parameters: Initial neighborhood radius $r_{init} \in R^+$ with decrease rate $\delta r \in (0,1)$, initial intensity $c_{init} \in R^+$ with decrease rate $\delta c \in (0,1)$.

$r = r_{init}$;
$c = c_{init}$;
Initialize competitive neuron vectors with random or predetermined values;
while terminating condition does not hold true **do**
    Randomly choose a vector v_{in} from the set of all input vectors;
    Determine neuron n_{win} whose vector v_{win} has the smallest Euclidean distance $d_v(v_{win}, v_{in})$ to v_{in};
    for all competitive neurons n with vector v **do**
      $f = c \cdot f_n(d_n(v, v_{win})/r)$;
      $v = v + f \cdot (v_{in} - v)$
    end for
    $r = r \cdot \delta r$;
    $c = c \cdot \delta c$
end while

From the function defined above, we get an activation value $f_n(d_n(v, v')/r)$ for two neural weights v and v', such that the activation for two identical vectors is 1 and the activation for any two neurons with a distance greater than r is 0. The value of the activation function obviously decreases when the distance between the vectors v and v' increases. The activation value $f_n(d_n(v, v_{win})/r)$ controls the influence of the winning neuron with vector v_{win} on a neuron with vector v. The neighborhood radius r limits the affected network area, and the intensity factor c additionally controls the amount of shifting, so that a gradual stablilization of the network is obtained. Algorithm 1 describes the general self-organizing learning procedure used in this paper. There are a number of possible termination criteria. The learning procedure is often terminated when the network stabilizes, after a certain number of runs through the while-loop have been completed, or when an application-specific termination criterion is met.

3   Edit Distance Model

The Euclidean distance matches the intuitive perception of a distance measure very well. Therefore, the Euclidean metric is often used as a similarity measure for node and edge labels in a graph that represent point coordinates or other positional entities in a multidimensional real space. The cost of substituting a node or an edge with another is defined as the Euclidean distance between the two corresponding label vectors. Moreover, insertion and deletion costs are defined to be constant or are computed by means of some application dependent function. Examples of such an approach have been reported in [11,12]. For a given application, however, it may be hard to determine optimal weights other than in an exhaustive search, and this approach may become computationally infeasible. Moreover, such a definition of edit distance may only inadequately be able to deal with certain distortions or transformations of labels, as it heavily relies on the fact that attributes are related in a Euclidean way. Some areas of the Euclidean space may be more important than others in the discrimination of different classes. But the Euclidean metric does not distinguish between areas of low importance and highly relevant areas. If these important areas were known, the Euclidean space could be stretched or shrunk to emphasize relevant areas in an attempt to improve the model's discrimination power. Such a distortion is the main idea on which the edit cost model proposed in this paper relies.

The n-dimensional Euclidean space can be represented by a sampling grid with equally spaced vertices. Such a grid may be interpreted as a mapping from the Euclidean space into itself, where points not coinciding with sampling points are linearly interpolated. A SOM with such a grid structure can be regarded as a sampling grid. Applying our learning algorithm, Algorithm 1, to a SOM, the sampling grid, and simultaneously the represented measuring space, is deformed. The map is modified such that grid points are attracted by input vectors. That is, the space is deformed so as to better represent the structure of the vectors occurring in the input patterns. If we measure, after the training phase, the distance between vectors with the standard Euclidean metric, then vectors that occur frequently will have a smaller distance than unfrequent ones.

To define an edit distance model, we consider node and edge distances independently of each other and provide one map for each type of edit operation, i.e. there is one map for node substitution, node deletion, node insertion, edge substitution, edge deletion, and edge insertion.

3.1   Distance Learning

Let us assume that two graphs, G and G', are given that are known to be similar or, equivalently, belong to the same class. The least expensive sequence of edit operations $(e_1, \ldots, e_l)$ transforming G into G' defines the edit distance $d(G, G')$. In order to reduce the transformation cost, we individually apply all involved edit operations, $e_1, \ldots, e_l$, to the respective SOMs. Since all node or edge labels have the same number of attributes, the map dimensions are clearly defined.

A substitution operation specifies the translation of one point in the map to another. For example, the substitution $\mathbf{x} \to \mathbf{y}$ can be interpreted as moving point $\mathbf{x} \in R^n$ to location $\mathbf{y} \in R^n$. If this operation occurs frequently, we can reduce its cost by drawing points $\mathbf{x}$ and $\mathbf{y}$ closer together. Given vectors $\mathbf{x}$ and $\mathbf{y}$, the grid points closest to either one of these vectors can easily be determined. The areas around these winner points are then moved in the other winner's direction. The neighborhood function f_n controls which nodes are affected. The learning parameters of the SOM are chosen to alter the map only slightly, so that the sampling grid neither collapses into a single point nor is torn apart. Algorithm 2 shows the details of this procedure for the node substitution map.

Algorithm 2 Node substitution distance learning for G and G'

Input: Node substitution label space SOM, a pair of graphs G and G'.
Output: Trained label space map.
Parameters: Initial neighborhood radius $r_{init} \in R^+$ with decrease rate $\delta r \in (0, 1)$, initial intensity $c_{init} \in R^+$ with decrease rate $\delta c \in (0, 1)$.

$r = r_{init}$
$c = c_{init}$
Initialize competitive neuron vectors with regular grid vertices;
while terminating condition does not hold true **do**
    Sequence $(e_1, \ldots, e_l)$ is the least costly transformation from G into G';
    for all node substitutions $e = \mathbf{x} \to \mathbf{y}$ in $(e_1, \ldots, e_l)$ **do**
        Vector $\mathbf{x}_{win}$ is the closest neuron to $\mathbf{x}$;
        for all competitive neurons $\mathbf{v}$ **do**
            $d_{grid} = d_n(\mathbf{v}, \mathbf{x}_{win})$;
            $f = c \cdot f_n(d_{grid}/r)$;
            $\mathbf{v} = \mathbf{v} + f \cdot (\mathbf{y} - \mathbf{v})$
        end for
        Vector $\mathbf{y}_{win}$ is the closest neuron to $\mathbf{y}$;
        for all competitive neurons $\mathbf{v}$ **do**
            $d_{grid} = d_n(\mathbf{v}, \mathbf{y}_{win})$
            $f = c \cdot f_n(d_{grid}/r)$
            $\mathbf{v} = \mathbf{v} + f \cdot (\mathbf{x} - \mathbf{v})$
        end for
        $r = r \cdot \delta r$
        $c = c \cdot \delta c$
    end for
end while

Deletion and insertion operations occurring in sequence $(e_1, \ldots, e_l)$ are treated in a similar way. There exists one individual SOM for each type of edit operation. For details the reader is referred to [13].

In order to actually apply the learning procedure sketched above, a sample set of graphs is needed. In the experiments described in Section 4, the class of each sample graph is known and pairs of graphs, G and G', from the same class are used in Algorithm 2.

3.2 Edit Operation Costs

Once a SOM for each type of edit operation has been trained, as described in Section 3.1, it completely encodes the costs of the corresponding edit operations. The cost of a substitution is defined to be proportional to the deformed distance in the substitution map. Similarly, the cost of an insertion (deletion) operation is proportional to the average deformed distance from the winner grid vertex to its connected neighbors. The proportion factors of all edit operations are initially set and remain constant during the learning. Due to the design of the learning procedure, distances between pairs of training graphs tend to decrease.

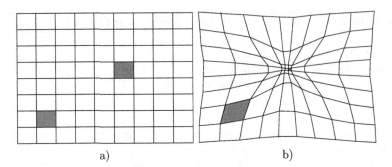

a)   b)

Fig. 1. Self-organizing map a) before and b) after learning

As an example, consider the SOM depicted in Fig. 1. Let the rectangles in Fig. 1a have width and height 1. The cost of a node substitution $\mathbf{x} \to \mathbf{y}$ from the lower left to the upper right corner of either gray rectangle is proportional to the diagonal distance, thus $c(\mathbf{x} \to \mathbf{y}) = 1.41\beta_{ns}$, where β_{ns} is the proportion factor of a node substitution. As indicated by Fig. 1b, the learning algorithm emphasizes the center area of the net. After learning, the area of the left gray polygon has evidently increased, whereas the right polygon collapsed into a tiny area. The same diagonal substitutions now cost $2.20\beta_{ns}$ for the left and $0.22\beta_{ns}$ for the right marked polygon. Thus the SOM has learned to assign low costs to substitutions in the center area.

4   Experimental Results

To evaluate the proposed self-organizing distance, two applications are considered. In the first application simulated data are used, while the second application is based on real data. First, we apply the learning procedure to graphs extracted from line drawings of capital letters. This application is similar to the one described in [12]. A drawing consisting of straight lines can be transformed into a graph by representing each end point of a line by a node (with a two-dimensional attribute specifying its position) and the line itself by an edge (without attributes). A corpus of sample graphs is generated by distorting copies of an ideal prototype. An example of a line drawing representing letter A is shown in Fig. 2. Notice that the degree of distortion is significantly increased as we go from Fig. 2a to b, c, d, and e. Not only line substitutions (i.e., change in length and angle of a line segment), but also deletions and insertions, which result in a change of the number of line segments, are applied.

To evaluate a given distance model, eight cluster validation indices (Calinski-Harabasz, C-index, Goodman-Kruskal, Dunn, Davies-Bouldin, Rand, Jaccard, and Fowlkes-Mallow, see [14]) are computed, normalized, and averaged. Each of these indices provides a measure of quality of a given clustering of a set of objects. In order to achieve improved robustness, the average of these indices is used. Choosing pairs of sample graphs, G and G', from the same class (i.e., cluster), their edit distance is minimized as described in Section 2. Hence the

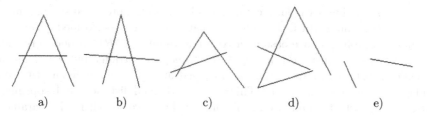

Fig. 2. Line drawing samples a) letter prototype and b)-e) distorted copies

application of the SOM learning procedure should lead to an improvement of the validation indices, in particular of the Average index. The Average index falls into the interval $[0, 1]$, where low values indicate a good clustering of the sample graphs.

The parameters of the self-organizing edit distance include the initial neighborhood radius r_{init} and decrease rate δr, the initial intensity c_{init} and decrease rate δc, and the weight factors $\beta_{ns}, \beta_{ni}, \beta_{nd}, \beta_{es}, \beta_{ei}, \beta_{ed}$ for each of the individual edit operations node substitution, node insertion, node deletion, edge substitution, edge insertion, and edge deletion. In the first experiment described in this paper all parameter values but c_{init} were fixed: $\beta_x = 1, r_{init} = 1, \delta_r = \delta_c = 0.9$. The intensity parameter c_{init} was varied and the Average index was measured over a certain number of iterations. The results of this experiment are shown in Fig. 3. The underlying sample set of graphs consists of distorted copies of letters L, I, and T. We note that for some values of c_{init} the Average index continuously decreases while for other values there is a local optimum. Hence the use of an independent validation set is advisable. Nevertheless, Fig. 3 clearly shows that during the SOM learning process, the edit costs are adapted so as to better model the given population of graphs.

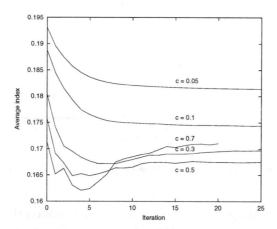

Fig. 3. Self-organizing distance learning with variable learning intensity

In the second experiment a comparison between the proposed self-organizing edit distance and a static Euclidean distance model is performed. The static Euclidean graph distance assigns a constant value $c_1 \in R^+$ to node insertions and deletions, a constant value $c_2 \in R^+$ to edge insertions and deletions, the Euclidean distance between the two involved attributes to node substitutions, and cost 0 to edge substitutions. This statically defined distance can be optimized for a sample set of graphs by means of an exhaustive search in the parameter space $R^+ \times R^+$. For instance, if node substitutions occur very frequently in the sample generation process, the node insertion and deletion weight c_1 is expected to tend to higher values for optimal clusterings, so that substitution costs are reduced. Note that this distance model is explicitly defined with the specific line drawing application in mind. The static Euclidean and the self-organizing model are applied to a sample set of graphs consisting of fifteen letters with three distorted copies and one prototype per class. The optimal parameters of the static Euclidean model are found to produce an Average index value of 0.275. The illustration of the self-organizing learning in Fig. 4 shows that most instances of the self-organizing model perform superior to the Euclidean model. In this experiment fixed parameter values $r_{init} = 1, c_{init} = 0.1, \delta_r = \delta_c = 0.9, \beta_{ns} = \beta_{ni} = \beta_{nd} = 1, \beta_{es} = 0$ are used, but $\beta_{ei} = \beta_{ed}$ is varied. Although the Euclidean model is particularly customized for the line drawing application and optimized for the given sample, the learning distance model achieves better Average indices for almost all configurations.

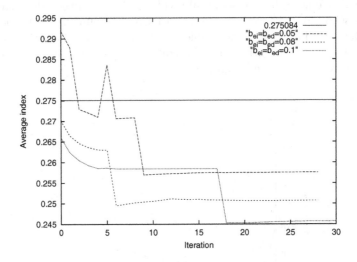

Fig. 4. Average index of the self-organizing distance learning

The static Euclidean edit distance crucially depends on the Euclidean relation between attributes, whereas the self-organizing distance is defined for more universal graphs with the capability to adapt the distance measure to a certain extent to some underlying specific distortion model. To verify the training effect, a homogeneous deterministic transformation is applied to the line drawings in the third experiment. We employ a shearing operation in the direction of the y-axis, defined by $f_\alpha(x, y) = (x, y + \alpha x)$, to deform a sample set of line drawings consisting of ten distorted copies of the three letter classes A, E, and X. An illustration of the shearing effect on an instance of letter A is provided in Fig. 5. The Average index for the Euclidean and the self-organizing models as a function of the shearing angle α are depicted in Fig. 6. For almost all distortion factors, the learning distance performs better. For higher values of α, the Euclidean distance fails to correctly model the distortions. The self-organizing distance is obviously less vulnerable to the considered type of distortions.

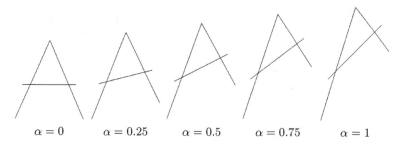

$\alpha = 0$ $\qquad$ $\alpha = 0.25$ $\qquad$ $\alpha = 0.5$ $\qquad$ $\alpha = 0.75$ $\qquad$ $\alpha = 1$

Fig. 5. Sheared letter A with shearing factor α

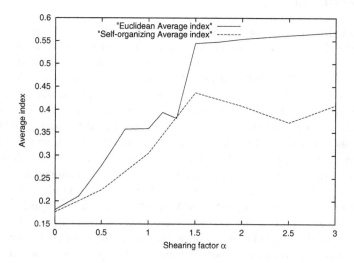

Fig. 6. Best Average index of the Euclidean and self-organizing distance models

To test the self-organizing graph edit distance on a real-world application, the learning model is applied to diatom graphs. Diatoms are unicellular algae occurring in water and humid places where light provides the basis for photosynthesis [15]. In an attempt to automatically identify diatoms, diatom images have been segmented into regions with homogeneous texture and then represented as numerically attributed graphs [11]. A subset of five diatom classes, including the two species with the highest error rates reported in [11], with five elements per class, are chosen. Part of the confusion matrix reported in [11] is presented in Table 1a. The results obtained with the self-organizing edit distance proposed in this paper are given in Table 1b. The learning procedure achieves to correct six out of eight misclassifications.

Table 1. Diatom recognition a) reported confusion matrix and b) self-organizing model confusion matrix (Cymbella hybrida, Cymbella subequalis, Gomphonema sp.1, Parlibellus delognei, and Stauroneis smithii)

Ch	Cs	Gs	Pd	Ss
5	–	–	–	–
–	5	–	–	–
3	1	1	–	–
–	–	–	5	–
1	–	–	3	1

a)

Ch	Cs	Gs	Pd	Ss
5	–	–	–	–
–	5	–	–	–
–	–	5	–	–
–	–	–	5	–
–	–	1	1	3

b)

5   Conclusions

Graph edit distance has been recognized as a flexible and powerful concept to measure the similarity, or distance, of graphs. However it has been suffering from the lack of suitable learning procedures that allow us to infer the costs of the individual edit operations automatically from a sample set of graphs. The present paper is, to the knowledge of the authors, the first attempt towards such an automatic learning procedure. This procedure is based on SOMs. The original space of the node and edge labels of the graphs is represented through a SOM where label distance corresponds to the cost of the associated edit operation. Through an iterative learning process, the structure of the label space is deformed, i.e. the locations of the node and edge labels in the SOM are changed in such a way that the costs of edit operations that occur frequently are gradually reduced.

The usefulness of the proposed learning scheme has been experimentally investigated on synthetically generated graphs that represent line drawings. Pairs of graphs from the same class have been chosen and the edit costs were changed so as to make the graphs more similar. In a first experiment, the feasibility of the proposed approach was verified by showing that the iterative SOM updating procedure really leads to a better separation of graphs from different classes. Next it was shown that the proposed SOM distance performs better than a Euclidean cost function that was optimized for the given sample set in an exhaustive search over all possible edit cost parameter values. Third, the influence of particular geometric distortions on the graph edit distance was studied. Again it was demonstrated that the edit costs obtained through SOM learning perform better than those of the Euclidean model. Finally, the proposed approach has been applied to the classification of diatoms. It was shown that on a small size sample set the errors committed by a graph matching based nearest neighbor classifier could be reduced from 32% to 8%.

The learning procedure presented in Algorithm 2 has been set up in such a way that only patterns from the same class are used. However, this procedure can be extended to also include pairs of graphs from different classes. In this case we wish to increase, rather than decrease, the corresponding edit distance. This can be accomplished in Algorithm 2 if we move a pair of label vectors away from each other, instead of making their distance smaller. A detailed investigation of this issue is left to future research. Another — perhaps more challenging — open question is the extension of the proposed learning procedure to graphs with symbolic node and edge labels.

Acknowledgment. This research was supported by the Swiss National Science Foundation NCCR program "Interactive Multimodal Information Management (IM)2" in the Individual Project "Multimedia Information Access and Content Protection".

References

1. IEEE Transactions on Pattern Analysis and Machine Intelligence: Special section on graph algorithms and computer vision. **23** (2001) 1040–1151
2. Pattern Recognition Letters: Special issue on graph based representations. **24** (2003) 1033–1122
3. Int. Journal of Pattern Recognition and Art. Intelligence: Special issue on graph matching in pattern recognition and computer vision. (In preparation)
4. Bunke, H., Shearer, K.: A graph distance metric based on the maximal common subgraph. Pattern Recognition Letters **19** (1998) 255–259
5. Wallis, W., Shoubridge, P., Kraetzl, M., Ray, D.: Graph distances using graph union. Pattern Recognition Letters **22** (2001) 701–704
6. Fernandez, M.L., Valiente, G.: A graph distance metric combining maximum common subgraph and minimum common supergraph. Pattern Recognition Letters **22** (2001) 753–758
7. Sanfeliu, A., Fu, K.: A distance measure between attributed relational graphs for pattern recognition. IEEE Transactions on Systems, Man, and Cybernetics **13** (1983) 353–363
8. Myers, R., Wilson, R., Hancock, E.: Bayesian graph edit distance. IEEE Transactions on Pattern Analysis and Machine Intelligence **22** (2000) 628–635
9. Messmer, B., Bunke, H.: A new algorithm for error-tolerant subgraph isomorphism detection. IEEE Transactions on Pattern Analysis and Machine Intelligence **20** (1998) 493–504
10. Kohonen, T.: Self-Organizing Maps. Springer (1995)
11. Ambauen, R., Fischer, S., Bunke, H.: Graph edit distance with node splitting and merging and its application to diatom identification (2003) Submitted to GbR 2003.

12. Günter, S., Bunke, H.: Self-organizing map for clustering in the graph domain. Pattern Recognition Letters **23** (2002) 405–417
13. Neuhaus, M.: Learning graph edit distance. Master's thesis, University of Bern, Switzerland (2002)
14. Günter, S., Bunke, H.: Validation indices for graph clustering. Pattern Recognition Letters **24** (2003) 1107–1113
15. du Buf, H., Bayer, M., eds.: Automatic Diatom Identification. World Scientific (2002)

Graph Edit Distance with Node Splitting and Merging, and Its Application to Diatom Identification

R. Ambauen, S. Fischer, and Horst Bunke

Department of Computer Science, University of Bern,
Neubrückstrasse 10, CH-3012 Bern, Switzerland
bunke@iam.unibe.ch

Abstract. A new graph matching scheme based on an extended set of edit operations, which include the splitting and merging of nodes, is proposed in this paper. This scheme is useful in applications where the nodes of the considered graphs represent regions extracted by some segmentation procedure from an image. To demonstrate the feasibility of the proposed method, its application to the automatic identification of diatoms (unicellular algae) is described.

1 Introduction

The area of graph matching has become a focus of intensive research recently [1], and a number of practically applicable algorithms have become available meanwhile [2,3,4]. Also a large number of graph matching applications have been reported in the literature. For a survey see [5]. Data from the real world is usually corrupted by noise and distortions. Consequently, error-correcting, or error-tolerant, graph matching has been proposed. A popular approach to error-correcting graph matching is graph edit distance [2,6]. In graph edit distance computation, one introduces a set of edit operations. Typically, three types of edit operations are considered, namely, insertion, deletion, and substitution. These edit operations can be applied to both nodes and edges in a graph. Depending on the underlying application, a weight, or cost, is sometimes assigned to each edit operation. The edit distance of two graphs is then defined as the minimum length (or cost) among all sequences of edit operations that transform one of the two graphs into the other.

It is well known that the classical set of edit operations, consisting of the deletion, insertion, and substitution of nodes and edges, is powerful enough to transform any two given graphs into each other. Despite their theoretical power, it can be argued that these edit operations are not perfectly suited for all problem domains. In the present paper we are concerned with an object classifications task that requires image segmentation. The underlying graph representation is such that nodes model elementary image regions and edges spatial relations between those regions. It is well known that image segmentation is an extremely difficult task that is far from being solved. Consequently, image segmentation

E. Hancock and M. Vento (Eds.): IAPR Workshop GbRPR 2003, LNCS 2726, pp. 95–106, 2003.

errors are inevitable. The two most common types of errors are over- and under-segmentation. To model these errors on the graph level, we propose two new edit operations in this paper, namely, node splitting and node merging. These two edit operations are intended to be at our disposal in addition to the normal edit operations, i.e. deletion, insertion and substitution of nodes and edges. The rationale behind node splitting (merging) is to provide a direct way to model the over(under)-segmentation of image regions. Clearly, the over(under)-segmentation of a region can also be modeled through node insertions (deletions), but it can be expected that the availability of separate edit operations allows us to discriminate between 'real' insertions (deletions) and those corresponding to over(under)-segmentations, which may lead to an overall enhancement of the modeling and discrimination ability of a graph matching scheme. Node splitting and merging has also been used in graph matching based object tracking [7], color image retrieval [8], and facial analysis [9].

To illustrate the feasibility of the proposed new graph edit distance measure, we consider the problem of automatic diatom identification in this paper. Diatoms are unicellular algae found in water and other places on earth where there is humidity and enough light for photosynthesis. Diatom identification has a number of applications in areas such as environment monitoring, climate research, and forensic medicine [10].

The remainder of this paper is organized as follows. In Section 2 we formally present our extended graph matching scheme including node splitting and merging. The task of diatom identification, the segmentation of diatom images, and the extraction of graphs from segmented diatom images are described in Section 3. Experimental results are reported in Section 4, and conclusions drawn in Section 5.

2   Graph Edit Distance with Node Splitting and Merging

In this paper we consider labeled and directed graphs of the form $g = (V E, \alpha, \beta)$ where V is a set of nodes, $E \subseteq V \times V$ is a set of edges, $\alpha : V \to L_V$ is the node labeling function, and $\beta : E \to L_E$ is the edge labeling function. L_V and L_E are the sets of node and edge labels, respectively. Given graphs $g_1 = (V_1, E_1, \alpha_1, \beta_1)$ and $g_2 = (V_2, E_2, \alpha_2, \beta_2)$ a graph isomorphism from g_1 to g_2 is a bijective mapping $f : V_1 \to V_2$ that preserves all edges and labels, i.e. (1) $\alpha_1(x) = \alpha_2(f(x))$ for all $x \in V_1$, (2) for any edge $e_1 = (x, y) \in E_1$ there exists an edge $e_2 = (f(x), f(y)) \in E_2$ such that $\beta_1(e_1) = \beta_2(e_2)$ and for any edge $e_3 = (u, v) \in E_2$ there exists an edge $e_4 = (f^{-1}(u), f^{-1}(v)) \in E_1$ such that $\beta_1(e_4) = \beta_2(e_3)$. Graphs g_1 and g_2 are called isomorphic if there exists a graph isomorphism from g_1 to g_2.

Let g be a graph. A classical edit operation δ is one of the following: (1) Node deletion, denoted by $v \to \varepsilon$, $v \in V$; (2) node insertion, denoted by $\varepsilon \to v$, $v \in V$; (3) node substitution, denoted by $\alpha(v) \to l$, $v \in V$, $l \in L_V$; (4) edge deletion, denoted by $e \to \varepsilon$, $e \in E$; (5) edge insertion, denoted by $\varepsilon \to e$, $e \in E$; (6) edge substitution, denoted by $\beta(e) \to l$, $e \in E$, $l \in L_E$.

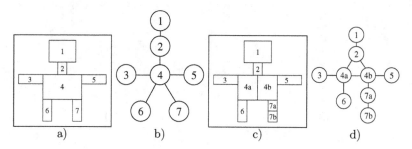

Fig. 1. An example of node splitting

As each edge $(x, y) \in E$ requires the existence of its two incident nodes $x, y \in V$, we need to impose the constraint that node deletions can only be applied to nodes that don't have any incident edges. Similarly the insertion of an edge $(x, y) \in E$ requires the existence of nodes $x, y \in V$. In this paper the list of classical edit operations given above is augmented by two additional edit operations: (7) Node splitting, denoted by $v \rightarrow (v_1, \ldots, v_n)$, $v \in V$; (8) node merging, denoted by $(v_1, \ldots, v_m) \rightarrow v$, $v_1, \ldots, v_m \in V$.

By means of node splitting any node $v \in V$ is split into n new nodes $v_1, \ldots, v_n$. This edit operation changes graph g by replacing node v by n new nodes $v_1, \ldots, v_n$. Moreover, edges between the new nodes or between new nodes and nodes that existed before the splitting operation may be inserted (the insertion of these new edges is usually application dependent). Node merging is the inverse operation to node splitting.

Figure 1 shows an example of node splitting. In this example we assume an image has been segmented into regions as shown in Fig. 1a. The corresponding graph is depicted in Fig. 1b. The graph representation has been chosen in such a way that each node represents a region, and edges represent region adjacency. The same image, but segmented differently, is shown in Fig. 1c. The corresponding graph is depicted in Fig. 1d. If we consider Fig. 1b being the perfect and Fig. 1d the distorted graph, then the distortions can be modeled through two node splitting operations, namely $4 \rightarrow (4a, 4b)$ and $7 \rightarrow (7a, 7b)$.

If certain edit operations are more likely to occur than others, it may be appropriate to assign an individual cost to each edit operation. Let δ be an edit operation and $c(\delta)$ its cost. Then the cost of a sequence of edit operations, $\Delta = \delta_1, \ldots, \delta_n$, is defined as $c(\Delta) = \sum_{i=1}^{n} c(\delta_i)$. Given two graphs g_1 and g_2 and our set of edit operations with corresponding costs, the edit distance of g_1 and g_2 is defined as $d(g_1, g_2) = \min_{\Delta} \{c(\Delta) | \Delta$ is a sequence of edit operations transforming g_1 into $g_2\}$.

Graph edit distance as defined above can be implemented in various ways. The greatest challenge in the implementation is the search for the minimum over all possible sequences of edit operations, which is computationally expensive. An alternative to searching in the space of edit sequences is searching for a minimum cost isomorphism from a subgraph of g_1 to a subgraph of g_2 [11] (which doesn't

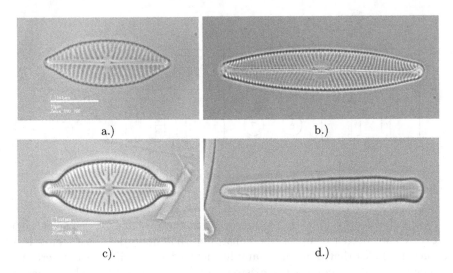

a.)   b.)

c).   d.)

Fig. 2. Example images of diatom valves

reduce the computational complexity, however). This formulation directly leads to a range of possible implementations using combinatorial search procedures such as best-first, breadth-first, or depth-first search. As the graphs considered in Section 3 are not excessively large, it was decided to base the implementation on such an algorithm. In particular the two new edit operations node splitting and merging can be seamlessly integrated into such a procedure. Details of our implementation are given in [12].

3   Diatom Identification by Means of Graph Matching

3.1   Problem Description

Examples of four different classes of diatoms are shown in Fig. 2. In previous papers the identification of diatoms based on their shape [13,14] and texture has been proposed. In [15] textural features of global nature have been used, while in [16] texture properties have been measured at the points of a regular grid overlaid on a diatom. In this paper we propose a more flexible approach where diatoms are segmented into homogeneous regions based on their texture. These homogeneous regions and their spatial relationships are represented by means of a graph. Graphs corresponding to unknown diatoms, which are to be classified, are matched to prototype graphs stored in a database, and a nearest-neighbor classification is performed, using the new graph edit distance measure introduced in Section 2.

3.2   Image Segmentation

The aim of image segmentation is to extract regions from the image of a diatom that are homogeneous with respect to texture. We assume that the underlying

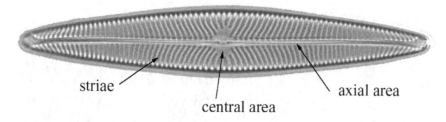

striae

central area

axial area

Fig. 3. Illustration of striae

image contains only one diatom the contour of which has been extracted before [17].

A large number of textural features have been proposed in the image processing literature and the segmentation of textured images is still a topic of ongoing research. In this paper we focus our attention on two special, problem-dependent textural features that have a high discriminatory power in diatom classification. These features are the direction and the density of the striae of a diatom's valve. Striae are stripes between the outer border of the valve and the inner axial area. Often the striation is orthogonal to the main axis of the diatom, but sometimes it is directed towards or away from the center of the valve. Furthermore the direction of the striation may change between the center and valve endings. A graphical illustration is provided in Fig. 3.

Before properties of the striae can be extracted, the vertical stripes have to be detected. This is accomplished by edge detection. Since the dominant striae direction is nearly orthogonal to the major axis, a unidirectional edge detector is used to extract edges inside a diatom. The detected edge points are connected to line elements by applying a line following and thinning algorithm, and their direction is determined.

In Fig. 4 this procedure is visualized for a region of the diatom shown in Fig. 3. The original image region is shown in Fig. 4a. In Fig. 4b the edge points, detected by a Canny-type edge detector, are represented in black and overlaid on the original image region. Finally, the connected and thinned line elements are shown in Fig. 4c.

After all line segments and their directions have been extracted from an image, a region growing procedure is started. The aim of this procedure is to segment a diatom into areas where the striae angle is approximately constant. In the region growing procedure we first assign to each pixel p inside the considered diatom the direction of the closest line segment l, as long as the distance between p and l is smaller than a threshold. The purpose of this threshold operation is to exclude pixels that don't belong to a textured area from the following region growing procedure. Then the pixels of the longest line segment are selected as seed region. Next the seed region is expanded by merging a pixel p inside the diatom with the seed region if p is adjacent to the seed region and the difference between the angle of p and the angle of the seed region is small (smaller than 20

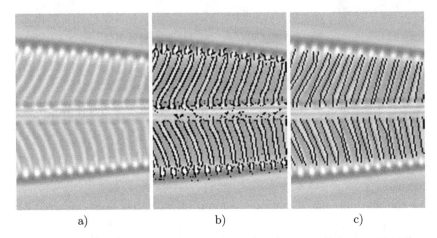

Fig. 4. Illustration of striae detection: a) a region from the image in Fig. 3, b) the detected edge points, c) the linked and thinned line segments

degree in the experiments described in Section 4). If the first seed region can't be grown any larger, the next seed region is selected (i.e. the largest line segment among the remaining lines segments). This process is repeated until no further line segments are available.

Finally a post-processing procedure is applied on the results of region growing. The purpose of this post-processing step is to eliminate small regions and to merge two adjacent regions if they have similar average striae direction. Elimination of small regions is accomplished by either merging them with larger regions or by deleting them. An example of the complete image segmentation procedure is given in Fig. 5.

3.3   Graph Representation

The regions extracted by the procedure described in the previous section are the basis for the graph representation of diatoms. In a diatom graph, the regions in the image are represented by nodes and spatial relations between regions as edges. The following attributes are assigned to each node:[1]

- Bounding box $B = (x_{min}, y_{min}, x_{max}, y_{max})$: The bounding box is the smallest rectangle which contains the complete region
- Relative size A: The relative size is the fraction of the number of pixels of the considered region and the complete valve
- Striation direction α: The striation direction is the average line segment direction inside the region

---

[1] Formally speaking each label $\alpha(v)$ attached to a node v is an n-tuple of attributes; see Section 2

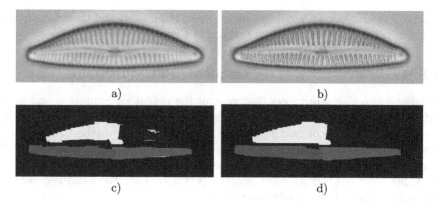

Fig. 5. Illustration of diatom image segmentation: a) original image, b) image with striae lines superimposed, c) results of region growing, d) final result after post-processing

- Striation density ρ: This attribute is the number of stripes per $1\mu m$ in the direction of the main axis[2]

To establish the edges of the graph, spatial relations between each pair of regions are analyzed. First of all, we define the relative neighborhood-degree of regions r_1 and r_2, $n(r_1, r_2)$, as being the relative percentage of pixels on the contour of r_1 for which there exists a pixel on the contour of r_2 such that the line connecting r_1 and r_2 doesn't intersect any region other than the background. Intuitively speaking if the line between a pixel p belonging to r_1 and a pixel q from r_2 doesn't intersect any region, we can conclude that p and q are neighbors. The relative neighborhood-degree indicates how many of such pixels exist on the contour of r_1 in relation to the total number of pixels on the contour of r_1. Note that this measure isn't symmetric in general, i.e. $n(r_1, r_2) \neq n(r_2, r_1)$. If $n(r_1, r_2) = 0$ or $n(r_2, r_1) = 0$ no edge will be inserted in the graph. Otherwise if $n(r_1, r_2) > 0$ and $n(r_2, r_1) > 0$ an edge between the nodes representing regions r_1 and r_2 will be created. This edge will be labelled with the average relative neighborhood degree $N(r_1, r_2) = 0.5 \cdot [n(r_1, r_2) + n(r_2, r_1)]$.

4   Experiments

4.1   Dataset

In the ADIAC project, several databases with images of diatoms have been acquired [18]. For the experiments described in this paper a subset of one of those databases was selected.[3] The dataset used in our experiments includes 110 images of diatoms belonging to 22 classes. Each diatom class is represented by

---

[2] Note that the absolute scale of each image is known

[3] At the time of the experiments, the full databases were not yet available.

exactly five samples. (A complete list of these 110 images and their taxa can be found in [12].) A few examples are shown in Figs. 2, 3, and 5a.

4.2 Parameterization of the Graph Matching Procedure

In addition to the parameters that control the region segmentation and graph extraction process (see Section 3), some parameters are needed for the graph similarity measure proposed in Section 2. First of all, to make the individual graph attributes compatible with each other, they are all normalized to the interval $[0, 1]$. Let x and y be nodes with normalized attributes $A(x)$, $B(x)$, $\alpha(x)$, $\rho(x)$ and $A(y)$, $B(y)$, $\alpha(y)$, $\rho(y)$, respectively. Then the following costs are assigned to the individual edit operations:

- node deletion and insertion: $c(x \to \varepsilon) = c(\varepsilon \to x) = c_i \cdot A(x)$
- node substitution:
  $$c(x \to y) = \sqrt{(A(x) - A(y))^2 + (\alpha(x) - \alpha(y))^2 + (\rho(x) - \rho(y))^2}$$
- node merging and splitting:
  $c((y_1, \ldots, y_n) \to x) = c(x \to (y_1, \ldots, y_n)) = 0$ where $y_1, \ldots, y_n$ are nodes

The parameter c_i controls the weight of a node deletion or insertion relative to a node subsitution. In the actual implementation it has been set equal to 5. The cost of a node deletion and insertion is proportional to the size of the corresponding region in the image, whereas the cost of a node substitution is equal to the Euclidean distance of the corresponding attribute vectors. Merging and splitting of nodes is motivated by attempting to recover from image segmentation errors, i.e. over- and under-segmentation. Therefore no extra charge is imposed on those edit operations. Note, however, that normally some substitution cost will occur when nodes resulting from a split or a merge operation in a prototype graph are assigned to the input graph. It is also to be noted that after a split or merge operation all node and edge attributes need to be recalculated.

Let $e_1 = (x_1, y_1)$ and $e_2 = (x_2, y_2)$ be edges with edge attributes $N_1(x_1, y_1)$ and $N_2(x_2, y_2)$, respectively. Edit operations on these edges have the following cost:

- edge deletion and insertion: $c(e_1 \to \varepsilon) = c(\varepsilon \to e_1) = c_e \cdot N_1(x_1, y_1)$
- edge substitution: $c(e_1 \to e_2) = c_e \cdot |N_1(x_1, y_1) - N_2(x_2, y_2)|$

The parameter c_e allows us to weight the importance of edit operations on the edges relative to those on the nodes. This parameter has been set to $1/3$ in the experiments.

Another important issue in the proposed graph matching procedure is the set of candidate nodes subject to splitting or merging. Clearly, if we allow many nodes to be split or merged, the flexibility of our graph matching procedure will increase, but also the computational complexity will go up. Basically the aim of node splitting and merging is to alter a prototype graph in such a way that it can be mapped to a given input graph at a lower cost. Therefore to generate the set of candidate nodes from the input into which node x of the prototype is potentially to be split, we first determine all nodes from the input graph with a corresponding region in the image that has a non-empty intersection with node x. Then all subsets of these nodes are considered that are adjacent to each other and consist of at least two nodes. Each of these subsets is a set of candidate nodes into which node x can be potentially split.[4] A similar procedure is used for node merging.

4.3   Results and Discussion

Given the dataset described in Section 4.1 a number of nearest-neighbor classifiers were applied under the leave-one-out method. That is, one out of the 110 diatoms from the database was used as a test sample, while all other 109 diatoms served as prototypes. This step was repeated 109 times such that each diatom was used once as test sample. The nearest-neighbor classifiers used in this work are described below.

- Nearest Neighbor (NN): This is the classical version of the nearest neighbor classifier. The test sample is assigned to the class of the prototype that has the smallest graph edit distance.
- k-Nearest Neighbor (kNN): The test sample is assigned to the class that is most often represented among the k nearest prototypes. In the experiments $k = 4$ was used, as there are four prototypes from the correct class for each input sample.
- Maximum rule (max): The prototype with maximum distance is determined for each class and the class with the smallest maximum distance is selected.
- Average rule (avg): the average distance between the unknown sample and the prototypes of each class is determined. The sample is assigned to the class with smallest average distance.
- k-Sum (sum): The sum of the k smallest distances is computed for each class, and the class with the smallest sum selected. Again $k = 4$ was used.

A graphical representation of the results is shown in Fig. 6, where up to 15 ranks are taken into consideration. The best result is obtained with the normal nearest-neighbor classifier. An explanation for the fact that the simplest classifier performs best is that the database contains only five prototypes per class. A closer analysis of the database reveals that the prototypes of some classes are quite different from each other. In fact there are classes where the within-class similarity is smaller than the inter-class similarity. A simple nearest neighbor classifier doesn't get confused by this situation, but classification performance is expected to deteriorate as soon as more than one prototype has an influence on the decision. This explanation is confirmed by the fact that the maximum rule has the least accuracy. Using the maximum rule, the decision of the classifier is based on the prototype from each class that has the largest distance to the input sample. Hence large in-class variability has a huge influence here. A similar effect

---

[4] For large graphs this procedure may become very time consuming. But in the application considered here the number of subsets to be considered is small.

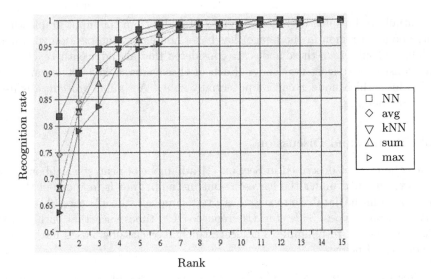

Fig. 6. Recognition rates obtained in the experiment

can be observed for the kNN, avg, and sum decision rule, where not only the most similar prototype from each class, but all samples have an impact on the decision.

When considering more than just the first rank, the performance of each classifier steadily improves, and for each classifier the correct class was always among the first 14 ranks. Regarding the top three ranks the NN classifier achieved a recognition rate of almost 95%.

5   Conclusions

In this paper we have proposed a new error-tolerant graph matching procedure with node splitting and merging. As an application of this new procedure the identification of diatoms was studied. On a complex database holding images of 22 different diatoms classes, a recognition rate of about 82% was achieved with a NN classifier. The recognition rate increases to almost 95% if the first three ranks are taken into account. Considering the difficulty of the underlying task, the recognition rates are quite impressive. Difficulties in the considered classification task arise from the rather small size of the database (just five samples per class), a high variance within some of the considered classes, and a high similarity between some classes.

There are some potential ways to improve the diatom classification method proposed in this paper. First of all, a higher recognition accuracy can be expected by enlarging the number of samples in the database. Secondly, one has to keep in mind that only texture, but no shape information, which is essential in the classification of diatoms, has been incorporated in the proposed graph matching scheme. Obviously, shape information can be utilized in various ways, for example, by designing a second independent classifier, similar to those described in [13,14], and combining it with the graph matching scheme. As an alternative, shape information can be directly integrated into the graph representation, by assigning moments, Fourier descriptors etc. as attributes to the nodes of a diatom graph. These issues are left to future research.

Acknowledgement. The work has been done in the framework of the EU-sponsored Marine Science and Technology Program (MAST-III), under contract no. MAS3-CT97-0122. Additional funding came from the Swiss Federal Office for Education and Science (BBW 98.00.48). We thank our project partners Micha Bayer and Stephen Droop from Royal Botanic Garden Edinburgh and Steve Juggins and co-workers at Newcastle University for preparing the images in the ADIAC image database and for useful discussions and hints.

References

1. IEEE Transactions on Pattern Analysis and Machine Intelligence: Special section on graph algorithms and computer vision. **23** (2001) 1040–1151
2. Messmer, B., Bunke, H.: A new algorithm for error-tolerant subgraph isomorphism detection. IEEE Trans. PAMI **20** (1998) 493–505
3. Luo, B., Hancock, E.: Structural graph matching using the EM algorithm and singular value decomposition. IEEE Trans. PAMI **23** (2001) 1120–1136
4. Hancock, E., Wilson, R.: Graph-based methods for vision: a Yorkish manifesto. In Caelli, T., Amin, A., Duin, R., Kamel, M., de Ridder, D., eds.: Structural, Syntactic, and Statistical Pattern Recognition. LNCS 2396. Springer (2002) 31–46
5. Bunke, H.: Recent developments in graph matching. In: Proc. 15th Int. Conf. on Pattern Recognition, Barcelona. (2000) 813–817
6. Sanfeliu, A., Fu, K.: A distance measure between attributed relational graphs for pattern recognition. IEEE Trans. SMC **13** (1983) 353–362
7. Gomila, C., Meyer, F.: Tracking objects by graph matching of image partition sequences. In: Proc. 3rd IAPR-TC15 Workshop on Graph-based Representations in Pattern Recognition. (2001) 1–11
8. Gregory, L., Kittler, J.: Using graph search techniques for contextual colour retrieval. In Caelli, T., Amin, A., Duin, R., Kamel, M., de Ridder, D., eds.: Structural, Syntactic, and Statistical Pattern Recognition. LNCS 2396. Springer (2002) 186–194
9. Cesar, R., Bengoetxea, E., Bloch, I.: Inexact graph matching using stochastic optimization techniques for facial feature recognition. In: Proc. 16th Int. Conf. on Pattern Recognition, Quebec-City. (2002) 465–468
10. Stoermer, E., Smol, J., eds.: The Diatoms: Applications for the Environmental and Earth Science. Cambridge University Press (1999)
11. Bunke, H.: On a relation between graph edit distance and maximum common subgraph. Pattern Recognition Letters **18** (1997) 689–694
12. Ambauen, R.: Identification of Diatoms based on Graph Matching. Master's thesis, University of Bern (2002) (in German).

13. Fischer, S., Binkert, M., Bunke, H.: Feature based retrieval of diatoms in an image database using decision trees. In: Proc. 2nd Int. Conf. on Advanced Concepts for Intelligent Vision Systems (ACIVS). (2000) 67–72
14. Fischer, S., Binkert, M., Bunke, H.: Symmetry based indexing of diatoms in an image database. In: Proc. 15th Int. Conf. on Pattern Recognition, Barcelona. (2000) 899–902
15. Fischer, S., Bunke, H.: Identification using classical and new features in combination with decision tree ensembles. In du Buf, H., Bayer, M., eds.: Automatic Diatom Identification. World Scientific (2002)
16. Fischer, S., Gilomen, K., Bunke, H.: Identification of diatoms by grid graph matching. In Caelli, T., Amin, A., Duin, R., Kamel, M., de Ridder, D., eds.: Structural, Syntactic, and Statistical Pattern Recognition. LNCS 2396. Springer (2002) 94–103
17. Fischer, S., et al.: Contour extraction. In du Buf, H., Bayer, M., eds.: Automatic Diatom Identification. World Scientific (2002)
18. Bayer, M., Juggins, S.: Adiac imaging techniques and databases. In du Buf, H., Bayer, M., eds.: Automatic Diatom Identification. World Scientific (2002)

Orthonormal Kernel Kronecker Product Graph Matching

Barend Jacobus van Wyk[1,2] and Michaël Antonie van Wyk[2]

[1] Kentron, a Division of Denel, Centurion, South Africa
ben.van.wyk@kentron.co.za
[2] Rand Afrikaans University, Johannesburg, South Africa
mavw@ing.rau.ac.za

Abstract. This paper introduces a novel algorithm for performing Attributed Graph Matching (AGM). The Orthonormal Kernel Kronecker Product Graph Matching (OKKPGM) algorithm is based on the recently introduced Kronecker Product Graph Matching (KPGM) formulation. However, unlike previous algorithms based on the KPGM formulation which avoided explicit compatibility calculations, the OKKPGM algorithm obtains an estimate to the Kronecker Match Matrix (KMM) using kernel function evaluations. A permutation sub-matrix (match matrix), is then inferred from the KMM using a summation procedure.

1 Introduction

An object can be described in terms of its parts, the properties of these parts and their mutual relationships. Representation of the structural descriptions of objects by attributed relational graphs reduces the problem of matching to an *Attributed Graph Matching* (AGM) problem.

Gold and Rangarajan [1] divided graph matching algorithms into two major classes of approach. In general, the first approach constructs a state-space, which is searched using heuristics to reduce complexity [2–11]. The second approach is based on transforming the discrete search space into a continuous state space which allows the use of optimization techniques such as Bayesian, linear-programming, continuation, eigen-decomposition, polynomial transform, genetic, neural network and relaxation-based methods [12–24].

Messmer and Bunke [11] perhaps more appropriately suggested that graph matching algorithms can be divided into search-based or random methods. According to this classification search-based methods in general finds optimal solutions, but require exponential time in the worst case. The computational time of random methods, on the other hand, is polynomially bounded, but in some cases may fail to find the optimal solution. The classification strategy of Messmer and Bunke [11] can be generalized to optimal and approximate methods. The Orthonormal Kernel Kronecker Product Graph Matching (OKKPGM) algorithm, which will be introduced in section 3, is an approximate method.

E. Hancock and M. Vento (Eds.): IAPR Workshop GbRPR 2003, LNCS 2726, pp. 107–117, 2003.

The focus of this paper is on matching fully-connected, undirected attributed graphs. Although the OKKPGM algorithm is also based on the Kronecker Product Graph Matching (KPGM) formulation introduced in [23–24], it differs from the Interpolator-Based Kronecker Product Graph Matching (IBKPGM) algorithm in the sense that it deviates from the idea of performing graph matching without the explicit calculation of compatibility values. An orthogonal kernel-based network is used to infer an approximation to the Kronecker Match Matrix (KMM), $\mathbf{\Phi}$, from which a permutation sub-matrix is inferred. The kernel evaluations, required by the OKKPGM algorithm to infer an approximation to the KMM, are in fact compatibility calculations. The OKKPGM algorithm circumvents a direct search for the largest common subgraph via an association graph, as required by the methods of Barrow and Burstall [9] and Horaud and Skordas [5].

The outline of the presentation is as follows: In section 2, we briefly review the concept of an attributed graph, formulate the *Attributed Graph Matching* (AGM) problem, and introduce the *Kronecker Product Graph Matching* KPGM formulation. The process to approximate $\mathbf{\Phi}$ is derived in section 3.1. A procedure to approximate the permutation sub-matrix is derived in section 4. The computational complexity of the OKKPGM algorithm is discussed in section 5 and some experimental results are reported in section 6.

2   Kronecker Product Graph Matching Formulation

The focus of this paper is on matching graphs where an input graph, say

$$G = \left(V, E, \{\mathbf{A}_i\}_{i=1}^r, \{\mathbf{B}_j\}_{j=1}^s\right) \tag{1}$$

is matched to a reference graph, say

$$G' = \left(V', E', \{\mathbf{A}_i'\}_{i=1}^r, \{\mathbf{B}_j'\}_{j=1}^s\right) \tag{2}$$

where $\mathbf{A}_i \in \mathbb{R}^{n \times n}$, $\mathbf{B}_j \in \mathbb{R}^{n \times 1}$, $\mathbf{A}_i' \in \mathbb{R}^{n' \times n'}$ and $\mathbf{B}_j' \in \mathbb{R}^{n' \times 1}$ represent the edge attribute adjacency matrices and vertex attribute vectors respectively. The reference and input graphs each have r edge attributes and s vertex attributes. The number of vertices of G' (respectively, G) is $n' = |V'|$ (respectively, $n = |V|$). Here we consider the general case of sub-graph matching. *Full-Graph Matching* (FGM) refers to matching two graphs having the same number of vertices (i.e. $n' = n$) while *Sub-Graph Matching* (SGM) refers to matching two graphs having a different number of vertices (i.e. $n' > n$).

We say that G is *matched* to some sub-graph of G' if there exists an $n \times n'$ permutation sub-matrix $\mathbf{P}$ such that $\mathbf{A}_i = \mathbf{P}\mathbf{A}_i'\mathbf{P}^T$ and $\mathbf{B}_j = \mathbf{P}\mathbf{B}_j'$ where $i = 1, ..., r$ and $j = 1, ..., s$.

We now observe that vertex attribute vectors converted to diagonal matrices, using the diag$(\cdot)$ operation in linear algebra, satisfy the same expression as edge attribute matrices do, namely diag $\mathbf{B}_j \approx \mathbf{P}$ diag$(\mathbf{B}_j')\,\mathbf{P}^T$, with exact equality holding for the ideal case (i.e. when G is just a permuted sub-graph of G').

This means that these converted vertex attribute vectors may be considered as additional edge attribute matrices. Observing that

$$\text{vec}(\mathbf{A}_i - \mathbf{P}\mathbf{A}'_i\mathbf{P}^T) \equiv \text{vec}\mathbf{A}_i - \boldsymbol{\Phi}\text{vec}\mathbf{A}'_i \tag{3}$$

where $\boldsymbol{\Phi} = \mathbf{P} \otimes \mathbf{P}$ and $\otimes$ denotes the *Kronecker Product* for matrices, the AGM problem can be expressed as

$$\min_{\boldsymbol{\Phi}} \left(\sum_{i=1}^{r+s} \|\text{vec}\mathbf{A}_i - \boldsymbol{\Phi}\,\text{vec}\mathbf{A}'_i\|^2 \right) \tag{4}$$

subject to $\boldsymbol{\Phi} = \mathbf{P} \otimes \mathbf{P}$, $\mathbf{P} \in Per(n, n')$ where $Per(n, n')$ is the set of all $n \times n'$ permutation sub-matrices. Here $\|\cdot\|$ denotes the Euclidean norm.

The following definitions and observation are used in the sequel:

Definition 1 *The matrix $\boldsymbol{\Phi} = \mathbf{P} \otimes \mathbf{P}$ minimizing equation 4, subject to $\mathbf{P} \in Per(n, n')$, is termed the Constrained* Kronecker Match Matrix (KMM).

Observation 1 *Given a Constrained Kronecker Match Matrix, $\boldsymbol{\Phi} = \mathbf{P} \otimes \mathbf{P}$, such that $\mathbf{P} \in Per(n, n')$, we can retrieve the unknown permutation sub-matrix $\mathbf{P} := (P_{ij})$ by*

$$P_{ij} = \frac{\sum_{k,l} \Phi_{kl}}{n}, \tag{5}$$

where $i = 1, ..., n$, $j = 1, ..., n'$, $k = (i-1)n + 1, ..., (i-1)n + n$ and $l = (j-1)n' + 1, ..., (j-1)n' + n'$.

3 The Orthonormal Kernel Kronecker Product Approach

The OKKPGM algorithm is summarized by the following steps:

1. Approximate the constrained KMM (section 3.1, equation 14).
2. Approximate the permutation sub-matrix $\mathbf{P}$ satisfying $\boldsymbol{\Phi} = \mathbf{P} \otimes \mathbf{P}$, $\mathbf{P} \in Per(n, n')$ (section 4).

3.1 Approximating $\boldsymbol{\Phi}$

The tractability of linear equations served as one of the motivations for devising interpolator-based algorithms [22–24] for matching attributed graphs. Unfortunately these algorithms sometimes fare poorly when performing subgraph matching. In addition they require the operation of matrix inversion. In order to avoid these two issues it is necessary to parameterize our kernels by the row vectors Π_i^T of the matrix $[\text{vec}\,\mathbf{A}_1, \ldots, \text{vec}\,\mathbf{A}_{r+s}]$ as opposed to the vectors $\text{vec}(A'_k)$, as in the case of the above-mentioned interpolator-based algorithms.

For the approximation of $\mathbf{\Phi}$ we consider the following cost function for the vector Π_i,

$$J(\Pi_i) = \left\| \Pi_i - \sum_{j=1}^{n'^2} \bar{c}_{ij} \Psi_i \left(\Pi'_j \right) \right\|^2, \tag{6}$$

where $i = 1, ..., n^2$ and $\Psi_i \left(\Pi'_j \right)$ is a kernel function, parameterized by Π_i, which takes Π'_j as an input. Here Π_i^T represents the i-th row of the matrix $\mathbf{A} = [\text{vec} \, \mathbf{A}_1, ..., \text{vec} \, \mathbf{A}_{r+s}]$, $\bar{c}_{ij}$ is a column vector with $r + s$ elements, Π'^T_j is the j-th row of the matrix $\mathbf{A}' = [\text{vec} \, \mathbf{A}'_1, ..., \text{vec} \, \mathbf{A}'_{r+s}]$ and $\|\cdot\|$ denotes the Euclidean norm. The cost function for each i is related to equation 3, which in component form can be expressed as,

$$\Phi_i \left(\Pi'_1, ..., \Pi'_{n'^2} \right)^T = \Pi_i^T, \tag{7}$$

where Φ_i is the i-th row of $\mathbf{\Phi}$, and $i = 1, ..., n^2$. Using the one-dimensional outputs of the orthonormal kernel functions, parameterized by Π_i where $j = 1, ..., n^2$, to approximate a row of $\mathbf{\Phi}$, leads to equation 6. The characteristics of the proposed network are summarized by the following definition and propositions:

Definition 2 *The network performance criterion is defined as the sum of the squared errors associated with each $J(\Pi_i)$, i.e.*

$$J(\mathbf{\Pi}) = \sum_{i=1}^{n^2} J(\Pi_i). \tag{8}$$

Proposition 1 *If the kernels $\Psi_i \left(\Pi'_j \right)$ are orthonormal, i.e.*

$$\Psi_i \left(\Pi'_j \right) = \begin{cases} 1 \ when \ \Pi_i = \Pi'_j \\ 0 \ otherwise \end{cases},$$

then the $\bar{c}_{ij}$ which minimize equation 8 are given by

$$\bar{c}_{ij} = \Pi_i \Psi_i \left(\Pi'_j \right). \tag{9}$$

Proof. From equation 6 we have

$$J(\Pi_i) = \Pi_i^T \Pi_i - 2\Pi_i^T \sum_{j=1}^{n'^2} \bar{c}_{ij} \Psi_i \left(\Pi'_j \right) + \left(\sum_{j=1}^{n'^2} \bar{c}_{ij} \Psi_i \left(\Pi'_j \right) \right)^T \left(\sum_{j=1}^{n'^2} \bar{c}_{ij} \Psi_i \left(\Pi'_j \right) \right), \tag{10}$$

and therefore

$$\frac{\partial J(\Pi_i)}{\partial c_{ij}} = -2\Pi_i \Psi_i \left(\Pi'_j \right) + 2\bar{c}_{ij}, \tag{11}$$

when the $\Psi_i \left(\Pi'_j \right)$ are orthonormal. By setting equation 11 equal to zero we obtain

$$\bar{c}_{ij} = \Pi_i \Psi_i \left(\Pi'_j \right). \tag{12}$$

which concludes the proof. $\quad\square$

A specific instance of the kernels Ψ_i will be considered in detail in section 3.2. The parameters $\bar{c}_{ij}$ of the network, described by equation 6, are only of theoretical importance. The following proposition constitutes the core of the OKKPGM algorithm:

Proposition 2 *If all Π'_j are distinct, no noise is present, and*

$$\Psi_i\left(\Pi'_j\right) = \begin{cases} 1, & when \ \Pi_i = \Pi'_j, \\ 0, & otherwise \end{cases}, \tag{13}$$

then

$$\overline{\Phi}_{ij} = \Psi_i\left(\Pi'_j\right), \tag{14}$$

where $\overline{\Phi} := (\overline{\Phi}_{ij})$ is the Constrained Kronecker Match Matrix which minimizes equation 4, $i = 1, ..., n^2$ and $j = 1, ..., (n')^2$.

Proof. If all Π'_j are distinct, then from $\mathbf{A}_k = \mathbf{P}\mathbf{A}'_k\mathbf{P}^T$ and $\mathbf{B}_l = \mathbf{P}\mathbf{B}'_l$ and the properties of the permutation sub-matrix $\mathbf{P}$, it follows that all Π_i are also distinct. This further implies that

$$\overline{\Phi}_{ij} = \begin{cases} 1, & when \ \Pi_i = \Pi'_j, \\ 0, & otherwise \end{cases}, \tag{15}$$

$$\sum_i \overline{\Phi}_{ij} = 1, \tag{16}$$

and

$$\sum_j \overline{\Phi}_{ij} \leq 1. \tag{17}$$

Since equations 15 to 17 describe the constraints associated with $\Phi = \mathbf{P} \otimes \mathbf{P}$ and satisfy

$$\left(\Pi_1, ..., \Pi_{n^2}\right)^T = \overline{\Phi}\left(\Pi'_1, ..., \Pi'_{n'^2}\right)^T, \tag{18}$$

$\overline{\Phi}$ will minimize equation 4. $\qquad\square$

3.2   Quadratic Kernel $\Psi_i\left(\Pi'_j\right)$

An easily implementable instance of the kernels $\Psi_i\left(\Pi'_j\right)$ is the quadratic kernel, given by

$$\Psi_i\left(\Pi'_j\right) = \frac{1}{\left(1 + \frac{(\Pi'_j - \Pi_i)^T(\Pi'_j - \Pi_i)}{v}\right)}. \tag{19}$$

The orthonormality conditions are fulfilled when $v \to 0$. When implementing the algorithm, care must be taken in choosing v. Setting $v = 1$ simplifies the kernel to

$$\Psi_i\left(\Pi'_j\right)_{v=1} = \frac{1}{\left(1 + (\Pi'_j - \Pi_i)^T(\Pi'_j - \Pi_i)\right)} \tag{20}$$

and yields an easily implementable algorithm but, as will be shown in section 4, may adversely affect performance. Obviously Gaussian or other kernels using the Mahalanobis distance can also be used, but the added complexity brought about by the inference of appropriate kernel parameters should be compared to the expected improvement in performance. The quadratic kernel given by equation 19 was found to be sufficient for our purposes.

4   Deriving an Approximation to P from $\overline{\overline{\Phi}}$

From equation 14 we see that our kernel evaluations yield an approximation to Φ, namely $\overline{\overline{\Phi}}$. Setting the element with the maximum value in each row equal to one, setting the rest of the elements to zero and applying observation 1 yield a permutation sub-matrix, say $\overline{P}$. When implementing the quadratic kernel given by equation 19, all Π'_j are distinct, and no noise is present, then

$$\Psi_i\left(\Pi'_j\right) \quad \begin{cases} = 1, & when \ \Pi_i = \Pi'_j \\ < 1, & otherwise \end{cases}. \tag{21}$$

Since $\overline{P} \otimes \overline{P}$ will minimize equation 4, $\overline{P}$ will be the desired match matrix (permutation sub-matrix) if all Π'_j are distinct and no noise is present. It is significant to note that this simple procedure circumvents a direct search for the largest common subgraph via an association graph (by searching for its maximum cliques).

Obviously all Π'_j are not distinct when dealing with undirected graphs. When matching undirected graphs in the absence of noise where there is no replication of edge attributes, all the elements of $\overline{P}$ will be equal to $\frac{1}{n}$, except for a single element in each row which will be equal to one. Setting the maximum value in each row equal to one and the rest of the elements equal to zero, will remedy the problem.

When the variable v in equation 19 is chosen small enough, equally desirable results can be obtained by applying observation 1 directly to $\overline{\overline{\Phi}}$ to obtain $\overline{P}$, where the maximum value in each row of $\overline{P}$ is set equal to one and the rest of the elements are set to zero. Although this strategy is simple to implement since it avoids thresholding the elements of $\overline{\overline{\Phi}}$, it causes the OKKPGM algorithm to become sensitive to the choice of v. Choosing v too small will adversely affect the performance of the algorithm in the presence of noise, since the kernel outputs will tend to zero even when $\Pi_i \approx \Pi'_j$. Choosing v too large will in turn cause the behaviour of the kernels to become less orthonormal and consequently the accuracy of $\overline{P}$ will deteriorate, even when no noise is present.

5   Complexity

Observation 2 *If* P *is a constrained permutation sub-matrix, then the elements of* $\Phi = P \otimes P \in \mathbb{R}^{n^2 \times (n')^2}$ *obey:*

1. $\Phi_{ij} = 0$ when $i = 1 + (qn + q)$, $j \neq 1 + (pn' + p)$, $q \in \{0, ..., n - 1\}$, $p \in \{0, ..., n' - 1\}$.

2. $\Phi_{ij} = 0$ when $i \neq 1 + (qn + q)$, $j = 1 + (pn' + p)$, $q \in \{0, ..., n - 1\}$, $p \in \{0, ..., n' - 1\}$.

Observation 2 indicates that we are not required to calculate $\Psi_i \left(\Pi'_j \right)$ when $i = 1 + (qn + q)$ and $j \neq 1 + (pn' + p)$ or when $i \neq 1 + (qn + q)$ and $j = 1 + (pn' + p)$ where $q \in \{0, ..., n - 1\}$ and $p \in \{0, ..., n' - 1\}$, as these values can be set to zero. In total we therefore only have to perform $\left(n^4 - 2n^3 + 2n^2 \right)$ kernel evaluations when $n = n'$. As a consequence, inferring $\overline{\mathbf{P}}$ from $\overline{\mathbf{\Phi}}$ requires $\left(n^4 - 2n^3 + 2n^2 \right)$ additions, and searching for the maximum value in each row of $\overline{\mathbf{P}}$ requires n^2 comparisons when $n = n'$. The resulting algorithm therefore has a complexity of $O(n^4)$ when $(r + s) \ll n$.

6   Experimental Results

Three line matching experiments were conducted to test the proposed graph matching algorithms on non-random graphs. For both experiments we used lines to represent the vertices of a graph. The binary relationships between lines were used as edge attributes, and unary line features as vertex attributes. Both experiments dealt with the recognition of a translated and rotated version of a reference line image. Figure 1 was used as a reference image.

To simulate anomalies associated with the line extraction process and endpoint errors, noise was added to the x and y coordinates of every line endpoint. Noise values were obtained by multiplying a random variable—uniformly distributed on the interval $[-1/2, 1/2]$—by the noise magnitude parameters specified in table 1. Although it might be argued that additive Gaussian noise is more

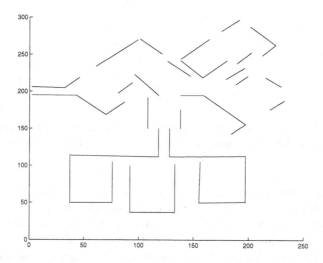

Fig. 1. Thirty-nine lines extracted from a non-occluded version of the original image without any noise present.

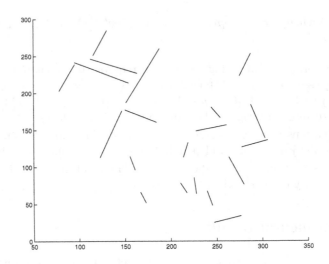

Fig. 2. Twenty lines extracted from a rotated and translated version of the original image with a noise factor of 10.

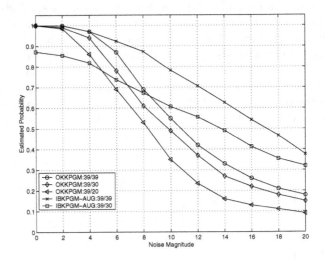

Fig. 3. Matching 39, 30 and 20 translated and rotated input lines to 39 reference lines: Estimated probability of a correct vertex-vertex assignment versus noise magnitude.

realistic, uniformly distributed noise was chosen since it affects the performance of the algorithms more adversely than Gaussian noise. Every line endpoint was uniformly displaced by a value in the interval $[0, \rho]$ in an arbitrary direction, where ρ is the maximum line endpoint shift given by table 1.

For the first experiment, all 39 reference lines from figure 1 were used to construct a reference graph. An input graph was constructed by randomly rotating, translating and adding noise to the reference lines. The estimated probability

Table 1. Noise magnitude versus maximum line endpoint shift.

Noise Magnitude	Maximum Endpoint Shift in Pixels (ρ)
2	1.0
4	2.8
6	4.2
8	5.7
10	7.0
12	8.5
14	9.9
16	11.3
18	12.7
20	14.1

of a correct vertex-vertex assignment was calculated for a given value of the noise magnitude parameter after every 100 trials. For the second and third experiments, we again used all 39 lines in figure 1 to construct a reference graph. Input graphs were constructed using a random selection of 30 and 20 rotated and translated lines from the reference list of lines as shown in figure 2. The lengths of the lines, in pixels, were used as vertex features. For edge features, the differences in orientation between lines, in degrees, line length ratios and distances in pixels between line midpoints were used.

For the first and second experiment the OKKPGM algorithm was compared to the IBKPGM algorithm described in [24] with polynomial feature augmentation. Unlike the OKKPGM algorithm the IBKPGM algorithm infers the KMM using an interpolator-based approach. Refer to figure 3 for the results obtained. Although the OKKPGM algorithm is more sensitive to noise than the IBKPGM algorithm, it's subgraph matching performance is superior for $\rho < 5.7$.

7   Conclusion

In this paper, the OKKPGM algorithm was derived. The OKKPGM algorithm uses the KPGM approach, but the way in which an approximation to Φ, the KMM was derived, differs significantly from the IBKPGM approach [23–24]. The OKKPGM algorithm is not based on the idea of performing graph matching without the explicit calculation of compatibility values since the kernel evaluations, required by the OKKPGM algorithm to infer an approximation to the constrained KMM, can be seen as compatibility calculations. A salient feature of the OKKPGM algorithm is the circumvention of a direct search for the largest common subgraph via an association graph.

References

1. Gold S. and Rangarajan A., A Graduated Assignment Algorithm for Graph Matching, IEEE Transactions on Pattern Analysis and Machine Intelligence, 18(4) (1996) 377–388.

2. Ullman J.R.: An Algorithm for Subgraph Isomorphism, Journal of the ACM, 23(1976)31–42.
3. Shapiro L.G., Haralick R.M.: Structural Descriptions and Inexact Matching, IEEE Transactions on Pattern Analysis and Machine Intelligence, 3(5)(1981)504–519.
4. Eshera M.A., Fu K-S.: A Graph Distance Measure for Image Analysis, IEEE Transactions on Systems Man and Cybernetics, 14(3)(1984)398–408.
5. Horaud R., Skordas T.: Stereo Correspondence Through Feature Grouping and Maximal Cliques, IEEE Transactions on Pattern Analysis and Machine Intelligence, 11(11)(1989)1168–1180.
6. Cinque L., Yashuda D., Shapiro L., Tanimoto S., Allen B.: An Improved Algorithm for Relational Distance Graph Matching, Pattern Recognition, 29(2)(1996)349–359.
7. Depiero F., Trived M., Serbin S.: Graph Matching using a Direct Classification of Node Attendance, Pattern Recognition, 29(6)(1996)1031–1048.
8. Shearer K., Bunke H., Venkatesh S.: Video Indexing and Similarity Retrieval by Largest Common Subgraph Detection Using Decision Trees, Pattern Recognition, 34(2001)1075–1091.
9. Barrow H.G., Burstall R.M.: Subgraph Isomorphism, Matching Relational structures and Maximal Cliques, Information Processing Letters, 4(4)(1976) 377–396.
10. Lazarescu M., Bunke H., Venkatesh S.: Graph Matching: Fast Candidate Elimination Using Machine Learning Techniques, in Lecture Notes in Computer Science Vol. 1876: Advances in Pattern Recognition, Ferri F.J., Iñesta J.M., Amin A. and Pudil P. (ed.), Springer-Verlag, 2000(236–245).
11. Messmer B.T., Bunke H.: A New Algorithm for Error-Tolerant Subgraph Isomorphism Detection, IEEE Transactions on Pattern Analysis and Machine Intelligence, 20(5)(1998) 493–503.
12. Christmas W.J., Kittler J., Petrou M.: Structural Matching in Computer Vision Using Probabilistic Relaxation, IEEE Transactions on Pattern Analysis and Machine Intelligence, 17(8)(1995)749–764.
13. Kittler J., Petrou M., Christmas W.J.: A Non-iterative Probabilistic Method for Contextual Correspondence Matching, Pattern Recognition, 31(1998)1455–1468.
14. Finch A.M., Wilson R.C., Hancock E.R.: Matching Delauney Triangulations by Probabilistic Relaxation, Proceedings CAIP, (1995)350–358.
15. Williams M.L., Wilson R.C., Hancock E.R.: Multiple Graph Matching with Bayesian Inference, Pattern Recognition Letters, 18(1997), 1275–1281.
16. Wilson R.C., Hancock E.R.: A Bayesian Compatibility Model for Graph Matching, Pattern Recognition Letters, 17 (1996)263–276.
17. Mjolsness E., Gindi G. , Anandan P.: Optimization in Model Matching and Perceptual Organization, Neural Computation, 1(1989)218–229.
18. Mjolsness E., Garrett C.: Algebraic Transformations of Objective Functions, Neural Networks, 3(1990)651–669.
19. Simić P.D.: Constrained Nets for Graph Matching and Other Quadratic Assignment Problems, Neural Computation, 3(1991)268–281.
20. Pelillo M.: Replicator Equations, maximal Cliques, and Graph Isomorphism, Neural Computation, 11(1999)1933–1955.
21. Van Wyk M.A., Clark J.: An Algorithm for Approximate Least-Squares Attributed Graph Matching, in Problems in Applied Mathematics and Computational Intelligence, N. Mastorakis (ed.), World Science and Engineering Society Press, 2000(67–72)

22. Van Wyk M.A., Durrani T.S., Van Wyk B.J.: A RKHS Interpolator-Based Graph Matching Algorithm, IEEE Transactions on Pattern Analysis and Machine Intelligence, 24(7)(2002)988–995.
23. Van Wyk B.J., Van Wyk M.A.: Non-Bayesian Graph Matching without Explicit Compatibility Calculations, in Lecture Notes in Computer Science Vol. 2396: Structural, Syntactic, and Statistical Pattern Recognition, Caelli T., Amin A., Duin R.P.W., Kamel M. and de Ridder D. (ed.), Springer-Verlag, 2002(74–82)
24. Van Wyk B.J., Van Wyk M.A.: Kronecker Product Graph Matching, Pattern Recognition Journal, In Press, (2003).

Theoretical Analysis and Experimental Comparison of Graph Matching Algorithms for Database Filtering

Christophe Irniger and Horst Bunke

Department of Computer Science, University of Bern
Neubrückstrasse 10, CH-3012 Bern, Switzerland
{bunke,irniger}@iam.unibe.ch

Abstract. In structural pattern recognition, an unknown pattern is often transformed into a graph that is matched against a database in order to find the most similar prototype in the database. Graph matching is a powerful yet computationally expensive procedure. If the sample graph is matched against a large database of model graphs, the size of the database is introduced as an additional factor into the overall complexity of the matching process. Database filtering procedures are used to reduce the impact of this additional factor. In this paper we report the results of a basic study on the relation between filtering efficiency and graph matching algorithm performance, using different graph matching algorithms for isomorphism and subgraph-isomorphism.

Keywords: graph isomorphism, subgraph isomorphism, database filtering

1 Introduction

Graphs play an important role in structural pattern recognition. Using graphs, the pattern recognition problem becomes a graph matching problem. In classification, a given sample pattern represented as a graph often needs to be matched against a database of model graphs [1,2,3]. Examples of applications of graph matching include document processing [4], image analysis [5], biometric identification [6] and video analysis [7].

Since graphs are a universal representation formalism, structural pattern recognition by means of graph matching is very attractive. The task of matching two graphs requires exponential time and memory resources in the number of nodes or edges involved. Furthermore, databases may be large and therefore introduce an additional factor in the overall complexity of the matching process. A variety of mechanisms have been proposed to reduce the complexity of graph matching when large databases are involved [3,8,9,10].

The present study is motivated by work on developing a filtering procedure based on machine learning methods [11,12]. This approach realizes database retrieval reducing the number of feasible candidates and then testing the remaining graphs with a conventional graph matching algorithm. In our previous work [11,

E. Hancock and M. Vento (Eds.): IAPR Workshop GbRPR 2003, LNCS 2726, pp. 118–129, 2003.

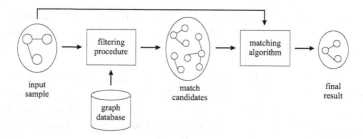

Fig. 1. Illustration of the filtering procedure in the database matching process.

12], it could very quickly be seen that the effectiveness of the overall scheme crucially depends on the underlying matching algorithm (and on the graphs of course). This led to the conjecture that the effectiveness of the proposed graph filtering approach could be increased if it was used in combination with suitable matching algorithms.

In this paper we will focus on graph and subgraph isomorphism. The problem of measuring the performance of various graph matching algorithms has been studied a number of times [13,14,15]. However the main objective in these studies was measuring and comparing the time needed to compare two "matching" graphs (i.e. isomorphic or subgraph-isomorphic graphs). Little attention was given to the performance of the algorithms when comparing "non-matching" graphs. This is, however, a crucial factor in graph database filtering methods as will be shown in this paper. The objective of this study is to examine the influence of graph matching algorithms when used in combination with graph database filtering methods, providing both theoretical and experimental analysis for the problem.

In the next section, theoretical foundations of the problem considered in this paper will be set. Then, in Section 3, the experimental setup including the underlying graph database will be presented, and results on matching and non-matching performance for various graph types and matching algorithms will be given and discussed. Finally, conclusions will be drawn in Section 4.

2   Theoretical Aspects

In this section, theoretical foundations for a performance evaluation of filtering procedures will be established. A filtering procedure in the context of this paper will be considered a method which allows the reduction of the initial graph database size with respect to a given input graph. The concept of such a procedure is shown in Figure 1. As can be seen, after the filtering procedure has been applied there are in general match candidates remaining that need to undergo a full matching procedure with the chosen matching algorithm, for example, Ullmann's algorithm [16] or VF 2 [17].

To measure the effectiveness of a filter it is best to compare it with a brute force approach. That is, filtering in combination with the chosen matcher is compared against sequential testing of the entire database with the same matcher (brute force approach). To formally describe the problem, we introduce the following variables:

- T^i_{filter}, total time needed to extract all matching graphs for a given input graph from the database using the filtering approach in conjunction with some graph matching algorithm i (for example, $i=$Ullmann's algorithm)
- $T^i_{\text{brute\_force}}$, total time needed to extract all matching graphs for a given input graph from the database applying the chosen graph matching algorithm i to each graph in the database in a brute force fashion.
- t^i_{match}, the time needed by the matching algorithm i to perform a comparison on two matching graphs (which are either isomorphic or subgraph-isomorphic to each other, depending on the considered problem).
- $t^i_{\text{non\_match}}$, the time needed by the matching algorithm i to perform a comparison on two non-matching graphs.
- t_{filter}, the time needed to perform the filtering procedure. Notice that this variable doesn't depend on the underlying graph matching algorithm.
- m_{db}, database size (the number of graphs stored in the database).
- m_{filter}, the size of candidates left for testing after the filtering procedure has been applied. The smaller this value, the more effective is the filter (m_{filter} is equal to the size of set 'match candidates' in Figure 1).
- s, the number of matching graphs for a given input sample. In case of isomorphism (subgraph-isomorphism), this number is equal to the number of graphs in the database that are isomorphic (supgraph-isomorphic) to the input. This means that the matching algorithm will have to execute s times a successful match.

Note that t^i_{match} and $t^i_{\text{non\_match}}$ are themselves depending on other parameters, such as the number of nodes and edges in a graph and the size of the alphabet of labels. Similarly, m_{filter} is highly dependent on the given input sample, the database as well as the filtering method. Note that the value of m_{filter} is bounded by s and m_{db}, such that $s \leq m_{\text{filter}} \leq m_{\text{db}}$.

We can now formally state the time needed to retrieve all candidate graphs of a given sample from the database. Assuming there is no filtering involved, the graphs in the database must sequentially be tested with the input sample using the matching algorithm. The total time needed to do this is composed of two terms, the time needed to process the matching graphs (left) and the time needed to process the non-matching graphs (right). Hence

$$T^i_{\text{brute\_force}} = \left(s \cdot t^i_{\text{match}}\right) + \left((m_{\text{db}} - s) \cdot t^i_{\text{non\_match}}\right). \qquad (1)$$

Assuming we are given a matching algorithm and a graph database, the behavior of brute force matching can be examined for varying values of s. There are two cases to be considered (note that s is limited to the interval $s \in [0, m_{\text{db}}]$):

1. s large, with a maximum of $s = m_{\text{db}}$: In this case, the time needed for comparing matching graphs t^i_{match} is weighted more whereas the weight of

the time needed for comparing non-matching graphs is reduced. Hence, the brute force performance will highly depend on t^i_{match}. (In case of $s = m_{db}$, $t^i_{non\_match}$ has no influence at all.)

2. s small, with a minimum of $s = 0$: Here, the weighting of the times is inverse to the case discussed before. Brute force performance will therefore mostly depend on $t^i_{non\_match}$ and the effect of t^i_{match} will be minimal.

Considering two graph matching algorithms $\{A,B\}$, using Equation (1) we can state

$$T^A_{brute\_force} = \left(s \cdot t^A_{match}\right) + \left((m_{db} - s) \cdot t^A_{non\_match}\right)$$
$$T^B_{brute\_force} = \left(s \cdot t^B_{match}\right) + \left((m_{db} - s) \cdot t^B_{non\_match}\right)$$

Defining δ_{match} and δ_{non_match} as

$$\delta_{match} = t^A_{match} - t^B_{match}$$
$$\delta_{non\_match} = t^A_{non\_match} - t^B_{non\_match},$$

then $T^B_{brute\_force}$ can be rewritten as

$$T^B_{brute\_force} = \left(s \cdot (t^A_{match} - \delta_{match})\right) + \left((m_{db} - s) \cdot (t^A_{non\_match} - \delta_{non\_match})\right). \quad (2)$$

Expressing $T^B_{brute\_force}$ as shown in Equation (2), we can now compare the performance of two different graph matching algorithms $\{A,B\}$ under brute force matching. Assuming $\delta_{match} \neq 0$ and $\delta_{non\_match} \neq 0$ (the case where these values are 0 is trivial), $T^A_{brute\_force}$ and $T^B_{brute\_force}$ will be equal if

$$\left(s \cdot t^A_{match}\right) + \left((m_{db} - s) \cdot t^A_{non\_match}\right) -$$
$$\left(s \cdot (t^A_{match} - \delta_{match})\right) + \left((m_{db} - s) \cdot (t^A_{non\_match} - \delta_{non\_match})\right) = 0.$$

This is equivalent to

$$\delta_{non\_match} m_{db} - \delta_{non\_match} s + \delta_{match} s = 0$$

which can be simplified to:

$$\frac{s}{m_{db}} = \frac{1}{1 - \dfrac{\delta_{match}}{\delta_{non\_match}}}. \quad (3)$$

Since $0 \leq s \leq m_{db}$, the value of the expression in the left-hand side of Equation (3) is obviously limited to the interval $[0, 1]$. Therefore, the value of the denominator in the right-hand side must be larger than or equal to 1. It follows that the value of $\delta_{match}/\delta_{non\_match}$ is limited to the interval $]-\infty, 0]$. This is particularly interesting because it shows that δ_{match} and δ_{non_match} are of inverse sign. Hence, if brute force matching with two different matching algorithms, A and B, takes equal time to match a sample with a given database and $\delta_{match} \neq 0$ as well as $\delta_{non\_match} \neq 0$, then if algorithm A's match performance is better than algorithm B's match performance, B's non-match performance must be better than A's non-match performance. In that case there must be an inverse relation between matching and non-matching performance for algorithms A and B. Furthermore, given a fixed value for $\delta_{\text{non\_match}}$ and assuming $T^{\text{A}}_{\text{brute\_force}} = T^{\text{B}}_{\text{brute\_force}}$, if the value of δ_{match} is increased, the ratio s/m_{db} must decrease, which means that either s becomes smaller or m_{db} grows larger. This means that if the performance of one matching algorithm drops when comparing matching graphs, either the database size must be increased (more non-matching graphs must be introduced) or the number of matching graphs must be reduced in order for the two brute force algorithms, A and B, to still perform equal. On the other hand, if we assume that $t^{\text{A}}_{\text{match}} \geq t^{\text{B}}_{\text{match}}$ and $t^{\text{A}}_{\text{non\_match}} \geq t^{\text{B}}_{\text{non\_match}}$, then $\delta_{\text{match}}/\delta_{\text{non\_match}}$ is always positive and it follows that the right-hand side of Equation (3) will never be within the interval $[0, 1]$ unless $\delta_{\text{match}}/\delta_{\text{non\_match}} = 0$ which is the case if $t^{\text{A}}_{\text{match}} = t^{\text{B}}_{\text{match}}$ and $t^{\text{A}}_{\text{non\_match}} = t^{\text{B}}_{\text{non\_match}}$ (which simply means that the two graph matching algorithms perform the same).

Introducing a filtering method, we can formulate its performance similarly to Equation (1) as:

$$T^i_{\text{filter}} = s \cdot t^i_{\text{match}} + (m_{\text{filter}} - s) \cdot t^i_{\text{non\_match}} + t_{\text{filter}} \qquad (4)$$

Note that due to the filter, the database size has been reduced from m_{db} to m_{filter}. On the other hand, an additional term, t_{filter}, has been added to the total time. For filtering to be effective, the relation $T^i_{\text{filter}} < T^i_{\text{brute\_force}}$ must hold. This is equivalent to

$$\left(s \cdot t^i_{\text{match}}\right) + \left((m_{\text{filter}} - s) \cdot t^i_{\text{non\_match}}\right) + t_{\text{filter}} < \left(s \cdot t^i_{\text{match}}\right) + \left((m_{\text{db}} - s) \cdot t^i_{\text{non\_match}}\right)$$

and can be simplified to

$$t_{\text{filter}} < t^i_{\text{non\_match}} \cdot (m_{\text{db}} - m_{\text{filter}}) \qquad (5)$$

When analyzing Equation (5) one can see that comparing the brute force approach with filtering, the time t^i_{match} needed for comparing two matching graphs is irrelevant. The inequality shows that the main factors for the performance of a filter approach are the time needed for comparing two non-matching graphs, $t^i_{\text{non\_match}}$, and the term m_{filter} which reflects the ability of the filter to reduce the database size m_{db}. If the matching algorithm performs well comparing non-matching graphs (i.e. if $t^i_{\text{non\_match}}$ is small), then the filter approach needs to be able to significantly reduce the database size in order to outperform the brute force approach (i.e. $m_{\text{db}} - m_{\text{filter}}$ needs to be large, see Equation (5)). On the other hand, if the matching algorithm performs poorly comparing two non-matching graphs (i.e. if $t^i_{\text{non\_match}}$ is large), then the reduction factor of the filter approach needs not to be quite as large in order for the filter to be faster than the brute force procedure (i.e. $m_{\text{db}} - m_{\text{filter}}$ can be rather small).

3   Experimental Comparison and Discussion

The primary goal of the experiments described in this section was to measure match and non-match performance of graph matching algorithms on different graph types. While matching performance has been studied in other papers before [13,14,15], we will include an investigation of the non-matching performance, motivated by the analysis given in the last section. The aim was to gain some understanding of how matching and non-matching performance of different graph matching algorithms depends on the graph type. The following parameters were varied in the experimental setup:

- matching task: graph isomorphism and subgraph isomorphism
- graph type: random graphs, meshes, bounded valence graphs
- graph matcher: Ullmann [16] and VF 2 [17] graph matching algorithms

In this study, the types of graphs examined are random graphs, bounded valence graph (regular/irregular) and meshes (regular/irregular). The generators for the graphs were either provided by the VFLib (see [13]) or implemented from scratch.

- Random Graphs: The random graph database consists of graphs with different node and label alphabet size. The graph size was varied from 5 to 70 nodes. The number of edges in a graph was varied such that connected graphs were created. The generator was started with as many edges as there were nodes. Then edges were sequentially added until the graph was connected.
- Meshes / Hyper-Cuboids (regular, irregular): The hypercuboid-generator produces cuboids of varying dimension n. In the experiments, databases of graphs with dimension $n = 2$ (meshes), $n = 3$ (cuboids), $n = 4, 5$ (hypercuboids) were created. Since the generator produced hypercuboids, the dimension and the interval of possible values on each axis of the cuboids were parameters to be fixed. The values e_i $(i = 1 \ldots n)$ per dimension for a single graph were chosen at random from the given interval. The boundaries for the axis-intervals were set as follows:
  - 2d-mesh: $e_i \in [2, \ldots, 10]$, i.e. a mesh of size 3×9
  - 3d-cuboid: $e_i \in [2, \ldots, 8]$, i.e. a cuboid of size $4 \times 6 \times 2$
  - 4d-hypercuboid: $e_i \in [2, \ldots, 6]$, i.e. a hypercuboid of size $3 \times 5 \times 2 \times 4$
  - 5d-hypercuboid: $e_i \in [2, \ldots, 4]$, i.e. a hypercuboid of size $2 \times 3 \times 2 \times 4 \times 3$
  To create irregular meshes, edges were randomly added to nodes according to a given probability. This probability was varied from 0.0, 0.2, 0.4 to 0.6.
- Bounded Valence Graphs (regular, irregular): Bounded valence graphs (bvg-graphs) are graphs where the number of incident edges per node is limited by a valence value. The generator creates graphs of either a fixed valence (regular bounded valence graphs) or graphs where after creation, a certain number of edges are moved to different nodes, thus ensuring constant valence over the entire graph (irregular bounded valence graphs). In these experiments, the number of nodes in the graphs was varied from 5 to 100 and the valence was set to 30% of the number of nodes in the graph. The irregularity was introduced by moving 10% of a graph's edges.

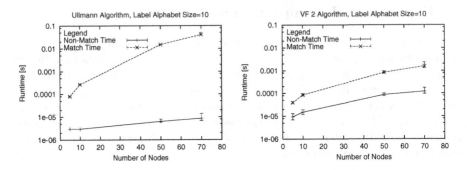

Fig. 2. Match times vs. non-match times for the Ullmann and the VF 2 algorithm on random graphs.

The label alphabet was varied, for all three graph types, from 5 to 100 labels. Labels were randomly assigned (based on a uniform distribution) to nodes. For all graph types and all parameter settings a database of 1000 graphs was created. During creation of the database, it was made sure that each graph was isomorphic only to itself. To measure non-match performance, each graph of the database was matched against the other 999 graphs. To measure match-performance, each graph was matched against itself 1000 times. To minimize the effect of outliers, 50% of the measured runtime values were discarded (25% of the smallest and 25% of the largest values).

In our study we used the VFlib (described in [13]) which provides implementations of several isomorphism and subgraph-isomorphism algorithms. We compared Ullmann and VF 2 against one another since these are the only fundamentally different algorithms contained in VFlib providing both graph isomorphism and subgraph isomorphism functionality.

The data was analyzed with the objectives of comparing matching performance versus non-matching performance for each of the selected matching algorithms on the classes of graphs described above. Figure 2 (left) shows match and non-match time of Ullmann's algorithm on random graphs as a function of graph size. The figure indicates very clearly that for Ullmann's algorithm the average time needed to match two graphs which are isomorphic to each other is much longer than the time needed to identify two graphs as being non-isomorphic. This is not just a phenomenon specific to Ullmann's algorithm, but can be observed for VF 2 as well, see Figure 2 (right). However, comparing the performance of Ullmann's algorithm and VF 2 it can also be seen that the gap between match-/non-match-performance is much smaller for the VF 2 algorithm than for Ullmann's method. Obviously VF 2 is faster than Ullmann's algorithm for matching graphs, but the inverse relation holds for non-matching graphs. Hence for the case of random graphs, Ullmann's method and VF 2 are instances of algorithms where parameters δ_{match} and $\delta_{\text{non\_match}}$, introduced in Section 2, have different sign. Therefore, which of the two algorithms is more efficient will eventually depend on the ratio s/m_{db} (see Equation (3)). If there are

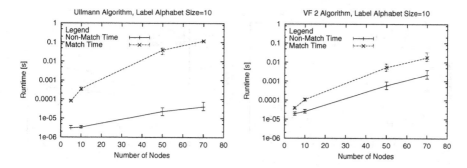

Fig. 3. Match times vs. non-match times for the Ullmann and the VF 2 algorithm on bounded valence graphs.

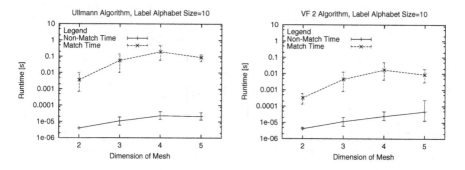

Fig. 4. Match times vs. non-match times for the Ullmann and the VF 2 algorithm on mesh graphs of varying dimension.

many matching graphs, then VF 2 will outperform Ullmann's method, while for the case where there are many non-matching graphs in the database, Ullmann's method will be faster.

The behavior of the two algorithms observed in Figure 2 can also be verified for other graph types. Figure 3 shows the performance for bounded valence graphs. In general, both algorithms are slower on graphs of this kind, but it can be seen again that Ullmann's algorithm is faster for the recognition of non-matching graphs whereas VF 2 performs better on matching graphs.

On mesh graphs there was again a large difference in runtime between the matching and the non-matching case (Figure 4). Again, VF 2 performs better than Ullmann in matching isomorphic graphs. However, the inverse relation between Ullmann and VF 2 concerning non-match performance that was observed on the bounded valence and random graphs no longer holds (here, both algorithms perform about the same).

The results obtained for meshes, bounded valence and random graphs are confirmed when looking at subgraph-isomorphism. To evaluate the performance of subgraph isomorphism detection, the graphs in the database were assumed

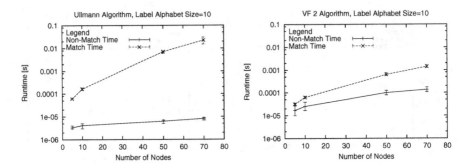

Fig. 5. Subgraph-Isomorphism match times vs. non-match times for the Ullmann and the VF 2 algorithm on random graphs.

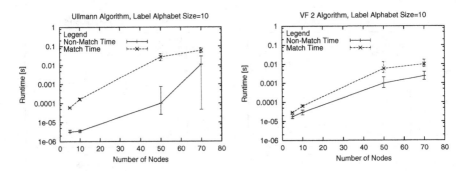

Fig. 6. Subgraph-Isomorphism match times vs. non-match times for the Ullmann and the VF 2 algorithm on bounded valence graphs.

to be supergraphs of the given input sample. The sample graph was initially defined to be one of the database graphs and then 30% of its nodes were cut. In Figure 5 we observe a behavior very similar to Figure 2. That is, Ullmann's approach is again very efficient compared to the VF algorithm on non-matching subgraph samples, yet it performs poorly on identifying subgraphs. Figures 6 and 7 confirm these observations for bounded valence and mesh graphs.

Now we give an example of how the results from Section 2 and 3 can be combined. Look at Figure 2 and assume the graphs in the database are of size 50. Then

$$\delta_{\text{match}} = t_{\text{match}}^{\text{Ullmann}} - t_{\text{match}}^{\text{VF\_2}} \approx 0.015\text{s} - 0.00083\text{s} = 0.014\text{s}$$

$$\delta_{\text{non\_match}} = t_{\text{non\_match}}^{\text{Ullmann}} - t_{\text{non\_match}}^{\text{VF\_2}} \approx 0.0000063\text{s} - 0.000087\text{s} = -0.00008\text{s}.$$

Using Equation (3) it follows that

$$\frac{s}{m_{\text{db}}} \approx \frac{1}{176}.$$

This means that for a database size of 176 graphs and 1 matching graph in the database, brute force matching using either Ullmann's algorithm or VF 2

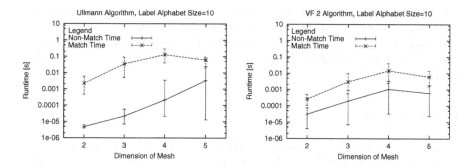

Fig. 7. Subgraph-Isomorphism match times vs. non-match times for the Ullmann and the VF 2 algorithm on mesh graphs.

will approximately perform the same. If the database is increased (for example to 1000 graphs) with constant value $s = 1$, then brute force matching using Ullmann's algorithm will be considerably faster than brute force matching using VF 2. If on the other hand there are more matching samples (s is increased) yet the database size stays constant, brute force matching will perform better if used in combination with VF 2.

The results derived in Section 2 also show that graph database filtering only improves the overall performance of database retrieval if the time needed to apply the filtering method is smaller than the time needed to compare, with the input sample, all non-matching graphs that are filtered out. For a given database of graphs and a graph matching algorithm, Equation (5) shows that filtering can be effective if either the filter method significantly reduces the number of candidates remaining for full-fledged graph matching or the non-match performance of the graph matching algorithm is very poor. Consider again Figure 2 for graphs consisting of 50 nodes. The average non-match runtime for Ullmann's algorithm is $t_{\text{non\_match}}^{\text{Ullmann}} \approx 0.0000063\text{s}$. The database itself consists of $m_{\text{db}} = 1000$ graphs. Assuming the filtering method reduces the database size to 2% ($m_{\text{filter}} = 20$), we obtain

$$t_{\text{filter}} < 0.0000063\text{s} \cdot 980 \approx 0.0062\text{s},$$

which means that a filtering method needs to be faster than 0.0062s in order for it to be more effective than a brute force approach using Ullmann's algorithm. Furthermore, the maximum performance gain to be expected from a filtering method is at most 0.0062s. With the average matching time $t_{\text{match}}^{\text{Ullmann}}$ for that database being $s \cdot t_{\text{match}}^{\text{Ullmann}} \approx 20 \cdot 0.015\text{s} = 0.3\text{s}$, the filtering effect for this database and graph matching algorithm is minimal. Considering VF 2's match and non-match performance on the other hand ($t_{\text{match}}^{\text{VF\_2}} \approx 0.00083\text{s}, t_{\text{non\_match}}^{\text{VF\_2}} \approx 0.000087\text{s}$ respectively), the maximum filter gain to be achieved is

$$t_{\text{filter}} < 0.000087\text{s} \cdot 980 \approx 0.085\text{s}.$$

Comparing this to the time needed for final matching $s \cdot t_{\text{match}}^{\text{VF\_2}} \approx 20 \cdot 0.00083\text{s} \approx 0.017\text{s}$, the effect of the filter will be significant. Note that, if database size m_{db} is increased further with constant matching sample value s and filter value m_{filter}, the effect of the filter will become even stronger.

It has to be pointed out that methods for database filtering and analysis of filtering efficiency are not the primary topic of the current paper. Such methods can be found in [11,12].

4   Conclusions

In this paper results of a theoretical and experimental study on graph database filtering and graph matching performance have been presented. In particular two popular graph matching algorithms and the relation between matching and non-matching performance on different graph types have been tested and compared. Furthermore, the algorithms' suitability for databases of various graph types has been evaluated. The results of the study may be useful for the design of optimal graph matching schemes that include a possibly large database of model graphs.

One particular aspect that has been widely neglected until now is the observation that the non-matching time of an algorithm has a crucial impact on the overall performance of a graph matching scheme. For large databases, a small number of expected matches, and no or only moderate filtering, the non-matching performance of the underlying graph matching algorithm is the dominant factor in the overall efficiency of a graph matching scheme. In addition to matching and non-matching time, we have identified a number of parameters, such as total database size, size of the database after filtering, number of matches in the database, and time needed for filtering. These parameters have a crucial influence on the behavior of a graph matching scheme. In particular, the number of matches in the database and database size can be used to predict the performance of given graph matching algorithms under both brute force matching and a filtering approach. Also, it has been shown that if two different matching algorithms perform equally on a given database, there exists an inverse relation between their matching and non-matching performance. The framework presented in the current paper can be extended, in a straightforward way, to other types of graphs, other (subgraph-) isomorphism algorithms and other graph matching tasks.

Acknowledgment. This research was supported by the Swiss National Science Foundation (Nr. 2100-066700). The authors thank the Foundation for the support.

References

1. Shapiro, L., Haralick, R.: Structural descriptions and inexact matching. In: IEEE Trans. on Pattern Analysis and Machine Intelligence. Volume 3. (1981) 504 – 519
2. Sanfeliu, A., Fu, K.: A distance measure between attributed relational graphs for pattern recognition. In: IEEE Trans. Systems, Man, and Cybernetics. Volume 13. (1983) 353 – 363

3. Messmer, B., Bunke, H.: A new algorithm for error–tolerant subgraph isomorphism detection. In: IEEE Trans. Pattern Analysis and Machine Intelligence. Volume 20. (1998) 493 – 505
4. Llados, J., Marti, E., Villanueva, J.: Symbol recognition by error-tolerant subgraph matching between region adjacency graphs. In: IEEE Transactions on Pattern Analysis and Machine Intelligence. Volume 23 – 10. (2001) 1137 – 1143
5. Torsella, A., Hancock, E.: Learning stuctural variations in shock trees. In: Proc. of the Joint IAPR International Workshops SSPR and SPR. (2002) 113 – 122
6. Tefas, A., Kotropoulos, C., Pitas, I.: Using support vector machines to enhance the performance of elastic graph matching for frontal face authentification. In: IEEE Transactions on Pattern Analysis and Machine Intelligence. Volume 23 – 7. (2001) 735 – 746
7. Chen, H., Lin, H., Liu, T.: Multi-object tracking using dynamical graph matching. In: Proc. of the 2001 IEEE Computer Society Conference on Computer Vision and Pattern Recognition. (2001) 210 – 217
8. Messmer, B., Bunke, H.: A decision tree approach to graph and subgraph isomorphism. In: Pattern Recognition. Volume 32. (1999) 1979 – 1998
9. Shapiro, L., Haralick, R.: Organization of relational models for scene analysis. In: IEEE Trans. Pattern Analysis and Machine Intelligence. Volume 3. (1982) 595 – 602
10. Sengupta, K., Boyer, K.: Organizing large structural modelbases. In: IEEE Trans. Pattern Analysis and Machine Intelligence. Volume 17. (1995)
11. Irniger, C., Bunke, H.: Graph matching: Filtering large databases of graphs using decision trees. In Jolion, J.M., Kropatsch, W., Vento, M., eds.: Graph-based Representations in Pattern Recognition, Cuen (2001) 239 – 249
12. Lazarescu, M., Bunke, H., Venkatesh, S.: Graph matching: Fast candidate elimination using machine learning techniques. In Ferri, F., ed.: Advances in Pattern Recognition, Springer Verlag (2000) 236 – 245
13. Foggia, P., Sansone, C., Vento, M.: A database of graphs for isomorphism and subgraph isomorphism. In Jolion, J.M., Kropatsch, W., Vento, M., eds.: Graph-based Representations in Pattern Recognition, Cuen (2001) 176 – 188
14. Foggia, P., Sansone, C., Vento, M.: A performance comparison of five algorithms for graph isomorphism. In Jolion, J.M., Kropatsch, W., Vento, M., eds.: Graph-based Representations in Pattern Recognition, Cuen (2001) 188 – 200
15. Kropatsch, W.: Benchmarking graph matching algorithms - a complementary view. In Jolion, J.M., Kropatsch, W., Vento, M., eds.: Graph-based Representations in Pattern Recognition, Cuen (2001) 210 – 217
16. Ullmann, J.: An algorithm for subgraph isomorphism. In: JACM. Volume 23. (1976) 31 – 42
17. Cordella, L., Foggia, P., Sansone, C., Vento, M.: An improved algorithm for matching large graphs. In Jolion, J.M., Kropatsch, W., Vento, M., eds.: Graph-based Representations in Pattern Recognition, Cuen (2001) 149 – 159

A Comparison of Three Maximum Common Subgraph Algorithms on a Large Database of Labeled Graphs

D. Conte[1], C. Guidobaldi[2], and C. Sansone[2]

[1] Dipartimento di Ingegneria dell'Informazione ed Ingegneria Elettrica,
Università di Salerno, via P.te Don Melillo, I-84084 Fisciano (SA), Italy
dconte@unisa.it
[2] Dipartimento di Informatica e Sistemistica,
Università di Napoli "Federico II", Via Claudio, 21 I-80125 Napoli (Italy)
{cguidoba,carlosan}@unina.it

Abstract. A graph g is called a *maximum common subgraph* of two graphs, g_1 and g_2, if there exists no other common subgraph of g_1 and g_2 that has more nodes than g. For the maximum common subgraph problem, exact and inexact algorithms are known from the literature. Nevertheless, until now no effort has been done for characterizing their performance, mainly for the lack of a large database of graphs. In this paper, three exact and well-known algorithms for maximum common subgraph detection are described. Moreover, a large database containing various categories of pairs of graphs (e.g. randomly connected graphs, meshes, bounded valence graphs...), having a maximum common subgraph of at least two nodes, is presented, and the performance of the three algorithms is evaluated on this database.

1 Introduction

Graphs are a powerful and versatile tool used in various fields of science and engineering. There are many applications where graphs are used for the representation of structured objects. When two such objects, as is often the case, are to be compared, some form of *graph matching* is used.

Among the different kinds of matching, graph isomorphism is useful to find out if two graphs have identical structure, while subgraph isomorphism can be used to check if one graph is part of another. A still more general form of comparison is the detection of the *Maximum Common Subgraph* (MCS), i.e. the largest graph that is a subgraph of both the graphs being compared.

The MCS detection problem has been widely investigated. Many of the proposed approaches compute the MCS by reducing it to the problem of finding the maximum *clique* (i.e. a completely connected subgraph) of a suitably constructed *association graph* [1]. Since it is known that both MCS and maximum clique detection are NP-complete problems, many approximate algorithms have also been developed [2].

Despite of the significant number of MCS detection algorithms proposed in the literature, until now no effort has been taken for characterizing their performance. The authors of each novel algorithm usually provide experimental results supporting the claim that, under suitable hypothesis, their method can outperform the previous ones. Nevertheless, it is almost always very difficult for a user to choose the algorithm which is best suited for dealing with the problem at hand. In fact, a report of the

E. Hancock and M. Vento (Eds.): IAPR Workshop GbRPR 2003, LNCS 2726, pp. 130–141, 2003.

algorithm performance on a specific test case can often provide no useful clues about what will happen in a different domain.

Unfortunately, only a few papers face the problem of an extensive comparison among graph matching algorithms, in terms of key performance indices [3,4]. One problem is that for such algorithms, a purely theoretical analysis can only provide lower and upper bounds that are not sufficiently tight to be of practical use, except for very special cases. So it is important to perform a good experimental comparison on a large database of graphs where different kinds of graph types are represented.

The creation of a graph database suitable for such a comparison is definitely not a simple task, since several issues have to be taken into account: for instance, the origin of the graphs; the kinds of graphs; their size, and the values of other parameters that can characterize the graphs; the kind of semantic attributes attached to nodes and edges. These questions do not have a definite answer valid for everyone. Nevertheless, it is our opinion that it is important to start building and sharing graph data sets, to provide a common ground for comparing the algorithms performance.

Hence, we present a synthetically generated large database containing various categories of pairs of attributed graph, i.e. randomly connected graphs, 2D, 3D and 4D regular and irregular meshes, regular and irregular bounded valence graphs, in which each pair has a known maximum common subgraph. Since graphs used in pattern recognition applications have usually attributes on nodes and edges, we have also introduced such attributes, trying to define them in a way that will not limit the generality of the database.

We have used this database to compare three well-known MCS algorithms [5,6,7], in order to highlight under which conditions each algorithm performs better than its competitors. All the three algorithms selected are optimal, in the sense that they will always find the best MCS; also, they are all based on some form of depth-first search with heuristic pruning of the search space. This choice will make their performance more directly comparable. Two of the algorithms are based on the reduction of MCS to the maximum clique problem, while the third relies on a State Space Representation formulated directly in terms of the MCS problem.

The remainder of the paper is organized as follows. In Section 2, basic terminology and concepts are introduced. In Section 3 the three selected MCS algorithms are described. Next, in Section 4 the database of graphs is presented, while experimental results are reported in Section 5. Finally, some conclusions are drawn in Section 6.

2 Basic Definitions

An *attributed graph* (hereinafter simply called *graph*) is a 4-tuple $g = (V, E, \alpha, \beta)$, where V is the finite set of vertices (also called nodes), $E \subseteq V \times V$ is the set of edges, $\alpha : V \rightarrow L$ is a function assigning labels to the vertices and $\beta : E \rightarrow L$ is a function assigning labels to the edges. L is the label alphabet; the same alphabet is used for both node and edge labels for notational convenience.

Given a graph $g = (V, E, \alpha, \beta)$, we say that $g' = (V', E', \alpha', \beta')$, is a *subgraph* of g (denoted as $g' \subseteq g$), if $V' \subseteq V$, $\alpha(v) = \alpha'(v)$ for all $v \in V'$, $E' = E \cap(V' \times V')$ and $\beta(e) = \beta'(e)$ for all $e \in E'$. From the above definition it follows that, given a graph $g = (V, E, \alpha, \beta)$, any subset $V' \subseteq V$ of its vertices uniquely defines a subgraph, called the subgraph *induced* by V'.

A matching process between two graphs g and g' consists in the determination of a mapping M which associates nodes of the graph g with nodes of g'.

A *graph isomorphism* between g and g' is a bijective mapping $f : V \rightarrow V'$ such that $\alpha(v) = \alpha'(f(v))$ for all $v \in V$, and for every edge $e = (u,v) \in E$ there exists an edge $e' = (f(u), f(v)) \in E'$ such that $\beta(e) = \beta'(e')$; furthermore, the inverse relation must hold for every edge of g'. If $f : V \rightarrow V'$ is a graph isomorphism between a graph g and a subgraph of another graph g', then f is called a *subgraph isomorphism* from g to g'.

A *common subgraph* of g_1 and g_2, is a graph g such that a subgraph isomorphism from g to g_1 and from g to g_2 exists. We call g a *maximum common subgraph* of g_1 and g_2, denoted as $mcs(g_1,g_2)$, if it is a common subgraph and there is no other common subgraph having more nodes than g. Notice that, according to this definition, $mcs(g_1,g_2)$ is not necessarily unique for two given graphs.

3   The Selected MCS Algorithms

In this section we will provide a brief description of the three algorithms that will be used for our experimental comparison. These algorithms are quite similar under several respects. They all belong to the category of exact, or optimal, matching algorithms, as opposed to approximate or suboptimal ones, in the sense that they always find the correct MCS, and not an approximate solution to the problem. Since MCS is a NP-complete problem, their worst case time complexity is exponential (more precisely, factorial) with respect to the number of nodes in the graphs. Also, as we will see in the next subsections, their structure is quite similar: they perform a depth-first search, with the help of some heuristic for pruning unfruitful search paths.

The differences among the three algorithms actually lie only in the information used to represent each state of the search space (that is reflected in their different space complexity), and in the kind of heuristic adopted.

The choice of three similar algorithms has been made for the very purpose of enabling a more effective interpretation of the experimental results: by reducing to a minimum the possible causes of the measured performance diversity, it will be easier to find a convincing explanation.

One final remark before introducing the algorithms: the original formulation provided by the authors of the three algorithms was not directly applicable to our needs, either because they used a slightly different definition of the MCS problem (e.g., requiring a maximization of the number of edges, instead of the number of nodes), or because of some minor limitations (e.g. the need of working on connected graphs). So we have modified the definition of the algorithms in a way that does not alter significantly their structure and their performance, but allows us to compare directly their behavior.

3.1   The McGregor Algorithm

The first algorithm proposed here has been derived from the one described in [5]. It can be suitably described through a State Space Representation [8]. Each state s represents a common subgraph of the two graphs under construction. This common subgraph is part of the MCS to be eventually formed. The first state is the empty state, in which two null-nodes are analyzed and an empty solution is stored. In each state a pair of nodes not yet analyzed, the first belonging to the first graph and the second belonging to the second graph, is selected (whenever it exists). The selected pair of nodes is analyzed to check whether it is possible to extend the common subgraph represented by the actual state by means of this pair of nodes. If this extension is possible, then the current partial solution is extended by adding the pair (n_1,n_2); if the current partial solution is larger then the stored solution, than it becomes the new stored solution. After that, if the current state s is not a leaf of the search tree, a new state, descending from the current one, is generated and the analysis of this new state is started. After the new state has been analyzed, a backtrack procedure is used to restore the common subgraph of the previous state and to choose a different new state. Using this search strategy, whenever a branch is chosen, it will be followed as deeply as possible in the search tree. The research along a branch is terminated when a leaf is reached or when the number of nodes from the current state to the leaf is not enough to construct a common subgraph larger than the stored common subgraph (*pruning condition*). It is noteworthy that every branch of the search tree has to be followed, because - except for trivial examples - is not possible to foresee if a better solution exists in a branch that has not yet been explored. It is also noteworthy that, whenever a state is not useful anymore, it is removed from the memory, thus during the exploration of a node, only the states of the branch ending with that node are stored.

A pseudo-code description of the maximum common subgraph detection algorithm is shown in Fig. 1. If N_1 and N_2 are the number of nodes of the starting graphs, with $N_1 \leq N_2$, it can be proved that each branch of the search tree is no more than N_1 levels deep. Since at each level the space requirement is constant, the total space requirement of the algorithm is $O(N_1)$.

```
procedure McGregor_MCS(s)
  while (NextPair(s,n1,n2))
    if (IsFeasiblePair(s,n1,n2)) then
      s' = AddPair(s,n1,n2);
      if (size(s') > CurrentMCSSize) then
        SaveCurrentMCS(s'); CurrentMCSSize = size(s');
      end if
      if (!LeafOfSearchTree(s') and !PruningCondition(s')) then
        McGregor_MCS(s');
      end if
      BackTrack(s');
    end if
  end while
end procedure
```

Fig. 1. A sketch of the space state search for the McGregor algorithm.

3.2 The Durand-Pasari Algorithm

The Durand-Pasari algorithm is based on the well-known reduction of the search of the MCS to the problem of finding a maximal clique in a graph [6]. The first step of the algorithm is the construction of the association graph, whose vertices corresponds to pair of vertices of the two starting graphs having the same label. The edges of the association graph (that are undirected) represent the compatibility of the pair of vertices to be included. That is, a node corresponding to the pair (n_1,n_2) is connected to a node corresponding to (m_1,m_2) if and only if the mapping of n_1 to n_2 does not preclude the mapping of m_1 to m_2 and vice versa. This condition can be easily checked by looking at the edges between n_1 and m_1 and between n_2 and m_2 in the two starting graphs; edge attributes, if present, must also be taken into account. It can been easily demonstrated that each clique in the association graph corresponds to a common subgraph and vice versa; hence, the MCS can be obtained by finding the maximal clique in the association graph. The algorithm for maximum clique detection generates a list of vertices that represents a clique of the association graph using a depth-first search strategy on a search tree, by systematically selecting one node at a time from successive levels, until it is not possible to add further vertices to the list. A sketch of the algorithm is in Fig. 2.

```
procedure DurandPasari_MC(s)
   while (NextNode(s,n))
      if (IsLegalNode(s,n) && !PruningCondition(s)) then
         s' = AddNode(s,n);
         if (size(s')>CurrentMCSize) then
            SaveCurrentMC(s'); CurrentMCSize = size(s');
         end if
         if(!LeafOfSearchTree(s')) then DurandPasari_MC(s');
         end if
         BackTrack(s');
      end if
   end while
end procedure
```

Fig. 2. A sketch of the Durand-Pasari algorithm.

When a node is being considered, the forward search part of the algorithm first checks whether this node is a *legal* node (i.e. a node connected to all the nodes in the current clique); in this case the algorithm verifies if the size of the clique obtained by adding the selected node to the current clique, is larger than the largest clique found so far. In this case the new clique is stored, becoming the current largest clique.

At each level k, the choice of the nodes to consider is limited to the ones which correspond to pairs (n_1,n_2) having $n_1 = k$. In this way the algorithm ensures that the search space is actually a tree, i.e. it will never consider twice the same list of nodes. After considering all the nodes for level k, a special node, called the null_node, is added to the list. This node is always legal, and can be added more than once to the list. The null_node is used to carry the information that no mapping is associated to a particular node of the first graph being matched. When all possible nodes (including the null_node) have been considered, the algorithm backtracks and explores a different branch of the search tree. The longest list of nodes (excluding null_nodes) is stored and updated, when needed.

If N_1 and N_2 are the number of nodes of the two starting graphs, with $N_1 \leq N_2$, it can be proved that the maximum depth of the search tree is N_1. Since at each level the space requirement is constant (the list of nodes can be shared across levels, since it is accessed in a stack-like fashion), the total space requirement of the algorithm is $O(N_1)$. To this, the space needed to represent the association graph must be added. In the worst case the association graph can be a complete graph of $N_1 \cdot N_2$ nodes.

3.3 The Balas-Yu Algorithm

Also the Balas-Yu algorithm [7] is based on the transformation of the MCS problem in maximum clique detection. The main difference with the other two selected algorithms is a more sophisticated heuristic function for pruning the search tree. This heuristic is based on graph coloring and graph triangulation. Some basic definitions are needed to describe how this algorithm works.

Let G an undirected graph, let V and E respectively the number of nodes and edges of G, and let $\omega(G)$ the size of a maximum clique. A *coloring* of G assigns colors to the nodes of G in such a way that no two adjacent nodes get the same color. The cardinality of a minimum nodes coloring is called the *chromatic number* $\chi(G)$ of G. It has been proved that $\chi(G)$ is an upper bound for $\omega(G)$ and that the coloring problem has an O(V+E) complexity.

A graph G_T is *triangulated* (or *chordal*) if every cycle of G_T, whose length is at least 4, has a chord. Let the graph LTS(G_T) be the largest triangulated subgraph of G. It has been proved that finding out LTS(G_T), has a O(V+E) complexity. A maximum clique of LTS(G_T), $\varphi(G)$ can be found as a byproduct during the search of LTS(G_T) and it is a lower bound for $\omega(G)$. Thus having: $|\varphi(G)| \leq |\omega(G)| \leq |\chi(G)|$.

```
procedure BalasYu_MC(s)
  while (SelectSubProblem(s,n))
    SolveSelectedSubProblem(s,n);
    if (SizeSelectedSubProblem(s)>CurrentMCSize) then
      SaveCurrentMC(s);
      CurrentMCSize = SizeSelectedSubProblem(s);
    end if
    while(ExistNewSubProblem(s))
      s'= update(s);
      BalasYu_MC(s');
    end while
    BackTrack(s');
  end while
end procedure
```

Fig. 3. A sketch of the Balas-Yu algorithm.

The algorithm proposed by Balas and Yu can be suitably described through a State Space Representation [8]. Each state *s* is associated to the subproblem of finding a maximum clique in a subgraph of the starting graph. Each subproblem is characterized by the size *k* of the current maximum clique and a partition of the nodes of the starting graph into 3 sets: included nodes *I* (i.e. those nodes that are forcibly included into the subproblem), excluded nodes *Ex* (i.e. those nodes that are forcibly excluded from the subproblem), unclassified nodes *S* (i.e. all the other nodes). Starting from a node *n* chosen into *S*, the solution of the subproblem consists of finding how many nodes can be colored in a graph G', whose nodes are the nodes of *S*, except *n*, and whose edges are the edges of the starting graph connecting the nodes of *S*, fixing a priori the number $c = |\varphi(G)|$ of colors. The uncolored nodes are stored in a set *W*. A byproduct of this coloring procedure is a clique, whose size is, in the best case, the sum of |*I*| and *c*. If this clique is larger than the current one, it is stored as the current maximum clique. If the set *W* is empty, then the node *n* is excluded, and a new subproblem is defined on the modified Ex and S sets; otherwise, for each node v_i in the set W, a new subproblem, in which v_i is inserted in I, is defined. If it is not possible to choose a further node n, an empty state is obtained and the exploration of the current branch is terminated. A sketch of the algorithm is shown in Fig. 3.

4   The Database

In general, two main approaches can be followed for generating a database. The easier way is that of collecting graphs obtained from real data; otherwise, a database can be obtained synthetically, by a suitably designed computer program.

Since the synthetic generation of graphs allows us to control several critical parameters of the underlying graph population (the average number of nodes, the average number of edges per node, the number of different attributes, and so on), we decided to synthetically generate a large database of pairs of graphs for benchmarking the above described MCS algorithms.

The categories of graphs included in the database have been chosen by considering the kinds of graphs more frequently used in the IAPR-TC15 community (see http://www.iapr-tc15.unisa.it/). A classification of the categories of graphs used within the Pattern Recognition field has been also proposed in [3].

The proposed database is structured in pairs of connected graphs; each pair of graphs has a MCS of at least two nodes, with a specified structure and size. A total of 81,400 pairs of graphs have been generated, as reported in Tab. 1. In particular, the following categories of MCS have been considered (see [9] for a detailed discussion about their properties and the parameters characterizing them):

- Randomly connected graphs, with different values of the edge density η (0.01, 0.05, 0.1, 0.2);
- Regular Meshes, with different dimensionalities (2D, 3D and 4D);
- Irregular Meshes, with different dimensionalities (2D, 3D and 4D) and degrees of irregularity ρ (0.2, 0.4, 0.6);
- Regular and irregular bounded valence graphs, with different values of the valence v (3, 6, 9).

For each of the above-mentioned categories, pairs of attributed graphs having different sizes N have been included in the database. In particular thirteen values of N have been considered (i.e., 10, 15, 20, 25, 30, 35, 40, 50, 60, 70, 80, 90 and 100). Moreover, for each value of N, five different sizes of the MCS (namely, 10%, 30%, 50%, 70% and 90% of N) have been taken into account and 100 pairs of graphs have been generated for each size. In this way, for each value of N, 500 pairs of graphs having an MCS belonging to a specific category have been included into the database. However, as it can be noted from Tab. 1, in some cases the number of generated pairs for a specific MCS size was less than 1300 (that is equal to the product of thirteen, i.e. the number of the different values of N, by one hundred, i.e. the number of the generated pairs per each value of N). This is due to the fact that, for some values of the parameters characterizing a specific category, it was not possible to have connected graphs without adding a significant number of extra edges.

As regards the labeling, our idea is to generate random values for the attributes, since any other choice would imply assumptions about an application dependent model of the represented graphs.

Table 1. The database of graphs for benchmarking MCS algorithms.

	Graph Category	Parameter value	MCS size				
			10%	30%	50%	70%	90%
DATABASE ORGANIZATION (81,400 pairs of graphs)	Randomly connected graphs *18600 pairs*	$\eta = 0.01$	100	100	100	100	100
		$\eta = 0.05$	1100	1100	1100	1100	1100
		$\eta = 0.1$	1100	1300	1300	1300	1300
		$\eta = 0.2$	1100	1300	1300	1300	1300
	Regular Meshes *8100 pairs*	2D	700	700	1000	1000	1000
		3D	500	700	400	400	400
		4D	100	500	300	100	300
	Irregular Meshes *23900 pairs*	2D $\rho = 0.2$	400	700	900	1000	1000
		2D $\rho = 0.4$	700	700	1000	1000	1000
		2D $\rho = 0.6$	700	700	1000	1000	1000
		3D $\rho = 0.2$	500	700	400	400	400
		3D $\rho = 0.4$	500	700	400	400	400
		3D $\rho = 0.6$	500	700	400	400	400
		4D $\rho = 0.2$	100	500	300	100	300
		4D $\rho = 0.4$	100	500	300	100	300
		4D $\rho = 0.6$	100	500	300	100	300
	Regular Bounded Valence Graphs *15400 pairs*	$v = 3$	700	1200	1300	1300	1300
		$v = 6$	400	1000	1200	1300	1300
		$v = 9$	100	800	1100	1200	1200
	Irregular Bounded Valence Graphs *15400 pairs*	$v = 3$	700	1200	1300	1300	1300
		$v = 6$	400	1000	1200	1300	1300
		$v = 9$	100	800	1100	1200	1200

If we choose a uniform distribution of the values, it is possible to assume, without any loss of generality, that attributes are represented through integer numbers. In fact, the purpose of the labeling in our tests is simply to restrict the possible node or edge pairings with respect to the unlabeled case. Moreover, the use of floating point values can be somewhat more problematic, because: *i*) usually it makes little sense to compare floating-point numbers for strict equality; *ii*) there is no universally usable binary format for storing binary numbers.

These disadvantages are not repaid for by significant advantages, since also integer attributes can be used to perform arithmetic tasks, e.g. distance evaluation.

An important parameter for characterizing the computational cost of the matching problem is the number M of different values of attributes (i.e. the size of the alphabet): obviously the higher is M, the lower is the cost of the matching.

It is important to have in the database different values of M: in such case it is possible to decide either to measure the matching time keeping M constant and varying the size of the graphs, or to increase M as the size of the graphs increases; both choices make sense for estimating the performance of different types of applications.

In order to avoid the need to have several copies of the database with different values of M, each attribute is generated as a 16-bit value, using a random number generation algorithm ensuring that each bit is random. Then, it is possible to choose any value of the form 2^k, with k not greater than 15, just by using, in the attributes comparison function, only the first k bits of the attribute. Furthermore, for values of M that are not powers of 2, attribute value module M can be used, if M is sufficiently smaller than 2^{16}, without introducing any undesired bias in the distribution of the values. Using this technique an attributed graph database has been built. It is enough general for experimenting with many different attribute cardinalities, avoiding the explosion of the size required to store a different database for each set of attributes.

5 Experimental Results

In order to make an unbiased comparison, a C++ implementation of all the three algorithms has been developed. In our implementation each algorithm ends when it finds out the first solution of the MCS problem.

In order to better compare the algorithms as the different parameters vary, we report, for each category of graphs, a table called the *winner map*. The columns report the number of nodes of the graphs to be matched, while the rows report the value of the parameter characterizing the considered graph category (for instance, the edge density η in case of randomly connected graphs). For each parameter value, different sizes M of the attribute cardinality have been considered. We have chosen to employ values of M proportional to the number of nodes in the graphs being matched, in order to keep the running times within a reasonable limit. In particular, we have tested each graph pair in the database with three values of M, equal to 33%, 50% and 75% of the number of nodes respectively. As regards the size N of the graph pairs, only the values ranging from 10 to 40 have been considered.

In the winner map, each cell represents the result of a comparison among the matching times of the three algorithms. A color is associated to each algorithm; thus, just observing the color of the cell it is possible to realize which algorithm performs better than the others for the values of the triple {parameter value, M, N} that corresponds to that cell. Furthermore, the number reported in each cell represents the order of magnitude of the ratio between the matching times of the winning algorithm and of the runner-up.

5.1 Randomly Connected Graphs

Tab. 2 shows the behavior of the selected algorithms with reference to the randomly connected graphs. The McGregor algorithm always performs better on sparse graphs (i.e. for $\eta=0.05$). The reason is that the McGregor algorithm solves the matching problem without using the association graph. When the graph density is low, the association graph is large and dense; thus, the transformation needed to obtain it is not so convenient. When the density increases (i.e. for $\eta=0.1$), the McGregor algorithm is still the winner when the graphs are small or the value of M is high. In the other cases the matching problem is more efficiently solved by using the association graph and the Durand-Pasari algorithm.

In some cases, in particular on large graphs with middle or high densities, the Balas-Yu algorithm performs better than the other two. This is due to the fact that the computation of the heuristic of the Balas-Yu algorithm is very expensive. For small graphs, the time needed for this computation overcomes the advantage produced by the heuristic itself. On the contrary, the use of a complex heuristic gives good results on large graphs.

Table 2. The winner map for randomly connected graphs: η represents the graph density, while M is the size of the alphabet of attributes with respect to the number of nodes.

η	M \\ N	10	15	20	25	30	35	40
0.05	33%			1	2	1	1	0
	50%			2	2	3	3	3
	75%			1	2	2	3	3
0.1	33%	0	1	0	0	0	0	1
	50%	0	0	0	0	0	0	0
	75%	0	0	0	1	0	0	0
0.2	33%	1	1	1	1	1	1	1
	50%	1	1	1	1	0	0	0
	75%	0	0	1	1	0	0	0

0 → DP
1 → BY
2 → MG

Table 3. The winner map for 2D and 3D meshes: ρ measures the degree of the mesh irregularity, while M is the size of the alphabet of attributes with respect to the number of nodes.

ρ	M \\ N	2D							3D						
		10	15	20	25	30	35	40	10	15	20	25	30	35	40
0	33%	0	1	0	1	1	0	2	0	1	0	0	0	0	2
	50%	0	0	1	2	3	2	3	0	0	0	1	1	0	2
	75%	0	1	1	2	2	2	3	0	0	1	2	2	1	2
0.2	33%	0	1	0	0	0	0	0	0	1	1	0	0	2	0
	50%	0	0	1	1	2	1	1	0	0	0	0	0	0	1
	75%	0	0	1	1	2	2	2	0	0	1	1	1	1	2
0.4	33%	0	1	1	0	0	1	0	0	1	1	1	1	2	1
	50%	0	0	0	0	1	0	2	0	0	0	0	0	0	1
	75%	0	0	1	1	2	2	2	0	0	1	1	1	0	2
0.6	33%	0	1	1	0	1	1	1	1	1	1	1	1	2	1
	50%	0	0	0	0	1	0	2	0	0	0	0	0	1	1
	75%	0	0	0	1	2	1	2	0	0	1	1	1	1	2

0 → DP
1 → BY
2 → MG

5.2 Regular and Irregular Meshes

In Tab. 3 the winner map on regular and irregular meshes is shown.
The behavior of all the algorithms is similar for both 2D, 3D and 4D meshes (this is the reason why the winner map for 4D meshes is not shown). We can see that the McGregor algorithm performs better in most cases. This can be explained by the fact that the mesh graphs are not very dense. Then, in most cases, the association graph is large and dense and therefore it is not particularly convenient to solve the problem of

finding the maximum clique. On the other hand, it is evident that the Durand-Pasari algorithm performs better on small graphs, especially when the irregularity of the meshes (i.e., the value of the parameter ρ) increases. Finally, for large graphs that have a high irregularity and for a small value of M, Balas-Yu is the best algorithm.

Table 4. The winner map for Bounded Valence Graphs (Regular and Irregular): v is the valence, while M is the size of the alphabet of attributes with respect to the number of nodes.

v	M \ N	Regular							Irregular							
		10	15	20	25	30	35	40	10	15	20	25	30	35	40	
3	33%	0	1	1	1	2	3	4	0	1	0	0	1	0	0	DP
	50%	0	0	1	2	3	3	4	0	1	1	0	0	3	0	
	75%	0	0	1	2	2	3	3	0	0	1	2	2	3	3	
6	33%	1	1	1	1	1	0	0	1	1	1	1	1	0	0	BY
	50%	0	0	0	0	0	0	0	0	1	1	0	0	0	0	
	75%	0	0	1	1	1	1	2	0	0	0	0	0	0	0	MG
9	33%	0	1	1	1	1	2	1		0	0	0	0	0	0	
	50%	0	0	0	0	0	0	1		0	0	0	0	0	0	
	75%	0	0	1	1	1	0	2		0	0	0	0	0	0	

5.3 Regular and Irregular Bounded Valence Graphs

In Tab. 4 the behavior of the selected algorithms on regular bounded valence graphs and irregular bounded valence graphs is shown.

For regular graphs, when the valence value is equal to 3, the McGregor algorithm performs better. Once again, the reason can be found in the fact that the number of edges per nodes is quite small, i.e. the graphs are not very dense. When the value of the valence is 6 or 9 the association graph becomes less dense, so making convenient the use of algorithms that find the maximum clique. In these cases, if the value of M is small, the Duran-Pasari algorithm performs better on smaller graphs, while the Balas-Yu algorithm is more convenient on larger graphs. Finally, when the value of M increases, the matching problem becomes simpler, and then the McGregor algorithm is always the winner because it does not pay the further cost of constructing the association graph.

For irregular bounded valence graphs, Durand-Pasari is the best algorithm in most cases. The only cases in which there is a different behavior is when the value of v is 3 and the size of the alphabet of the attributes is middle or large. In these cases, in fact, the association graphs are large and dense and then it is not convenient to solve the problem of finding the maximum clique. When the value of v is 6 or 9, the association graph becomes less dense and thus its use becomes more convenient. In these cases, the Duran-Pasari algorithm gives better results on smaller graphs while the Balas-Yu algorithms performs better on larger graphs due the its more refined heuristic.

6   Discussion and Conclusions

In this paper we presented a benchmarking activity for assessing the performance of some widely used exact MCS algorithms. The comparison has been carried out on a large database of synthetically generated labeled graphs, which will be made publicly available to provide a common reference data set for further benchmarking activities.

As it could be expected, it does not exist an algorithm that is definitively better than the others. In particular, for randomly connected graphs, the McGregor algorithm is the best one if the graphs are quite sparse and/or of quite small size. For larger and quite dense graphs, on the contrary, the Durand-Pasari algorithm performs better.

If the MCS has a more regular structure, i.e., on mesh-like MCS, the McGregor algorithm is in most cases the best algorithm; the Durand-Pasari algorithm performs better only for small and dense graphs. On the contrary, when the irregularity degree grows up, the Balas-Yu algorithm performs better for large and dense graphs.

In case of bounded valence graphs, the McGregor algorithm is the best one if the valence is small, while for larger values of the valence the Durand-Pasari algorithm is more convenient when the graphs to be matched are dense. If the graphs are both dense and large, the Balas-Yu algorithm is the best one. Finally, when an irregularity degree is added to the bounded valence graphs, the Durand-Pasari algorithm performs better in most cases, even if the Balas-Yu algorithm is still winning for large graphs.

Future steps of this benchmarking activity will involve the comparison of other MCS algorithms. We are also planning to extend the database with other graph categories and to add an indexing facility (based on several graph parameters), for making its use easier and more convenient to other researchers.

References

[1] C. Bron and J. Kerbosch, "Finding All the Cliques in an Undirected Graph", Communication of the ACM 16(1973), 575–577.

[2] I. M. Bomze, M. Budinich, P. M. Pardalos, and M. Pelillo, "The Maximum Clique Problem", Handbook of Combinatorial Optimization, vol. 4, Kluwer Academic Publisher, Boston Pattern MA, 1999.

[3] H. Bunke, M.Vento, "Benchmarking of Graph Matching Algorithms", Proc. 2nd IAPR TC-15 GbR Workshop, Haindorf, pp. 109–114, 1999.

[4] M. De Santo, P. Foggia, C. Sansone, M. Vento, "A Large Database of Graphs and its use for Benchmarking Graph Isomorphism Algorithms" Pattern Recognition Letters, vol. 24, no. 8, pp. 1067–1079, 2003.

[5] J.J. McGregor, "Backtrack Search Algorithms and the Maximal Common Subgraph Problem", Software Practice and Experience, Vol. 12, pp. 23–34, 1982.

[6] P. J. Durand, R. Pasari, J. W. Baker, and C.-C. Tsai, "An Efficient Algorithm for Similarity Analysis of Molecules", Internet Journal of Chemistry, vol. 2, 1999.

[7] E. Balas, C. S. Yu, "Finding a Maximum Clique in an Arbitrary Graph", SIAM J. Computing, Vol. 15, No 4, 1986.

[8] N. J. Nilsson, "Principles of Artificial Intelligence", Springer-Verlag, 1982.

[9] P.Foggia, C.Sansone, M.Vento, "A Database of Graphs for Isomorphism and Sub-Graph Isomorphism Benchmarking", Proc. 3rd IAPR TC-15 GbR Workshop, Italy, pp. 176–187, 2001.

Swap Strategies for Graph Matching[*]

Paolo Fosser[1], Roland Glantz[1], Marco Locatelli[2], and Marcello Pelillo[1]

[1] Dipartimento di Informatica
Università Ca' Foscari di Venezia
Via Torino 155, I-30172 Venezia Mestre, Italy
{fosser,glantz,pelillo}@dsi.unive.it
[2] Dipartimento di Informatica
Università di Torino
Corso Svizzera 185, I-10149 Torino, Italy
Locatelli@.di.unito.it

Abstract. The problem of graph matching is usually approached via explicit search in state-space or via energy minimization. In this paper we deal with a class of heuristics coming from a combination of both approaches. Combinatorially, the basic heuristic of the class can be interpreted as a greedy algorithm to form maximal cliques in an association graph. To avoid one of the main drawbacks of greedy strategies, i.e. that they are easily fooled by poor local optima, we propose a modification which allows for vertex swaps during the formation of a clique. Experiments on random graphs show the effectiveness of the proposed heuristics both in terms of quality of solutions and speed.

1 Introduction

Graph matching is a fundamental problem in computer vision and pattern recognition, and a great deal of effort has been devoted over the past decades to devise efficient and robust algorithms for it (see [7] for an update on recent developments). Basically, two radically distinct approaches have emerged, a distinction which reflects the well-known dichotomy originated in the artificial intelligence field between "symbolic" and "numeric" methods. The first approach views the matching problem as one of explicit search in state-space (see, e.g., [14,20,21]). The pioneering work of Ambler et al. [1] falls into this class. Their approach is based on the idea that graph matching is equivalent to the problem of finding maximal cliques in the so-called association graph, an auxiliary graph derived from the structures being matched. This framework is attractive because it casts the matching problem in terms of a pure graph-theoretic problem, for which a solid theory and powerful algorithms have been developed [4].

In the second approach, the relational matching problem is viewed as one of energy minimization. In this case, an energy (or objective) function is sought whose minimizers correspond to the solutions of the original problem, and a

---

[*] This work is supported by the Austrian Science Foundation (FWF) under grant P14445-MAT and by MURST under grant MM09308497.

E. Hancock and M. Vento (Eds.): IAPR Workshop GbRPR 2003, LNCS 2726, pp. 142–153, 2003.

dynamical system, usually embedded into a parallel relaxation network, is used to minimize it [9,10,22]. Typically, these methods do not solve the problem exactly, but only in approximation terms. Energy minimization algorithms are attractive because they are amenable to parallel hardware implementation and also offer the advantage of biological plausibility.

In [16], we have developed a new framework for graph matching which does unify the two approaches just described, thereby inheriting the attractive features of both. The approach is centered around a remarkable result in graph theory which allows us to map the maximum clique problem onto the problem of extremizing a quadratic form over a linearly constrained domain (i.e., the standard simplex in Euclidean space) [15]. Local gradient-based search methods such as *replicator dynamics* have proven to be remarkably effective when we restrict ourselves to simple versions of the problem, such as tree matching [19,18] or graph isomorphism [17]. For more difficult versions of the problem a pivoting-based heuristic (PBH), which is based on a linear complementarity formulation [8], has recently proven to be quite effective [13,12].

In [11] it has been shown that PBH is essentially equivalent to a greedy combinatorial heuristic and that modifying the combinatorial counterpart of PBH, the quality of the results may be improved, at the cost of moderately increasing the computation time. Basically, the modification consists of allowing for vertex swaps during the clique-construction process. In this paper we apply these heuristics to graph matching problems. Experiments on random graphs with various levels of corruption are presented. We found that the purely greedy heuristic performed as well as PBH but was dramatically faster, and the swap strategy always improved the results. Both algorithms were also found to perform better than graduated assignment [9], a state-of-the-art representative of energy minimization methods.

The plan of the paper is the following. In Section 2 we formulate the graph matching problem as a maximum clique problem on a derived association graph. The interpretation of PBH in terms of a greedy combinatorial heuristic is given in Section 3 and Section 4 is devoted to vertex swaps. Experimental results are provided in Section 5.

2   Graph Matching and Maximal Cliques

Given two graphs $G_1 = (V_1, E_1)$ and $G_2 = (V_2, E_2)$, an *isomorphism* between them is any bijection $\phi : V_1 \rightarrow V_2$ such that $(i,j) \in E_1 \Leftrightarrow (\phi(i), \phi(j)) \in E_2$, for all $i, j \in V_1$. Two graphs are said to be isomorphic if there exists an isomorphism between them. The maximum common subgraph problem consists of finding the largest isomorphic subgraphs of G_1 and G_2. A simpler version of this problem is to find a maximal common subgraph, i.e., an isomorphism between subgraphs which is not included in any larger subgraph isomorphism.

The *association graph* derived from $G_1 = (V_1, E_1)$ and $G_2 = (V_2, E_2)$ is the undirected graph $G = (V, E)$ defined as follows:

$$V = V_1 \times V_2$$

and

$$E = \{((i,h),(j,k)) \in V \times V \ : \ i \neq j, \ h \neq k, \ \text{and} \ (i,j) \in E_1 \Leftrightarrow (h,k) \in E_2\} \ .$$

Given an arbitrary undirected graph $G = (V, E)$, a subset of vertices C is called a *clique* if all its vertices are mutually adjacent, i.e., for all $i, j \in C$, with $i \neq j$, we have $(i, j) \in E$. A clique is said to be *maximal* if it is not contained in any larger clique, and *maximum* if it is the largest clique in the graph. The *clique number*, denoted by $\omega(G)$, is defined as the cardinality of the maximum clique.

The following result establishes an equivalence between the graph matching problem and the maximum clique problem (see, e.g., [2]).

Theorem 1. *Let $G_1 = (V_1, E_1)$ and $G_2 = (V_2, E_2)$ be two graphs, and let G be the corresponding association graph. Then, all maximal (maximum) cliques in G are in one-to-one correspondence with maximal (maximum) common subgraph isomorphisms between G_1 and G_2.*

The problem of finding a maximum clique in a graph is known to be NP-hard for arbitrary graphs and so is the problem of approximating it within a constant factor. However, several powerful heuristics have been developed in the last few years. We refer to [4] for a recent review concerning algorithms, applications and complexity issues of this important problem.

3   Combinatorics of Pivoting

Recently a powerful pivoting-based heuristic (PBH) has been introduced for attacking the maximum clique problem, based on the linear complementarity formulation of an equivalent (standard) quadratic program [13]. In [11] it has been shown that the continuous approach PBH for the solution of the maximum clique problem is equivalent to a combinatorial heuristic. Such a heuristic turns out to belong to the class of greedy heuristics denoted in [5] by SM^i ($i = 1, 2, \ldots$), which are described in what follows.

Greedy heuristic SM^i
Step 1 Given a graph $G = (V, E)$, let $\mathcal{Q}$ be the set of all cliques of graph G with cardinality i. Set $\mathcal{K} = \emptyset$, $\mathcal{K}^* = \emptyset$ and $max = 0$.
Step 2 If $\mathcal{Q} \neq \emptyset$, select a clique $H \in \mathcal{Q}$ and update $\mathcal{Q} = \mathcal{Q} \setminus \{H\}$. Put $K = H$ and go to Step 3; otherwise, return max and $\mathcal{K}^*$.
Step 3 If K is a maximal clique, then update $\mathcal{K} = \mathcal{K} \cup \{K\}$ and go to Step 4; otherwise go to Step 5.
Step 4 If $|K| > max$, set $max = |K|$ and $\mathcal{K}^* = \mathcal{K}$. Go back to Step 2.
Step 5 Let N_j denote the neighborhood of vertex j, i.e.

$$N_j = \{k \in V \ : \ (j, k) \in E\}. \tag{1}$$

Select a vertex

$$\ell \in \arg \max_{j \in C_0(K)} |C_0(K) \cap N_j| \tag{2}$$

at random, where

$$C_0(K) = \{j \in V \setminus K : (j,k) \in E \ \forall \, k \in K\} = \bigcap_{k \in K} N_k \qquad (3)$$

is the set of all vertices outside K that are adjacent to each vertex in the current clique K.

Step 6 Set $K = K \cup \{\ell\}$ and go back to Step 3.

We notice that the combinatorial version of PBH is equivalent to SM^1, except for the minimal difference of the random selection of ℓ in (2)—in PBH ℓ is selected as the node with the lowest index in $C_0(K)$. Thus, we established a new connection between the technique of pivoting and well-known heuristics for finding large cliques. To our knowledge, the SM^i heuristics have never been applied to association graphs.

Although the heuristic SM^1 frequently returns results of good quality, the heuristic SM^2 is often superior to SM^1. This was shown in [11], where numerous tests were made using DIMACS graphs. On the other hand, as expected, the computation times for SM^2 are much larger than those for SM^1 [11]. In the next section we shall modify the heuristic SM^1 in such a way that the modification is closer to SM^2 than to SM^1 from the point of view of the quality of the results, but is much closer to SM^1 than to SM^2 from the point of view of the computation times.

4 Vertex Swaps

The heuristics SM^i start with a clique of cardinality i and add a new vertex at each iteration until a maximal clique is reached. Some other approaches do the same but do not stop once a maximal clique is reached, allowing for the removal of vertices when this situation occurs (see e.g. the plateau search in [3]). The modification proposed here does not always add a new vertex to the current clique but also allows to swap vertices. If K is the current clique, let $C_0(K)$ be defined as in (3) and let

$$C_1(K) = \{j \in V \setminus K : |N_j \cap K| = |K| - 1\},$$

i.e. $C_1(K)$ is the set of vertices outside K which are connected to all vertices in K but exactly one. If we select a vertex $\ell \in C_0(K)$, then we can add it to K and update K as follows

$$K = K \cup \{\ell\}.$$

This is exactly what heuristics SM^i do with the selection (2). But if we select a vertex $\ell \in C_1(K)$ we can not simply add it to K; if we still want to have a clique we need to swap vertex ℓ with the unique vertex

$$k_\ell \in K \qquad \text{such that} \quad (\ell, k_\ell) \notin E, \qquad (4)$$

i.e. the new clique will be

$$K = K \cup \{\ell\} \setminus \{k_\ell\}.$$

Therefore, if we select a vertex in $C_1(K)$ we change the current clique but without increasing the cardinality, i.e. the method is not strictly monotone. In the modification proposed here we decide whether to increase the cardinality, by selecting a vertex in $C_0(K)$, or to swap vertices, by selecting a vertex in $C_1(K)$. Note that $C_0(K) = \emptyset$ if K is a maximal clique. Similarly, some authors call a maximal clique K *strictly maximal*, if $C_1(K) = \emptyset$, i.e., if every swap destroys the clique property.

Now, selection of a vertex in $C_0(K) \cup C_1(K)$ is done according to a rule similar to (2) but with the computation of the maximum extended also to vertices in $C_1(K)$, i.e.

$$\ell \in \arg \max_{j \in C_0(K) \cup C_1(K)} |C_0(K) \cap N_j|. \tag{5}$$

We underline that in some cases we force the heuristic to select the next vertex according to (2), namely:

- during a fixed number $(START\_SWAP)$ of initial iterations;
- when the number of swaps performed is a fixed number T times the cardinality of the current clique $|K|$;
- when the vertex selected in $C_1(K)$ is the same as the last one removed by the last swap. Without this condition, which can be interpreted as a very limited application of taboo rules, it has been observed that the heuristic is much less efficient (worse results and larger computation times), apparently because it happens that the same two vertices are swapped in and out until the maximum allowed number of swaps is reached and then the heuristic behaves exactly as the standard greedy one. Experimentally, we found that longer lists of forbidden vertices, corresponding to higher order taboo rules, have no effect on the quality of the results, i.e. on the sizes of the cliques.

The heuristic with the proposed modification, denoted with SM^1_SWAP is described in what follows.

Heuristic SM^1_SWAP

Step 1 Given a graph $G = (V, E)$, let $W = V$, $K^* = \emptyset$ and $max = 0$. Select a value for the parameters $START\_SWAP$ and T.

Step 2 If $W = \emptyset$ return K^* and max. Otherwise select a vertex $h \in W$, set $W = W \setminus \{h\}$ and set $K = \{h\}$. Set $n\_swaps = 0$ and $last\_swap = \emptyset$.

Step 3 If K is a maximal clique, then set $\mathcal{K} = \mathcal{K} \cup \{K\}$ and go to Step 4; otherwise go to Step 5.

Step 4 If $|K| > max$, set $max = |K|$ and $K^* = K$. Go back to Step 2.

Step 5 If $n\_swaps \leq T|K|$ and $|K| \geq START\_SWAP$, then go to Step 6, otherwise go to Step 7.

Step 6 Select randomly $\ell \in V$ according to (5). If $\ell = last\_swap$, then go to Step 7, otherwise go to Step 8.

Step 7 Select $\ell \in V$ according to (2).

Step 8 If $\ell \in C_1(K)$, then update

$$n\_swaps = n\_swaps + 1 \quad \text{and} \quad last\_swap = k_\ell$$

where k_ℓ is defined in (4), and

$$K = K \cup \{\ell\} \setminus \{k_\ell\}.$$

Otherwise set

$$K = K \cup \{\ell\}.$$

Go back to Step 3.

5   Experimental Results

In this section we test the performance of SM^1 and SM^1_SWAP on graph matching problems, and compare them with a well-known representative of energy minimization methods, i.e., Gold and Rangarajan's graduated assignment algorithm [9] (abbreviated as GR in what follows). All experiments are performed on random graphs. Random structures represent a useful benchmark not only because they are not constrained to any particular application, but also because it is simple to replicate experiments and hence to make comparisons with other algorithms.

The following protocol is standard and was already used in [12]. We generated random 50-node graphs using edge-probabilities (i.e. densities) ranging from 0.1 to 0.9. For each density value, 20 graphs were constructed so that, overall, 180 graphs were used in the experiments. Each graph had its vertices permuted and was then subject to a corruption process which consisted of randomly deleting a fraction of its nodes. In so doing we obtained a graph isomorphic to a proper subgraph of the original one. Various levels of corruption (i.e., percentage of node deletion) were used, namely 0% (the pure isomorphism case), 10%, 20% and 30%. In other words, the order of the corrupted graphs ranged from 50 to 35. Each algorithm was applied on each pair of graphs thus constructed and the percentage of matched nodes was recorded. All experiments were performed on a 1.4 GHz AMD Athlon processor.

Since PBH is essentially equivalent to SM^1, the results obtained by the two heuristics were almost identical, as expected. However, the computation time differed dramatically. For example, in case of pure isomorphism (i.e., 0% level of corruption) and a density of 0.5, PBH required more than 4 hours, whereas SM^1 terminated after only 14 seconds. Similar differences were also found at other densities and levels of corruption. We note that in [13] it was shown that PBH compares well with state-of-the-art clique finding algorithms.

Before proceeding with the experimental results, we note that the GR algorithm may yield "inconsistent" results. Formally, let the two graphs be denoted by G_1 and G_2. Furthermore let U_1 [U_2] denote the set of vertices in G_1 [G_2] that have been matched to a vertex in G_2 [G_1]. Then, the subgraph of G_1 induced by U_1 and the subgraph of G_2 induced by U_2 are not necessarily isomorphic. In other words, the matches found by GR do not necessarily induce a clique in the association graph of G_1 and G_2 (see Section 2). To find a maximum consistent "kernel" of the results from GR, we applied an exact algorithm for the maximum clique problem, i.e. Bron and Kerbosch [6], to the subgraph of the association graph induced by the matches. The following results on GR refer to the output of the Bron and Kerbosch algorithm. The computation times, however, do not include the time needed by the Bron and Kerbosch algorithm.

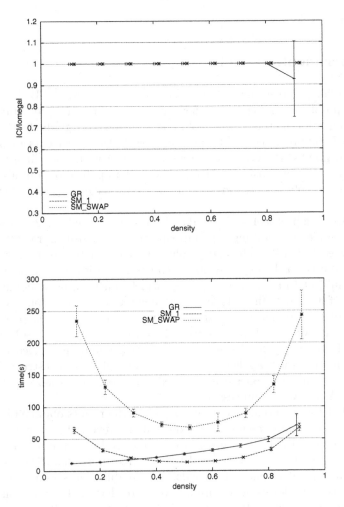

Fig. 1. Results of matching 50-node random graphs with varying densities, and 0% corruption. Top: Portion of correct matches. Bottom: Average computational time taken by the algorithms (in seconds).

Figures 1 to 4 show the results obtained at various levels of corruption. Each figure shows two plots. One concerns the quality of the solutions found and the other the CPU time. For each plot, the x-axis represents the (approximate) density of the matched graphs. The y-axis of the top plots represents the portion of the correct matches. Here, ω is the size of the maximum clique of the association graph, i.e., the size of the maximum isomorphism, and $|C|$ is the size of

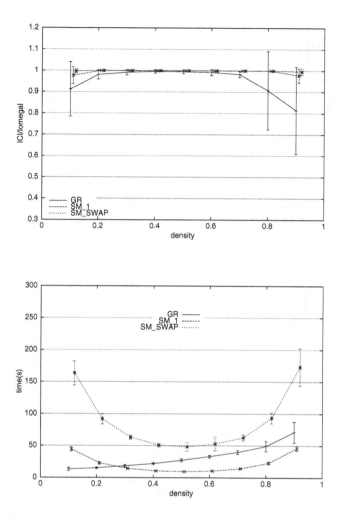

Fig. 2. Results of matching 50-node random graphs with varying densities, and 10% corruption. Top: Portion of correct matches. Bottom: Average computational time taken by the algorithms (in seconds).

the isomorphism returned by the algorithms, i.e. the size of the maximal clique found (in the case of GR, this corresponds to the size of the maximum consistent kernel, as described before). Finally, y-axis of the bottom plots represents CPU time expressed in seconds. All curves represent average over 20 trials.

Let us first focus on the quality of the results, i.e. on the top part of the figures. As corruption increases, the advantage of SM^1 and SM^1_SWAP over GR becomes more and more evident. Moreover, except for the the pure isomorphism case in which SM^1 and SM^1_SWAP both yielded always the optimal result, SM^1_SWAP is always superior to SM^1.

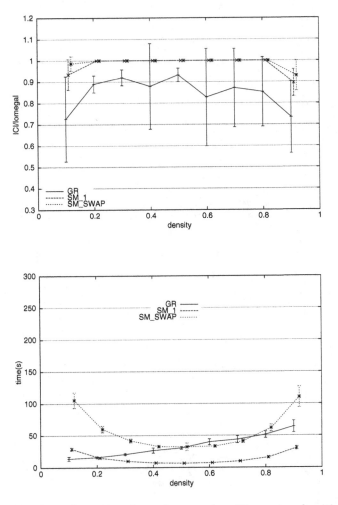

Fig. 3. Results of matching 50-node random graphs with varying densities, and 20% corruption. Top: Portion of correct matches. Bottom: Average computational time taken by the algorithms (in seconds).

Looking at the computation times, see the bottom part of the figures, one can easily distinguish the bowl shapes of SM^1 and SM^1_SWAP from the monotonic behavior of GR. The "bowls" are due to the fact that the computation times of SM^1 and SM^1_SWAP depend mainly on the number of edges in the association graph and that the density of the latter becomes high, if G_1 and G_2 are sparse or dense.

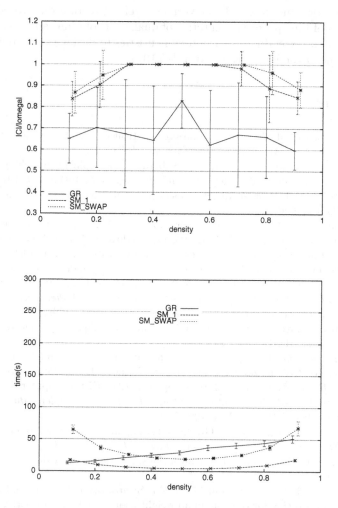

Fig. 4. Results of matching 50-node random graphs with varying densities, and 30% corruption. Top: Portion of correct matches. Bottom: Average computational time taken by the algorithms (in seconds).

6   Conclusions

In this paper, we have proposed a class of combinatorial heuristics for graph matching, based on the well-known idea of finding maximal cliques in the association graph. The basic heuristic of the class, equivalent to a recently introduced pivoting-based algorithm, was shown to fall into the class of greedy algorithms. Trying to avoid one of the classical drawbacks of greedy strategies (i.e., the ease of being trapped in poor local optima), we proposed a modification of the basic algorithm allowing for vertex swaps during the clique-construction process.

Experiments on random graphs at various levels of corruption have shown the effectiveness of the approaches in terms of quality of solutions and speed. Future work will aim at extending the proposed heuristics to deal with attributed graph matching, along the lines suggested in [16,19].

References

1. A. P. Ambler et al. A versatile computer-controlled assembly system. In *Proc. of 3rd Int. J. Conf. Art. Intell.*, pages 298–307, 1973.
2. H. G. Barrow and R. M. Burstall. Subgraph isomorphism, matching relational structures and maximal cliques. *Inform. Process. Lett.*, 4(4):83–84, 1976.
3. R. Battiti and M. Protasi. Reactive local search for the maximum clique problem. *Algorithmica*, 29:610–637, 2001.
4. I. M. Bomze, M. Budinich, M. P. Pardalos, and M. Pelillo. The maximum clique problem. In D.-Z. Du and P. M. Pardalos, editors, *Handbook of Combinatorial Optimization (Suppl. Vol. A)*, pages 1–74. Kluwer, Boston, MA, 1999.
5. M. Brockington and J. C. Culberson. Camouflaging independent sets in quasi-random graphs. In D. Johnson and M. A. Trick, editors, *Cliques, Coloring and Satisfiability*, volume 26 of *DIMACS series in Discrete Mathematics and Theoretical Computer Science*, pages 75–88. Americ. Math. Soc., 1996.
6. C. Bron and J. Kerbosch. Algorithm457: Finding all cliques of an undirected graph. *Comm. ACM*, 16:575–577, 1973.
7. H. Bunke. Recent developments in graph matching. In A. Sanfeliu, J. Villanueva, M. Vanrell, R. Alquezar, A. Jain, and J. Kittler, editors, *Proc. 15th Int. Conf. Pattern Recognition*, volume 2, pages pp. 117–124. IEEE Computer Society, 2000.
8. R. W. Cottle, J.-S. Pang, and R. E. Stone. *The Linear Complementarity Problem*. Accademic Press, Boston, MA, 1992.
9. S. Gold and A. Rangarajan. A Graduated Assignment Algorithm for Graph Matching. *IEEE Trans. Pattern Anal. and Machine Intell.*, 18(4):377–388, 1996.
10. S. Z. Li. Matching: invariant to translations, rotations and scale changes. *Pattern Recogognition*, 25(6):583–594, 1992.
11. M. Locatelli, I. M. Bomze, and M. Pelillo. Swaps, diversification, and the combinatorics of pivoting for the maximum weight clique. Technical Report CS-2002-12, Dipartimento di Informatica, Università Ca' Foscari di Venezia, 30172 Venezia Mestre, Italy, 2002.
12. A. Massaro and M. Pelillo. Matching graphs by pivoting. *Pattern Recognition Letters*, 24:1099–1106, 2003.
13. A. Massaro, M. Pelillo, and I. M. Bomze. A complementary pivoting approach to the maximum weight clique problem. *SIAM Journal on Optimization*, 12(4):928–948, 2002.
14. B. Messmer and H. Bunke. A New Algorithm for Error-Tolerant Subgraph Isomorphism Detection. *IEEE Trans. on Pattern Analysis and Machine Intelligence*, 20(7):493–504, 1998.
15. T. S. Motzkin and E. G. Straus. Maxima for graphs and a new proof of a theorem of Turán. *Cand. J. Math.*, 17:533–540, 1965.
16. M. Pelillo. A unifying framework for relational structure matching. In A. K. Jain, S. Venkatesh, and B. C. Lovell, editors, *Proc. 14th Int. Conf. Pattern Recognition*, pages 1316–1319. IEEE-Computer Society Press, 1998.

17. M. Pelillo. Replicator equations, maximal cliques, and graph isomorphism. *Neural Computation*, 11:1933–1955, 1999.
18. M. Pelillo. Matching free trees, maximal cliques, and monotone game dynamics. *IEEE Trans. Pattern Anal. and Machine Intell.*, 24(11):1535–1541, 2002.
19. M. Pelillo, K. Siddiqi, and S. W. Zucker. Matching Hierarchical Structures Using Association Graphs. *IEEE Trans. Pattern Anal. and Machine Intell.*, 21(11):1105–1120, 1999.
20. L. G. Shapiro and R. M. Haralick. Structural descriptions and inexact matching. *IEEE Trans. Pattern Anal. and Machine Intell.*, 3:504–519, 1981.
21. W. H. Tsai and K. S. Fu. Subgraph error-correcting isomorphisms for syntactic pattern recognition. *IEEE Trans. Syst. Man Cybern.*, 13:48–62, 1983.
22. R. C. Wilson and E. R. Hancock. Structural Matching by Discrete Relaxation. *Trans. Pattern Anal. Machince Intell.*, 19(6):634–648, 1997.

Graph Matching Using Spectral Seriation and String Edit Distance

Antonio Robles-Kelly* and Edwin R. Hancock

Department of Computer Science
University of York
York, Y01 5DD, UK
{arobkell,erh}@cs.york.ac.uk

Abstract. This paper is concerned with computing graph edit distance. One of the criticisms that can be leveled at existing methods for computing graph edit distance is that it lacks the formality and rigour of the computation of string edit distance. Hence, our aim is to convert graphs to string sequences so that string matching techniques can be used. To do this we use graph spectral seriation method to convert the adjacency matrix into a string or sequence order. We show how the serial ordering can be established using the leading eigenvector of the graph adjacency matrix. We pose the problem of graph-matching as maximum a posteriori probability alignment of the seriation sequences for pairs of graphs. This treatment leads to an expression in which for edit cost is the negative logarithm of the a posteriori sequence alignment probability. To compute the string alignment probability we provide models of the edge compatibility error and the probability of individual node correspondences.

1   Introduction

One of the classical approaches to graph matching is to extend the concept of edit distance from strings to graphs [15,3]. One of the criticisms that can be aimed at the early work on graph edit distance is that it lacks the formal rigour of the corresponding work on string edit distance. However, recently, Bunke and his co-workers have returned to the problem and have shown the relationship between graph edit distance and the size of the maximum common subgraph [2]. An alternative approach to the problem is to convert graphs to string sequences and to use the existing theory of string edit distance. In recent work, we have shown how the steady state random walk on a graph can be used to establish a string order. Using the theory of Markov chains on graphs, we have shown how the leading eigenvector of the normalised adjacency matrix can be used to establish a string order sequence from the steady state random walk [14]. However, one of the shortcomings of the method is that it does not guarantee or even necessarily encourage edge connectivity in the sequence order. In this

* Supported by CONACYT, under grant No. 146475/151752.

E. Hancock and M. Vento (Eds.): IAPR Workshop GbRPR 2003, LNCS 2726, pp. 154–165, 2003.

paper, we turn to recent work in bioinformatics to provide an improved solution to the problem.

In bioinformatics the problem of converting a graph to a sequence is known as seriation. Stated succinctly, it is the task of ordering the set of nodes in a graph in a sequence such that strongly correlated nodes are placed next to one another. The seriation problem can be approached in a number of ways. Clearly the problem of searching for a serial ordering of the nodes, which maximally preserves the edge ordering is one of exponential complexity. As a result approximate solution methods have been employed. These involve casting the problem in an optimisation setting. Hence techniques such as simulated annealing and mean field annealing have been applied to the problem. It may also be formulated using semidefinite programming, which is a technique closely akin to spectral graph theory since it relies on eigenvector methods. However, recently Atkins, Boman and Hendrikson [1] have shown how to use an eigenvector of the Laplacian matrix to sequence relational data. There is an obvious parallel between this method and the use of eigenvector methods to locate steady state random walks on graphs.

The aim in this paper is to exploit this seriation method to develop a spectral method for computing graph edit distance. We pose the recovery of the seriation as that of finding an optimal permutation order subject to edge connectivty constraints. By using the Perron-Frobenius theorem, we show that the optimal permuatation order is given by the leading eigenvector of a matrix derived from the adjacency matrix. The seriation path may be obtained from this eigenvector using a simple edge-filtering technique. By using the spectral seriation method, we are able to convert the graph into a string. This opens up the possibility of performing graph matching by performing string alignment. We pose the problem as maximum a posteriori probability estimation of of an optimal alignment sequence on an edit lattice. The edit distance is the negative log-likelihood of the a posteriori alignment probability. To compute the alignment probability, we require two model ingredients. The first of these is an edge compatibility measure for the edit sequence. We use a simple error model to show how these compatibilities may be computed using the edge densities for the graphs under match. The second model ingredient is a set of node correspondence probabilities. We model these using a Gaussian distribution on the components of the leading eigenvectors of the adjacency matrices. by minimising the Levenshtein or edit distance [4,18]. We can follow Wagner and use dynamic programming to evaluate the edit distance between strings and hence recover correspondences [18]. It is worth stressing that although there been attempts to extend the string edit idea to trees and graphs [19,12,15,16], there is considerable current effort aimed at putting the underlying methodology on a rigourous footing.

2   Graph Seriation

Consider the graph $G = (V, E)$ with node index-set V and edge-set $E = \{(i, j) | (i, j) \in V \times V, i \neq j\}$. Associated with the graph is an adjacency matrix A whose elements are defined as follows

$$A(i,j) = \begin{cases} 1 & \text{if } (i,j) \in E \\ 0 & \text{otherwise} \end{cases} \tag{1}$$

Our aim is to assign the nodes of the graph to a sequence order which preserves the edge ordering of the nodes. This sequence can be viewed as an edge connected path on the graph. Let the path commence at the node j_1 and proceed via the sequence of edge-connected nodes $X = \{j_1, j_2, j_3, ...\}$ where $(j_i, j_{i-1}) \in E$. With these ingredients, the problem of finding the path can be viewed as one of seriation, subject to edge connectivity constraints.

As noted by Atkins, Boman and Hendrikson [1], many applied computational problems, such as sparse matrix envelope reduction, graph partitioning and genomic sequencing, involve ordering a set according to a permutation $\pi = \{\pi(j_1), \pi(j_2), \ldots, \pi(j_{|V|})\}$ so that strongly related tokens are placed next to one another. The seriation problem is that of finding the permutation π that satisfies the condition

$$\pi(j_i) < \pi(j_k) < \pi(j_l) \Rightarrow \{A(i,k) \geq A(i,l) \wedge A(k,l) \geq A(i,l)\}$$

This task has been posed as a combinatorial optmisisation problem which involves minimising the penalty function

$$g(\pi) = \sum_{i=1}^{|V|} \sum_{k=1}^{|V|} A(i,k) \big(\pi(j_i) - \pi(j_k)\big)^2$$

for a real symmetric adjacency matrix A.

Unfortunately, the penalty function $g(\pi)$, as given above, does not impose edge connectivity constraints on the ordering computed during the minimisation process. Furthermore, it implies no directionality in the transition from the node indexed j_i to the one indexed j_{i+1}. To overcome these shortcomings, we turn our attention instead to the penalty function

$$g(\pi) = \sum_{i=1}^{|V|-1} A(i, i+1)\big(\pi(j_i) - \pi(j_{i+1})\big)^2 \tag{2}$$

where the nodes indexed j_i and j_{i+1} are edge connected. After some algebra, it is straightforward to show that

$$g(\pi) = \sum_{i=1}^{|V|-1} A(i, i+1)(\pi(j_i)^2 + \pi(j_{i+1})^2) - 2 \sum_{i=1}^{|V|-1} A(i, i+1)\pi(j_i)\pi(j_{i+1}) \tag{3}$$

It is important to note that $g(\pi)$ does not have a unique minimiser. The reason for this is that its value remains unchanged if we add a constant amount to each of the co-efficients of π. We also note that it is desirable that the minimiser of $g(\pi)$ is defined up to a constant λ whose solutions are polynomials in the elements of A. Therefore, we subject the minimisation problem to the constraints

$$\lambda \pi(j_i)^2 = \sum_{k=1}^{|V|} A(k, i)\pi(j_k)^2 \quad \text{and} \quad \sum_{k=1}^{|V|} \pi(j_k)^2 \neq 0 \tag{4}$$

Combining the constraint conditions given in Equation 4 with the definition of the penalty function given in Equation 3, it is straightforward to show that the permutation π satisfies the condition

$$\sum_{k=1}^{|V|} \sum_{i=1}^{|V|-1} \left(A(k,i) + A(k,i+1) \right) \pi(j_k)^2 = \lambda \sum_{i=1}^{|V|-1} \left(\pi(j_i)^2 + \pi(j_{i+1})^2 \right) \quad (5)$$

Using matrix notation, we can write the above equation in the more compact form

$$\Omega A \phi = \lambda \Omega \phi \quad (6)$$

where $\phi = \{\pi(j_1)^2, \pi(j_2)^2, \ldots, \pi(j_{|V|})^2\}^T$ and Ω is the $(N-1) \times N$ matrix

$$\Omega = \begin{bmatrix} 1 & 1 & 0 & \ldots & 0 \\ 0 & 1 & 1 & \ddots & \vdots \\ \vdots & \ddots & \ddots & \ddots & 0 \\ 0 & \ldots & 0 & 1 & 1 \end{bmatrix} \quad (7)$$

Hence it is clear that locating the permuation π that minimises $g(\pi)$ can be posed as an eigenvalue problem, and that ϕ is an eigenvector of A. This follows from the fact Equation 4 can be obtained by multiplying both sides of the eigenvector equation $A\phi = \lambda\phi$ by Ω. Furthermore, due to the norm condition of the eigenvector, the constraint $\sum_{k=1}^{|V|} \pi(j_k)^2 \neq 0$ is always satisfied. Taking this analysis one step further, we can premultiply both sides of Equation 6 by ϕ^T to obtain the matrix equation $\phi^T \Omega A \phi = \lambda \phi^T \Omega \phi$. As a result, it follows that

$$\lambda = \frac{\phi^T \Omega A \phi}{\phi^T \Omega \phi} \quad (8)$$

We note that the elements of the permutation π are required to be real. Consequently, the co-efficients of the eigenvector ϕ are always non-negative. Since the elements of the matrices Ω and A are positive, it follows that the quantities $\phi^T \Omega A \phi$ and $\phi^T \Omega \phi$ are positive. Hence, the set of solutions reduces itself to those that are determined up to a constant $\lambda > 0$. As a result, the co-efficients of the eigenvector ϕ are linearly independent of the all-ones vector $\mathbf{e} = (1, 1...., 1)^T$.

With these observations in mind, we focus on proving the existence of a permutation that minimises $g(\pi)$ subject to the constraints in Equation 4, and demonstrating that this permutation is unique. To this end we use the Perron-Frobenius theorem [17]. This concerns the proof of existence regarding the eigenvalue $\lambda^* = \max_{i=1,2,\ldots,|V|}\{\lambda_i\}$ of a primitive, real, non-negative, symmetric matrix A, and the uniqueness of the corresponding eigenvector ϕ^*. The Perron-Frobenius theorem states that the eigenvalue $\lambda^* > 0$ has multiplicity one. Moreover, the co-efficients of the corresponding eigenvector ϕ^* are all positive and the eigenvector is unique. As a result the remaining eigenvectors of A have at least one negative co-efficient and one positive co-efficient. Since, ϕ^* is also known to be linearly independent of the all-ones vector $\mathbf{e}$, the leading eigenvector of A is the minimiser of $g(\pi)$.

The elements of the leading eigenvector ϕ^* can be used to construct a sertial ordering of the nodes in the graph. We commence from the node associated with the largest component of ϕ^*. We then sort the elements of the leading eigenvector such that they are both in the decreasing magnitude order of the co-efficients of the eigenvector, and satisfy edge connectivity constraints on the graph.The procedure is a recursive one that proceeds as follows. At each iteration, we maintain a list of nodes visited. At iteration k let the list of nodes be denoted by $\mathcal{L}_k$. Initially, $\mathcal{L}_0 = j_o$ where $j_0 = \arg\max_j \phi^*(j)$, i.e. j_0 is the component of ϕ^* with the largest magnitude. Next, we search through the set of first neighbours $\mathcal{N}_{j_0} = \{k | (j_0, k) \in E\}$ of j_o to find the node associated with the largest remaining component of ϕ^*. The second element in the list is $j_1 = \arg\max_{l \in \mathcal{N}_{j_0}} \phi^*(l)$. The node index j_1 is appended to the list of nodes visited and the result is $\mathcal{L}_1$. In the kth (general) step of the algorithm we are at the node indexed j_k and the list of nodes visited by the path so far is $\mathcal{L}_k$. We search through those first-neighbours of j_k that have not already been traversed by the path. The set of nodes is $C_k = \{l | l \in \mathcal{N}_{j_k} \wedge l \notin \mathcal{L}_k\}$. The next site to be appended to the path list is therefore $j_{k+1} = \arg\max_{l \in C_k} \phi^*(l)$. This process is repeated until no further moves can be made. This occurs when $C_k = \emptyset$ and we denote the index of the termination of the path by T. The serial ordering of the nodes of the graph X is given by the ordered list or string of nodes indices $\mathcal{L}_T$.

In practice we will be intrested in finding the edit distance for a pair of graphs $G_M = (V_M, E_M)$ and $G_D = (V_D, E_D)$. The leading eigenvectors of the corresponding adjacency matrices A_M and A_D are respectivlty ϕ_M^* and ϕ_D^*. The string representations of the graph G_M is denoted by X and that for the graph G_D by Y.

There are similarities between the use of the leading eigenvector for seriation and the use of spectral methods to find the steady state random walk on a graph. There are more detailed discussions of the problem of locating the steady state random walk on a graph in the reviews by Lovasz [5] and Mohar [10]. An important result described in these papers, is that if we visit the nodes of the graph in the order defined by the magnitudes of the co-efficients of the leading eigenvector of the transition probability matrix, then the path is the steady state Markov chain. However, this path is not guaranteed to be edge connected. Hence, it can not ne used to impose a string ordering on the nodes of a graph. The seriation approach adopted in this paper does impose edge connectivity constraints and can hence be used to convert graphs to strings in a manner which is suitable for computing edit distance.

3   Probabilistic Framework

We are interested in computing the edit distance between the graphs $G_M = (V_M, E_M)$ referred to as the model graph and the graph $G_D = (V_D, E_D)$ referred to as the data-graph. The seriations of the two graphs are denoted by $X = \{x_1, x_2, ..., x_{|V_M|}\}$ for the model graph and $Y = \{y_1, y_2, ..., y_{|V_D|}\}$ for the data graph. These two strings are used to index the rows and column of an edit lattice. The rows of the lattice are indexed using the data-graph string, while the columns are indexed using the model-graph string. To allow for differences in the sizes of the graphs we introduce a null symbol ϵ which can be used to pad the strings. We pose the problem of computing the edit distance as that of finding a path $\Gamma =< \gamma_1, \gamma_2, ...\gamma_k, ..., \gamma_L >$ through the lattice. Each element $\gamma_k \in (V_D \cup \epsilon) \times (V_M \cup \epsilon)$ of the edit path is a Cartesian pair. We constrain the path to be connected on the edit lattice. In particular, the transition on the edit lattice from the state γ_k to to the state γ_{k+1} is constrained to move in a direction that is increasing and connected in the horizontal, vertical or diagonal direction on the lattice. The diagonal transition corresponds to the match of an edge of the data graph to an edge of the model graph. A horizontal transition means that the data-graph index is not incremented, and this corresponds to the case where the traversed nodes of the model graph are null-matched. Similarly when a vertical transition is made, then the traversed nodes of the data-graph are null-matched.

Suppose that $\gamma_k = (a, b)$ and $\gamma_{k+1} = (c, d)$ represent adjacent states in the edit path between the seriations X and Y. According to the classical approach, the cost of the edit path is given by

$$d(X, Y) = C(\Gamma) = \sum_{\gamma_k \in \Gamma} \eta(\gamma_k \to \gamma_{k+1}) \tag{9}$$

where $\eta(\gamma_k \to \gamma_{k+1})$ is the cost of the transition between the states $\gamma_k = (a, b)$ and $\gamma_{k+1} = (c, d)$. The optimal edit path is the one that minimises the edit distance between string, and satisfies the condition $\Gamma^* = \arg\min_\Gamma C(\Gamma)$ and hence the edit distance is $d(X, Y) = C(\Gamma^*)$. Classically, the optimal edit sequence may be found using Dijkstra's algorithm or by using the quadratic programming method of Wagner and Fisher [18].

However, in this paper we adopt a different approach. We aim to find the edit path that has maximum probability given the available leading eigenvectors of the data-graph and model-graph adjacency matrices. Hence, the optimal path is the one that satisfies the condition

$$\Gamma^* = \arg\max_\Gamma P(\Gamma|\phi_X^*, \phi_Y^*) \tag{10}$$

Under assumptions concerning conditional dependence on the edit lattice it is a straightforward matter to simplify the *a posteriori* probability to show that

$$\Gamma^* = \arg\max_{\gamma_1, \gamma_2 \cdots, \gamma_L} \left\{ \prod_{k=1}^L P(\gamma_k|\phi_X^*(a), \phi_Y^*(b)) \frac{P(\gamma_k, \gamma_{k+1})}{P(\gamma_k)P(\gamma_{k+1})} \right\} \tag{11}$$

The information concerning the structure of the edit path on the lattice is captured by the compatibility measure

$$R_{k,k+1} = \frac{P(\gamma_k, \gamma_{k+1})}{P(\gamma_k)P(\gamma_{k+1})} \tag{12}$$

To establish a link with the classical edit distance picture presented in Section 3, we can re-write the optmimal edit path as a minmisation problem involving the negative logarithm of the *a posteriori* path probability. The optimal path is the one that satisfies the condition

$$\Gamma^* = \arg \min_{\gamma_1, \gamma_2 \dots, \gamma_L} \left\{ \sum_{k=1}^{L} \left[-P(\gamma_k | \phi_X^*(a), \phi_Y^*(b)) - \ln R_{k,k+1} \right] \right\} \tag{13}$$

As a result the elementary edit cost $\eta(\gamma_k \to \gamma_{k+1}) = \eta((a,b) \to (c,d))$ from the site (a,b) and the edit lattice to the site (c,d) is

$$\eta(\gamma_k \to \gamma_{k+1}) = -\left(\ln P(\gamma_k | \phi_X^*(a), \phi_Y^*(b)) + \ln P(\gamma_{k+1} | \phi_*^X(c), \phi_*^Y(d)) + \ln R_{k,k+1} \right) \tag{14}$$

We have recently reported a methodology which lends itself to the modelling of the edge compatibility quantity $R_{k,k+1}$ [11]. It is based on the idea of constraint corruption through the action of a label-error process.

The model leads to an expression for the compatibility that is devoid of free parameters and which depends on the edge-set in the graphs being matched. Details of the derivation of the model are omitted from this paper for reasons of brevity. The compatibility contingency table is

$$R_{k,k+1} = \begin{cases} 1 - \rho_M \rho_D & \text{if } \gamma_k \to \gamma_{k+1} \text{ if } (a,c) \in E_D \text{ and } (b,d) \in E_M \\ 1 - \rho_M & \text{if } \gamma_k \to \gamma_{k+1} \text{ if } (a,c) \in E_D \text{ and } (b = \epsilon \vee d = \epsilon) \\ 1 - \rho_D & \text{if } \gamma_k \to \gamma_{k+1} \text{ if } (a = \epsilon \vee c = \epsilon) \text{ and } (b,d) \in E_M \\ 0 & \text{if } (a = \epsilon \vee c = \epsilon) \text{ and } (b = \epsilon \vee d = \epsilon) \end{cases} \tag{15}$$

The second model ingredient is the *a posteriori* probability of visiting a site on the lattice. Here we assume that the differences in the components of the adjacency matrix leading eigenvectors are drawn from a Guassian distribution. Hence,

$$P(\gamma_k | \phi_X^*(a), \phi_Y^*(b)) = \begin{cases} \frac{1}{\sqrt{2\pi}\sigma} \exp\left\{ -\frac{1}{2\sigma^2}(\phi_X^*(a) - \phi_Y^*(b))^2 \right\} & \text{if } a \neq \epsilon \text{ and } b \neq \epsilon \\ \alpha & \text{if } a = \epsilon \text{ or } b = \epsilon \end{cases} \tag{16}$$

Once the edit costs are computed, we proceed to find the path that yields the minimum edit distance. Following Levenshtein [4] we compute a transition-cost matrix. The minimum cost path on the matrix is located using Dijkstra's algorithm.

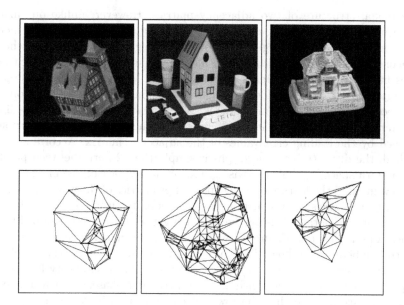

Fig. 1. Example frames and graphs of the sequences used for the experiments.

4 Experiments

We have experimented with our new matching method on an application involving clustering a database containing different perspective views of a number of 3D objects. In doing this, we aim to determine which distance measure results in a cluster assignment which best corresponds to the three image sequences. The objects used in our study are model houses. The different views are obtained as the camera circumscribes the object. The three object sequences used in our experiments are the CMU-VASC sequence, the INRIA MOVI sequence and a sequence of views of a model Swiss chalet, that we collected in-house. In our experiments we use ten images from each of the sequences. To construct graphs for the purposes of matching, we have first extracted corners from the images using the corner detector of Luo, Cross and Hancock [6]. The graphs used in our experiments are the Delaunay triangulations of these corner points. In the top row of Figure 1, we show the first frame of the sequences used in our experiments. The corresponding graphs are displayed in the bottom row. For the 30 graphs contained in the database, we have computed the complete set of 30 × 29 =870 distances between each of the distinct pairs of graphs.

Next, we compare our results with those obtained when using the distances computed using two alternative methods. The first of these, is the negative log-likelihood function computed using the EM algorithm reported by Luo and Hancock [7]. This similarity measure uses purely structural information. The method shares with our edit distance framework the use of a statistical method to compare modal structure of two adjacency matrics. However, unlike our method which uses only the leading eigenvector of the adjacency matrix, this method uses the full pattern of singular vectors of a weighted correlation of the two adjacency matrices. In addition, the method is an iterative one which alternates between computing singular vectors in the M or maximisation step and re-computing the correspondence weight matrix in the E or expectation step. The second of the distance measures is computed using a spectral emdeding of the graphs [8]. The method involves embedding the graphs in an pattern space spanned by the leading eigenvalues of the adjacency matrix. According to this method, the distance between graphs is simply the L2 norm between points in this pattern space [9]. This pairwise graph similarity measure again shares with our new method the feature of using spectral properties of the adjacency matrix.

In Figure 2a we show the distance matrix computed using our algorithm. The distance matrix obtained using the negative log-likelihood measure of Luo and Hancock [8] is shown in Figure 2b, while the distance matrix computed from the spectral embedding is shown in Figure 2c. In each distance-matrix the element with row column indices i, j corresponds to the pairwise similarity between the graph indexed i and the graph indexed j in the data-base. In each matrix the graphs are arranged so that the row and column index increase monotonically with viewing angle. The blocks of views for the different objects follow one another. From the matrices in Figure 2, it is clear that as the difference in viewing direction increases, then the graphs corresponding to different views of the same object are more similar than the graphs for different objects. The different objects appear as distinct blocks in the matrices. Within each block, there is substructure (sub-blocks) which correspond to different characteristic views of the object.

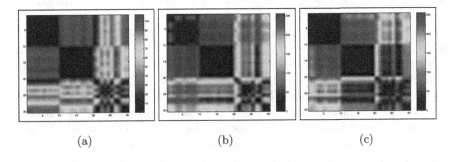

(a)   (b)   (c)

Fig. 2. Edit distance matrices

We have applied the pairwise clustering clustering algorithm of Robles-Kelly and Hancock [13] to the similarity data for the complete data-base of graphs. Our aim is to investigate which distance measure results in the best set of graph-clusters. The pairwise clustering algorithm requires distances to be represented by a matrix of pairwise affinity weights. Ideally, the smaller the distance, the

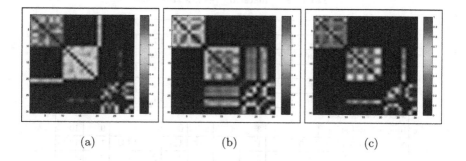

(a)   (b)   (c)

Fig. 3. Final similarity matrices associated to the clustering process.

stronger the weight, and hence the mutual affinity to a cluster. The affinity weights are required to be in the interval $[0, 1]$. Hence, for the pair of graphs indexed $i1$ and $i2$ the affinity weight is taken to be

$$W_{i1,i2} = \exp\left(-k\frac{d_{i1,i2}}{\max(D)}\right) \qquad (17)$$

where k is a constant and $d_{i1,i2}$ is the edit distance between the graph indexed $i1$ and the graph indexed $i2$. The clustering process is an iterative one that maintains two sets of variables. The first of these is a set of cluster membership indicators $s_{i\omega}^{(n)}$ which measures the affinity of the graph indexed i to the cluster indexed ω at iteration n of the algorithm. The second, is an estimate of the affinity matrix based on the current cluster-membership indicators $W^{(n)}$. These two sets of variables are estimated using interleaved updates steps, which are formulated to maximise a likelihood function for the pairwise cluster configuration.

The initial similarity matrices generated by the clustering process for each of the distance measures have already been shown in Figure 2. In Figure 3, we show the final similarity matrices after 10 iterations of the clustering process. The left-hand panel is for the edit-distance reported in this paper, the middle panel is for the negative log-likelihood function of Luo and Hancock, and the right-hand panel is for the spectral feature vectors.

We summarise the clustering results for the three similarity measures in Table 1. In contrast with the results obtained when using the structural graph matching algorithm, Only two images of the Chalet sequence have been misassigned when the graph edit distance is used. This is an improvement on the results obtained with the log-likelihood function of Luo and Hancock, and comparable to the results obtained with the spectral feature vectors. The errors are due to the fact that the graphs in question are morphologically more similar to the graphs in the CMU and MOVI sequence than to those in the Chalet sequence. Nevertheless a success rate of 93.3% is obtained when edit distance is used for the affinity matrix computation. Our method compares favourably when computational cost

Table 1. Clustering results statistics.

CMU Sequence										
Image Number	1	2	3	4	5	6	7	8	9	10
Number of Nodes	30	32	32	30	30	32	30	30	30	31
Cluster (SGM)	1	1	1	1	1	1	1	1	1	1
Cluster (MDS)	1	1	1	1	1	1	1	1	1	1
Cluster (ED)	1	1	1	1	1	1	1	1	1	1
MOVI Sequence										
Image Number	1	2	3	4	5	6	7	8	9	10
Number of Nodes	140	134	130	136	137	131	139	141	133	136
Cluster (SGM)	2	2	2	2	2	2	2	2	2	2
Cluster (MDS)	2	2	2	2	2	2	2	2	2	2
Cluster (ED)	2	2	2	2	2	2	2	2	2	2
Chalet Sequence										
Image Number	1	2	3	4	5	6	7	8	9	10
Number of Nodes	40	57	92	78	90	64	113	100	67	59
Cluster (SGM)	3	3	2	2	2	3	2	2	3	3
Cluster (MDS)	4	3	3	3	3	3	2	3	3	3
Cluster (ED)	1	3	3	3	3	3	2	3	3	3

is taken into account. This is because our method only requires the computation of the leading eigenvector while the spectral feature vectors require the complete eigenstructure to be computed. In consequence, the edit distance takes in average 2.6 times less time to compute than the spectral feature vectors.

5   Conclusions

The work reported in this paper provides a synthesis of ideas from spectral graph-theory and structural pattern recognition. We use a graph spectral seriation method based on the leading eigenvector of the adjacency matrix to convert graphs to strings. We match the resulting string representations by minimising edit distance. The edit costs needed are computed using a simple probabilistic model of the edit transitions which is designed to preserve the edge order on the correspondences. The minimum cost edit sequence may be used to locate correspondences between nodes in the graphs under study.

We have demonstrated that both the edit sequence and the associated distance are of practical use. The edit sequence delivers correspondences that are robust to structural error, and compare favourably with those delivered by a number of alternative graph-matching algorithms. The edit distances can be used to cluster graphs into meaningful classes.

References

1. J. E. Atkins, E. G. Roman, and B. Hendrickson. A spectral algorithm for seriation and the consecutive ones problem. *SIAM Journal on Computing*, 28(1):297–310, 1998.

2. H. Bunke. On a relation between graph edit distance and maximum common subgraph. *Pattern Recognition Letters*, 18(8):689–694, 1997.
3. M. A. Eshera and K. S. Fu. A graph distance measure for image analysis. *IEEE Transactions on Systems, Man and Cybernetics*, 14:398–407, 1984.
4. V. I. Levenshtein. Binary codes capable of correcting deletions, insertions and reversals. *Sov. Phys. Dokl.*, 6:707–710, 1966.
5. L. Lovász. Random walks on graphs: a survey. *Bolyai Society Mathematical Studies*, 2(2):1–46, 1993.
6. Bin Luo and E. R. Hancock. Procrustes alignment with the EM Algorithm. In *8th International Conference on Computer Analysis of Images and Image Patterns*, pages 623–631, 1999.
7. Bin Luo and E. R. Hancock. Structural graph matching using the EM algorithm and singular value decomposition. *IEEE Trans. on Pattern Analysis and Machine Intelligence*, 23(10):1120–1136, 2001.
8. Bin Luo, R. Wilson, and E. Hancock. Eigenspaces for graphs. *International Journal of Image and Graphics*, 2(2):247–268, 2002.
9. Bin Luo, R. C. Wilson, and E. R. Hancock. Spectral feature vectors for graph clustering. In *S+SSPR 2002*, pages 82–90, 2002.
10. B. Mohar. Some applications of laplace eigenvalues of graphs. In G. Hahn and G. Sabidussi, editors, *Graph Symmetry: Algebraic Methods and Applications*, NATO ASI Series C, pages 227–275, 1997.
11. R. Myers, R. C. Wilson, and E. R. Hancock. Bayesian graph edit distance. *PAMI*, 22(6):628–635, June 2000.
12. B. J. Oommen and K. Zhang. The normalized string editing problem revisited. *PAMI*, 18(6):669–672, June 1996.
13. A. Robles-Kelly and E. R. Hancock. A maximum likelihood framework for iterative eigendecomposition. In *Proc. of the IEEE International Conference on Conputer Vision*, pages 654–661, 2001.
14. A. Robles-Kelly and E. R. Hancock. An expectation-maximisation framework for segmentation and grouping. *Image and Vision Computing*, 20(9-10):725–738, 2002.
15. A. Sanfeliu and K. S. Fu. A distance measure between attributed relational graphs for pattern recognition. *IEEE Transactions on Systems, Man and Cybernetics*, 13:353–362, 1983.
16. L. G. Shapiro and R. M. Haralick. Relational models for scene analysis. *IEEE Transactions on Pattern Analysis and Machine Intelligence*, 4:595–602, 1982.
17. R. S. Varga. *Matrix Iterative Analysis*. Springer, second edition, 2000.
18. R. A. Wagner and M. J. Fisher. The string-to-string correction problem. *Journal of the ACM*, 21(1):168–173, 1974.
19. J. T. L. Wang, B. A. Shapiro, D. Shasha, K. Zhang, and K. M. Currey. An algorithm for finding the largest approximately common substructures of two trees. *PAMI*, 20(8):889–895, August 1998.

Graph Polynomials, Principal Pivoting, and Maximum Independent Sets[*]

Roland Glantz and Marcello Pelillo

Dipartimento di Informatica
Università Ca' Foscari di Venezia
Via Torino 155, 30172 Venezia Mestre, Italy
{glantz,pelillo}@dsi.unive.it

Abstract. The maximum independent set problem (or its equivalent formulation, which asks for maximum cliques) is a well-known difficult combinatorial optimization problem that is frequently encountered in computer vision and pattern recognition. Recently, motivated by a linear complementarity formulation, standard pivoting operations on matrices have proven to be effective in attacking this and related problems. An intriguing connection between the maximum independent set problem and pivoting has also been recently studied by Arratia, Bollobás and Sorkin who introduced the *interlace polynomial*, a graph polynomial defined in terms of a new pivoting operation on undirected, unweighted graphs. Specifically, they proved that the degree of this polynomial is an upper bound on the independence number of a graph. The first contribution of this paper is to interpret their work in terms of standard matrix pivoting. We show that Arratia *et al.*'s pivoting operation on a graph is equivalent to a principal pivoting transform on a corresponding adjacency matrix, provided that all calculations are performed in the Galois field $\mathbb{F}_2$. We then extend Arratia *et al.*'s pivoting operation to fields other than $\mathbb{F}_2$, thereby allowing us to apply their polynomial to the class of *gain* graphs, namely bidirected edge-weighted graphs whereby reversed edges carry weights that differ only by their sign. Finally, we introduce a new graph polynomial for undirected graphs. Its recursive calculation can be done such that all ends of the recursion correspond to independent sets and its degree equals the independence number. However, the new graph polynomial is different from the independence polynomial.

1   Introduction

Many problems in computer vision and pattern recognition, such as graph matching, clustering and segmentation, image sequence analysis, etc., can be formulated in terms of finding an independent set (i.e., a subset of mutually non-adjacent vertices) of a given graph, having maximal cardinality.[1] This problem

---

[*] This work is supported by the Austrian Science Foundation (FWF) under grant P14445-MAT and by MURST under grant MM09308497.

[1] Note that determining a maximum independent set in a graph G is equivalent to finding a maximum clique (namely, a subset of mutually adjacent vertices) in the complement of G.

is known to be NP-hard, and so is that of approximating it within a constant factor (see [3] for a recent review concerning algorithms, applications, and complexity issues related to this problem).

Motivated by a linear complementarity formulation, standard pivoting operations on matrices have recently proven to be effective in attacking this and related problems [7,6]. Another intriguing connection between the maximum independent set problem and pivoting has also been recently studied by Arratia, Bollobás and Sorkin [1,2] who introduced the *interlace polynomial*, a graph polynomial defined in terms of a new pivoting operation on undirected, unweighted graphs. Specifically, they proved that the degree of this polynomial is an upper bound on the independence number of a graph.

In this paper, we make three theoretical contributions. Firstly, we show that pivoting on an edge $e = \{u, v\}$ of an undirected graph G, as defined in [1,2], is equivalent to a sequence of two classical pivoting steps on an adjacency matrix of G, provided that the calculations are performed in the Galois field $\mathbb{F}_2$. If a and b denote the row and column numbers of u and v in an adjacency matrix A of G, pivoting on e is equivalent to classical matrix pivoting on $A(a, b)$, followed by classical matrix pivoting on $A(b, a)$. This sequence of pivoting operations, in turn, is equivalent to a principal pivoting transform of A on the non-singular 2×2 principal submatrix $A[\{a, b\}]$, followed by swapping row a with row b and column a with column b. Here, $A[\{a, b\}]$ denotes the principal submatrix of A given by the rows and columns a, b. The principal pivoting transform can be understood as a generalization of classical pivoting, where pivoting is performed on non-singular principal submatrices instead of single non-zero entries [9].

Our second contribution comes from the observation that the principal pivoting transformation on $A[\{a, b\}]$ preserves antisymmetry in *all* fields. Note also that A is antisymmetric in $\mathbb{F}_2$. Hence, by extending the pivoting operation from $\mathbb{F}_2$ to arbitrary fields, we extend the interlace polynomial to gain graphs, with gains from an arbitrary field, i.e. bidirected graphs, the edges of which have attributes $g(\cdot)$ such that $g(r, s) = -g(s, r) \neq 0$ for all edges (r, s).

The final part of the paper is devoted to finding a variant p of q that again takes undirected unweighted graphs G. The plan is to define $p(G) := q(G')$, where G' is from a class of gain graphs that have the same vertices and the same edges (apart from the directions) as G. In addition, the class of gain graphs is restricted by a non-singularity condition on the gains. The latter also ensures that $q(G')$ can be computed such that none of the edges in the original gain graph is removed by pivoting. Thus, the vertex sets of the edge-less graphs arising from selected calculations of the extended interlace polynomial are always independent sets of the original gain graphs. It turns out to be crucial that the gains come from an infinite field. The new graph polynomial is called *pivoting polynomial*. Its degree equals the independence number, whereas the degree of $q(G)$ is merely an upper bound of the independence number.

2   Double Pivots of Antisymmetric Matrices

Pivoting is a standard method in linear and quadratic optimization. Given a field $\mathbb{F}$, consider q, x, $x' \in \mathbb{F}^n$ and $M = (M(i,j))_{i,j=1}^n \in \mathbb{F}^{n \times n}$. The *simple pivot* on $M(a,b) \neq 0$ of the system $Mx = q$ is the system

$$M^{(a,b)}x' = q^{(a,b)},$$

where $q^{(a,b)} \in \mathbb{F}^n$ and $M^{(a,b)} \in \mathbb{F}^{n \times n}$ are defined as follows [4]

$$q_a^{(a,b)} = -q_a/M(a,b)$$
$$q_i^{(a,b)} = q_i - (M(i,b)/M(a,b))q_a \quad i \neq a$$
$$M^{(a,b)}(a,b) = 1/M(a,b)$$
$$M^{(a,b)}(i,b) = M(i,b)/M(a,b) \quad i \neq a$$
$$M^{(a,b)}(a,j) = -M(a,j)/M(a,b) \quad j \neq b$$
$$M^{(a,b)}(i,j) = M(i,j) - (M(i,b)/M(a,b))M(a,j) \quad i \neq a, j \neq b.$$

From now on we require M to be antisymmetric, i.e. $M(i,j) = -M(j,i)$ and $M(i,i) = 0$ for all i,j. The former requirement implies the latter whenever $\mathbb{F} \neq \mathbb{F}_2$. From $M(a,b) \neq 0$ it follows $a \neq b$, and

$$M^{(a,b)}(b,j) = M(b,j) \quad \forall j,$$
$$M^{(a,b)}(i,a) = M(i,a) \quad \forall i.$$

In particular, $M^{(a,b)}(b,a) = M(b,a) = -M(a,b) \neq 0$ and thus $(M^{a,b})^{(b,a)}$ is well defined. By symmetry, $(M^{(b,a)})^{(a,b)}$ is also well defined. In the following we write $M^{(a,b)(c,d)}$ instead of $(M^{(a,b)})^{(c,d)}$. The straight forward proof of the following proposition can be found in [5].

Proposition 1 (Double pivot, antisymmetry). *Let $M \in \mathbb{F}^{n \times n}$ be antisymmetric with $M(a,b) \neq 0$. Then, $M^{(a,b)(b,a)} = M^{(b,a)(a,b)}$. Moreover, $M^{(a,b)(b,a)}$ is again antisymmetric.*

Definition 1 (Double pivot $M^{\{a,b\}}$). *Let $M \in \mathbb{F}^{n \times n}$ be antisymmetric and let $M(a,b) \neq 0$. Then the antisymmetric matrix*

$$M^{\{a,b\}} := M^{(a,b)(b,a)} \tag{1}$$

is called the double pivot of M on $\{a,b\}$.

3   Double Pivoting in Terms of PPT

For the double pivot we will derive the analogue of Lemma 10(ii) in [1], the key to prove the existence of the interlace polynomial. The plan is to express the double pivot in terms of the principal pivot transform. Using well known theorems about iterated principal pivot transforms we will then be able to derive the analogue of Lemma 10(ii).

In the following, n denotes a positive integer and M is from $\mathbb{F}^{n \times n}$. The next definitions are from [9,4]:

- $\langle n \rangle := \{1, 2, \ldots, n\}$. For any $\alpha \subseteq \langle n \rangle$ the set $\langle n \rangle \setminus \alpha$ is denoted by $\overline{\alpha}$.
- $M[\alpha, \beta]$ is the submatrix of M whose rows and columns are indexed by α and β, respectively. The submatrix $M[\alpha, \alpha]$ of M is written as $M[\alpha]$. The matrix $M[\alpha]$ is a so called *principal* submatrix of M.
- The *Schur complement* $M \setminus M[\alpha]$ of a non-singular principal submatrix $M[\alpha]$ in M is defined by

$$M \setminus M[\alpha] := M[\overline{\alpha}] - M[\overline{\alpha}, \alpha] M[\alpha]^{-1} M[\alpha, \overline{\alpha}]. \tag{2}$$

Definition 2 (Principal pivot transform $ppt(M, \alpha)$). *Let $\alpha \subseteq \langle n \rangle$ be such that $M[\alpha]$ is non-singular. The principal pivot transform $ppt(M, \alpha)$ of $M \in \mathbb{F}^{n \times n}$ is obtained from M through replacing*
$M[\alpha]$ *by* $M[\alpha]^{-1}$,
$M[\overline{\alpha}]$ *by* $M \setminus M[\alpha]$,
$M[\alpha, \overline{\alpha}]$ *by* $-M[\alpha]^{-1} M[\alpha, \overline{\alpha}]$, *and*
$M[\overline{\alpha}, \alpha]$ *by* $M[\overline{\alpha}, \alpha] M[\alpha]^{-1}$.

Double and principal pivots turn out to differ only by a swap of two rows and the corresponding columns. Such a swap will be denoted by a subscript indicating the numbers of the rows (columns) being swapped.

Proposition 2 (Double and principal pivots). *Let $a \neq b \in \langle n \rangle$ be such that $M[\{a, b\}]$ is non-singular. Then $M^{\{a,b\}} = (ppt(M, \{a, b\}))_{ab}$.*

A straight forward proof can be found in [5]. The following theorems differ from Theorem 3.1 and Theorem 3.2 in [9] only in that they are formulated for arbitrary fields $\mathbb{F}$ and not just for the field of complex numbers. However, the proofs given in [9] depend merely on the axioms for fields. The two theorems will serve to derive rules for iterated double pivots.

Theorem 1. *Let $M \in \mathbb{F}^{n \times n}$ and $\alpha \subseteq \langle n \rangle$ such that $M[\alpha]$ is non-singular. For each $x, y \in \mathbb{F}^n$ define $u = u(x, y), v = v(x, y) \in \mathbb{F}^n$ by $u_i = y_i, v_i = x_i$ for all $i \in \alpha$, and $u_j = x_j, v_j = y_j$ for all $j \in \overline{\alpha}$. Then $N = ppt(M, \alpha)$ is the unique matrix with the property that*

$$Mx = y \quad \leftrightarrow \quad Nu = v. \tag{3}$$

Theorem 2. *Let $M \in \mathbb{F}^{n \times n}$ and let $\alpha = \bigcup_{i=1}^{k} \alpha_i \subseteq \langle n \rangle$ for some k such that $\alpha_i \cap \alpha_j = \emptyset$ for all $i \neq j$. If the sequence of matrices*

$$M_0 = M, \quad M_i = ppt(M_{i-1}, \alpha_i), \quad i = 1, \ldots, k$$

is well defined, i.e. if all $M_{i-1}[\alpha_i]$ are non-singular, then $ppt(M, \alpha) = M_k$.

In the following we write $M^{\{a,b\}\{a,c\}}$ instead of $(M^{\{a,b\}})^{\{a,c\}}$.

Lemma 1 (First analogue of Lemma 10(ii) in [1]). *Let $M \in \mathbb{F}^{n \times n}$ be antisymmetric with $M(a,b), M(a,c) \neq 0$. Then,*

$$M^{\{a,b\}\{a,c\}} = (M^{\{a,c\}})_{bc} . \tag{4}$$

Proof: First we prove that $ppt(ppt(M, \{a,b\}), \{b,c\}) = ppt(M, \{a,c\})$. Applying Theorem 1 first to principal pivoting on $\{a,b\}$ and then to principal pivoting on $\{b,c\}$, it follows that $Mx = y \leftrightarrow Nu = v$, where $N = ppt(ppt(M, \{a,b\}), \{b,c\})$ is unique and u, v are defined by

$$u_i = \begin{cases} y_i \text{ if } i \in \{a,c\}, \\ x_i \text{ otherwise} \end{cases} \qquad v_i = \begin{cases} x_i \text{ if } i \in \{a,c\}, \\ y_i \text{ otherwise.} \end{cases} \tag{5}$$

Note that x_b and y_b have been exchanged twice. Applying Theorem 1 to principal pivoting on $\{a,c\}$ only, it follows that $Mx = y \leftrightarrow N'u = v$, where $N' = ppt(M, \{a,c\})$ is unique and u, v are as above. Hence, $N = N'$.

Finally, using Prop. 2 and the result just proven, we have:

$$\begin{aligned} M^{\{a,b\}\{a,c\}} &= (ppt((ppt(M, \{a,b\}))_{ab}, \{a,c\}))_{ac} \\ &= (ppt(ppt(M, \{a,b\}), \{b,c\})_{ab})_{ac} = ((ppt(M, \{a,c\}))_{ab})_{ac} \\ &= (((M^{\{a,c\}})_{ac})_{ab})_{ac} = (M^{\{a,c\}})_{bc} \end{aligned}$$

$\square$

4   Double Pivots of Gain Graphs

For the rest of the paper, let $G = (V, E, g)$ denote a gain graph with non-zero gains. Formally, the edge set E is a subset of $(V \times V) \setminus \{(v,v) \mid v \in V\}$ such that $(u,v) \in E$ implies $(v,u) \in E$, and the gains are given by a mapping $g(\cdot)$ from E to $\mathbb{F} \setminus \{0\}$ such that $g(u,v) = -g(v,u)$ for all $(u,v) \in E$.

In this section we define double pivots of G by means of double pivots of an (antisymmetric) adjacency matrix of G. In order to show that the double pivots of G do not depend on the choice of the adjacency matrix, we first define adjacency matrices in terms of functions $l(\cdot)$ that specify the correspondence between the vertices of G and the rows (columns) of the adjacency matrices.

Definition 3 (Adjacency matrix M_l of G). *Let $l : V \to \{1, \ldots, |V|\}$ be bijective. The adjacency matrix M_l of G is defined by*

$$M_l(i,j) = \begin{cases} g(l^{-1}(i), l^{-1}(j)) \text{ if } (l^{-1}(i), l^{-1}(j)) \in E, \\ 0 \qquad\qquad\qquad otherwise. \end{cases} \tag{6}$$

Note that $M_l(l(u), l(v))$ is independent of the labeling function $l(\cdot)$. From the definition of $M^{\{a,b\}}$ it follows that $M_l^{\{l(r), l(s)\}}(l(u), l(v))$ is also independent of $l(\cdot)$. Finally, independence of $l(\cdot)$ extends to the following.

Definition 4 (Double pivot of G). *Let M_l denote an adjacency matrix of G and let $(r,s) \in E$. The graph $G^{\{r,s\}} := (V^{\{r,s\}}, E^{\{r,s\}}, g^{\{r,s\}})$ defined by*

- $V^{\{r,s\}} := V,$
- $E^{\{r,s\}} := \{(u,v) \mid u,v \in V,\ M_l^{\{l(r),l(s)\}}(l(u),l(v)) \neq 0\},$
- $g^{\{r,s\}}(u,v) := M_l^{\{l(r),l(s)\}}(l(u),l(v))$ for all $(u,v) \in E^{\{r,s\}}.$

is called the double pivot of G on r and s.

In particular, Lemma 1 extends to

$$G^{\{r,s\}\{r,t\}} = (G^{\{r,t\}})_{st} \quad \forall (r,s), (r,t) \in E, \tag{7}$$

where the subscript indicates that the vertex names r and s have been exchanged. In accordance with Prop. 2 we may define

$$ppt(G, \{r,s\}) := (G^{\{r,s\}})_{rs} \quad \forall (r,s) \in E. \tag{8}$$

If $\sum_{v \in V} g(u,v) = 0$ for all $u \in V$, then $g(\cdot)$ is a flow on G. Hence,

$$\sum_j M_l(i,j) = 0 \quad \forall i$$

holds for any adjacency matrix M_l of G. Let (r,s) be an edge of G and let $l(r) = a$, $l(s) = b$. Due to the definition of $M_l^{\{a,b\}}$ via simple pivoting it holds that

$$\sum_j M_l^{\{a,b\}}(i,j) = 0 \quad \forall i.$$

Since $M_l^{\{a,b\}}$ is antisymmetric and since $g^{\{a,b\}}(u,v) \neq 0$ for all $(u,v) \in E^{\{r,s\}}$, it follows that $g^{\{r,s\}}$ is a flow on $G^{\{r,s\}}$. In other words, Kirchhoff's conservation law is preserved by double pivoting.

5   Double Pivots in $\mathbb{F}_2$

In case of $\mathbb{F} = \mathbb{F}_2$, the Galois field containing only the numbers 0 and 1, the simple pivot of M_l on $M_l(a,b) \neq 0$, i.e. on $M_l(a,b) = 1$, takes the form

$$M_l^{(a,b)}(i,j) = \begin{cases} 1 - M_l(i,j) & \text{if } i \neq a,\ j \neq b \text{ and } M_l(i,b) = M_l(a,j) = 1, \\ M_l(i,j) & \text{otherwise.} \end{cases}$$

In Fig.1 the matrix $M_l^{(a,b)}$ is interpreted as a directed graph with self-loops. Double pivoting on $\{a,b\}$ (see again Fig. 1) yields $M_l^{\{a,b\}}(i,j) = M_l^{(a,b)(b,a)}(i,j)$

$$= \begin{cases} 1 - M_l^{(a,b)}(i,j) & \text{if } i \neq b,\ j \neq a \text{ and } M_l^{(a,b)}(i,a) = M_l^{(a,b)}(b,j) = 1, \\ M_l^{(a,b)}(i,j) & \text{otherwise,} \end{cases}$$

$$= \begin{cases} 1 - M_l^{(a,b)}(i,j) & \text{if } i \neq b,\ j \neq a \text{ and } M_l(i,a) = M_l(b,j) = 1, \\ M_l^{(a,b)}(i,j) & \text{otherwise.} \end{cases}$$

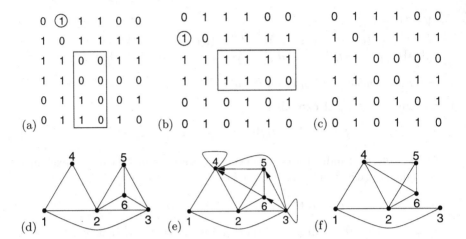

Fig. 1. (a) Adjacency Matrix M_l. The pivoting element $(1,2)$ is circuited. Changes will take place in the rectangle. (b) $M_l^{(1,2)}$. The new pivoting element is $(2,1)$. (c) $M_l^{\{1,2\}}$. (d), (e), (f) Graphs defined by (a), (b), and (c), respectively.

Comparing M_l with $M_l^{\{a,b\}}$, the entry at (i,j) has changed, if and only if the following boolean expression is true.

$$(i \neq a, j \neq b, M_l(i,b) = M_l(a,j) = 1) \ \text{Xor} \ (i \neq b, j \neq a, M_l(i,a) = M_l(b,j) = 1) \tag{9}$$

Let $G_u = (V, E_u)$ be an undirected simple graph, i.e. E_u is a collection of 2-element subsets of V. G_u corresponds to a bidirected graph $G = (V, E, g)$ with gains in $\mathbb{F}_2$ by $E = \{(u,v) \mid \{u,v\} \in E_u\}$ and $g(e) = 1$ for all $e \in E$. Let (r,s) be an edge of G and let M_l be an antisymmetric adjacency matrix of G with $l(r) = a$, $l(s) = b$. Partitioning $V \setminus \{r, s\}$ into the classes

1) $C_1^{r,s} := \{t \in V \setminus \{r,s\} \mid (t,r) \in E, (t,s) \notin E\}$,
2) $C_2^{r,s} := \{t \in V \setminus \{r,s\} \mid (t,s) \in E, (t,r) \notin E\}$,
3) $C_3^{r,s} := \{t \in V \setminus \{r,s\} \mid (t,r), (t,s) \in E\}$,
4) $C_4^{r,s} := \{t \in V \setminus \{r,s\} \mid (t,r), (t,s) \notin E\}$.

and evaluating the boolean expression (9) according to the membership of $l^{-1}(i)$ and $l^{-1}(j)$ to the classes $C_1^{r,s}$ to $C_4^{r,s}$ yields Tab. 1. Note that $G^{\{r,s\}}$ corresponds to another undirected simple graph. Thus, double pivoting on gain graphs with non-zero gains in $\mathbb{F}_2$ induces an operation on the corresponding undirected simple graphs which may be characterized as follows (see Tab. 1). An edge $\{u,v\}$ of G_u is toggled (between a non-edge and an edge), if and only if u and v belong to different classes other than class $C_4^{r,s}$. This characterization coincides with the definition of the pivoting operation in [1].

Table 1. Eq. (9) depending on the classes of the nodes $l^{-1}(i)$, $l^{-1}(j)$.

	$l^{-1}(j) \in C_1^{r,s}$	$l^{-1}(j) \in C_2^{r,s}$	$l^{-1}(j) \in C_3^{r,s}$	$l^{-1}(j) \in C_4^{r,s}$
$l^{-1}(i) \in C_1^{r,s}$	0 Xor 0	0 Xor 1	0 Xor 1	0 Xor 0
$l^{-1}(i) \in C_2^{r,s}$	1 Xor 0	0 Xor 0	1 Xor 0	0 Xor 0
$l^{-1}(i) \in C_3^{r,s}$	1 Xor 0	0 Xor 1	1 Xor 1	0 Xor 0
$l^{-1}(i) \in C_4^{r,s}$	0 Xor 0	0 Xor 0	0 Xor 0	0 Xor 0

6   The Extended Interlace Polynomial

Let $\mathcal{G}$ denote the class of gain graphs and let $G \in \mathcal{G}$. The order of G is written as $|G|$. The following is analogous to [1] (an undirected edge in [1] corresponds to a pair of reversed edges in the gain graph).

Theorem 3 (Interlace polynomial).
There is a unique map $q : \mathcal{G} \to \mathbf{Z}[x]$, $\mathbf{G} \mapsto \mathbf{q(G)}$, such that

$$q(G) = \begin{cases} q(G - r) + q(G^{\{r,s\}} - s) \text{ for any edge } (r, s) \text{ of } G, \\ x^{|G|} \qquad\qquad\qquad\qquad\qquad\quad \text{if } G \text{ has no edges} \end{cases} \tag{10}$$

Proof: Using Eq. (7) instead of Lemma 10(ii) in [1], the proof is the same as the one in [1].   □

Using Eq. (8) the recursive part of Eq. (10) can be rewritten as

$$q(G) = q(G - r) + q(ppt(G, \{r, s\}) - r). \tag{11}$$

7   The Pivoting Polynomial

The plan for the rest of the paper is to derive a graph polynomial for undirected simple graphs $G_u = (V, E_u)$ via the extended interlace polynomial of "corresponding" gain graphs. The term "corresponding" implies that the gain graphs take the form $G = (V, E, g(\cdot))$, where $E = \{(u, v) \mid \{u, v\} \in E_u\}$ and $g(e) \neq 0$ for all $e \in E$. We will have to make sure that from G and G' both "corresponding" to G_u, it always follows that $q(G) = q(G')$. The problem of finding "corresponding" gain graphs is approached via classes of adjacency matrices. The following is needed to relate the adjacency matrix of G to the adjacency matrix of G_u.

Definition 5 (Template $\overline{M}$, supremum of matrices from $\{0,1\}^{n \times n}$).
Let $M \in \mathbb{F}^{n \times n}$. The template $\overline{M} \in \{0,1\}^{n \times n}$ of M is defined by

$$\overline{M}(i,j) = \begin{cases} 1 \text{ if } M(i,j) \neq 0, \\ 0 \text{ otherwise.} \end{cases}$$

Let $\mathcal{M}$ be a collection of matrices from $\{0,1\}^{n \times n}$ and let $S \in \mathcal{M}$. S is called supremum of $\mathcal{M}$, if $S(i,j) = 0$ implies that $M(i,j) = 0$ for all $M \in \mathcal{M}$.

Given a symmetric adjacency matrix $A \in \{0,1\}^{n \times n}$ of G_u, we are looking for a class sup^A of antisymmetric adjacency matrices from $\mathbb{F}^{n \times n}$ such that

(1) $\overline{M} = A$ for all $M \in sup^A$.
(2) Let $M, M' \in sup^A$ be antisymmetric adjacency matrices of the gain graphs G and G'. Then, $q(G) = q(G')$.

To arrive at the definition of sup^A, we consider Schur complements $M \setminus M[\alpha]$. Varying the non-zero elements outside the non-singular principal submatrix $M[\alpha]$, we will maximize the number of non-zero entries in the Schur complement. Formally, we set

$$V_{M,\alpha} := \{N \in \mathbb{F}^{n \times n} \mid N \text{ is antisymmetric}, \overline{N} = \overline{M} \text{ and } N[\alpha] = M[\alpha]\} \quad (12)$$

and determine $S \in V_{M,\alpha}$ such that $\overline{S \setminus S[\alpha]}$ is the supremum of

$$T_{M,\alpha} := \{\overline{N \setminus N[\alpha]} \mid N \in V_{M,\alpha}\} . \quad (13)$$

In [5] it is shown that the supremum exists and that non-supremum Schur complement templates are singular cases. This suggests to define the following class of antisymmetric adjacency matrices.

Definition 6 (Matrix class $sup_{M,\alpha}$). *Let $\mathbb{F}$ be infinite, let $\alpha \in \langle n \rangle$ and let $M \in \mathbb{F}^{n \times n}$ be antisymmetric such that $M[\alpha]$ is non-singular. Set*

$$sup_{M,\alpha} := \{S \in V_{M,\alpha} \mid \overline{S \setminus S[\alpha]} \text{ is the supremum of } T_{M,\alpha}\} . \quad (14)$$

In the following we will specify those principal submatrices of a symmetric adjacency matrix A that have a non-singular counterpart in an antisymmetric adjacency matrix M with $\overline{M} = A$.

Definition 7 (Regular with respect to A, n_A). *Let $\mathbb{F}$ be infinite and let $A \in \{0,1\}^{n \times n}$ be symmetric with $A(i,i) = 0$ for all i. The set $\alpha \in \langle n \rangle$ is called regular with respect to A, if there exists antisymmetric $M \in \mathbb{F}^{n \times n}$ such that $\overline{M} = A$ and $M[\alpha]$ is non-singular. Set*

$$n_A := \{\alpha \in \langle n \rangle \mid \alpha \text{ is regular with respect to } A\} .$$

Finally, the matrix class sup^A with respect to $\mathbb{F}$ is defined as follows.

Definition 8 (Matrix class sup^A with respect to $\mathbb{F}$). *Let $\mathbb{F}$ be infinite and let $A \in \{0,1\}^{n \times n}$ be symmetric with $A(i,i) = 0$ for all i. The class sup^A of A is defined by*

$$sup^A := \bigcap_{\alpha \in n_A} \left(\bigcup (sup_{N,\alpha} \mid \overline{N} = A \text{ and } N[\alpha] \text{ is non-singular}) \right) .$$

The next proposition is proven in [5].

Proposition 3. *Let $M \in sup^A$ and let $M[\alpha]$ be a non-singular principal submatrix of M. Then, $(ppt(M,\alpha))(i,j) = 0$ implies $M(i,j) = 0$ for all $i,j \in \overline{\alpha}$.*

In the following we define the gain graphs "corresponding" to an undirected simple graph.

Definition 9 (Supremum gain graphs from G_u).
Let G_u be an undirected simple graph with vertex set V. Moreover, let $l(\cdot)$ be a bijective mapping from V to $\{1, \ldots, |V|\}$, and let A be the symmetric adjacency matrix of G_u with respect to l. A gain graph with vertex set V is called supremum gain graph from G_u, if its antisymmetric adjacency matrix with respect to $l(\cdot)$ is in $\sup^A$.

To see that the above definition is independent of $l(\cdot)$ consider an edge $\{u, v\}$ of G_u. It corresponds to the entries $A(l(u), l(v)) = A(l(v), l(u)) = 1$ of A, i.e. to the entries $M(l(u), l(v)) = -M(l(v), l(u)) \neq 0$ of *all* matrices $M \in \sup^A$. Those entries, in turn, correspond to the edges (u, v) and (v, u) of *all* supremum gain graphs from G_u.

In the following we will see that the extended interlace polynomials of supremum gain graphs from G_u are identical.

A gain graph G' is called *first-order-descendant* of G, if G has an edge (a, b) such that $G' = G - a$ or $G' = G^{\{a,b\}} - b$. A *descendant of order $n + 1$* is a first-order-descendant of an an n-th order descendant. It will turn out to be convenient, if G may be referred to as its descendant of order 0. G' is called a *descendant* of G, if G' is an n-th order descendant of G for some $n \in \mathbb{N}_0$. The vertex set of a descendant of $G = (V, E, g)$ is always a subset of V. A descendant of G is called *terminal*, if it has no edges (and thus no descendants).

Proposition 4 (Descendants of $q(G)$). *Let G be a gain graph. Then, the recursive computation of $q(G)$ can be done such that each descendant is a subgraph of G or a subgraph of some $G^{U_1 \ldots U_m} \setminus U$, where $U = \bigcup_{i=1}^m U_i \subset V$ and each U_i has cardinality 2.*

Proof: We may assume that G has at least one non-isolated vertex r_1. Consider the following scheme for computing $q(G)$.

- **Step $i = 1$.** As long as r_1 is a non-isolated vertex of D (note that $D = G$ in the beginning), continue the calculation of $q(D)$ via the descendants $D - r_1$ and $D^{\{r_1, s_1\}} - s_1$ for some s_1. The descendant $D - r_1$ has no further descendants at this stage of the scheme. If r_1 is not isolated in $D^{\{r_1, s_1\}} - s_1$, the scheme transforms $D^{\{r_1, s_1\}} - s_1$ into the descendants $D^{\{r_1, s_1\}} - r_1 - s_1$ and $(D^{\{r_1, s_1\}} - s_1)^{\{r_1, s_2\}} - s_2$ for some s_2. Again, the first descendant has no further descendants at this stage and, using Eq. (7), the second descendant can be written as $(D^{\{r_1, s_1\}} - s_1)^{\{r_1, s_2\}} - s_2 =$

$$(D^{\{r_1, s_1\}})^{\{r_1, s_2\}} - s_1 - s_2 = D^{\{r_1, s_2\}} - s_1 - s_2.$$

In the following let d denote the degree of r_1 in D. Continued expansion of each second descendant eventually leaves us with the descendant $D - r$, descendants of the form

$$D^{\{r_1, s_k\}} - r_1 - s_1 - s_2 - \ldots - s_k \quad (k < d), \tag{15}$$

and the descendant $D^{\{r_1,s_d\}} - s_1 - \ldots - s_d$ with isolated r_1. The expression of the latter descendant can be simplified as follows. Pivoting on $\{r_1, s_d\}$ has no effect on the neighborhood of r_1. Hence, $s_1, \ldots, s_d$ are precisely the neighbors of r_1 in D and s_d is the only neighbor of r_1 in $D - s_1 - \ldots - s_{d-1}$. Thus, $D^{\{r_1,s_d\}} - s_1 - \ldots - s_d = (D - s_1 - \ldots - s_{d-1})^{\{r_1,s_d\}} - s_d =$

$$D - s_1 - \ldots - s_d .$$

Hence, each descendant is a subset of D or it has the form (15).

- **Step $i > 1$.** We may assume that each descendant D is a subgraph of G or a subgraph of $G^{U_1,\ldots,U_j} - \bigcup_{k=1}^{j} U_k$ for some U_k, $1 \le k \le j \le i - 1$. If D has a non-isolated vertex r_i, we apply the first step of the scheme to r_i instead of r_1. All descendants still have the required form. □

Proposition 5 (Identical polynomials, independent sets).
Let $G = (V, E, g(\cdot))$ and G' be supremum gain graphs from G_u. Then $q(G) = q(G')$. Furthermore, for each term $a_k x^k$ of $q(G)$ with $a_k \ne 0$ there exist independent sets of G_u with cardinality k.

Proof: According to Prop. 4 the polynomial $q(G)$ can be computed such that every descendant is a subgraph of G or a subgraph of some $G^{U_1 \ldots U_m} \setminus U$, where $U = \bigcup_{i=1}^{m} U_i$ and each U_i has cardinality 2.

Let A be a the template of an antisymmetric adjacency matrix of G with respect to the bijective mapping $l : V \to \{1, \ldots |V|\}$. From Theorem 2 it follows that the adjacency matrix of $G^{U_1 \ldots U_m} \setminus U$ with respect to $l(\cdot)$ is $M \setminus M[\alpha]$ for some $M \in sup^A$, where $\alpha = \bigcup_{i=1}^{m} \alpha_i$, $\alpha_i = l(U_i)$ for all i.

From $\overline{M \setminus M[\alpha]}$ being unique it follows that the templates of the terminal descendants are also unique. Moreover, Prop. 3 implies that the edge set of $G^{U_1 \ldots U_m} \setminus U$ is a superset of the edge set of $G \setminus U$. Hence, the vertex sets of the (edge-less) terminal descendants of $q(G)$ are unique independent sets of G_u. □

Prop. 5 implies that the following is well defined.

Definition 10 (Pivoting polynomial $p(G_u)$). *Let G_u be an undirected simple graph and let G be a supremum gain graph from G_u. The pivoting polynomial $p(G_u)$ of G_u is defined by $p(G_u) := q(G)$.*

The following consideration is from [1]. Take a vertex a_1. If a_1 is an isolated vertex of G, then $q(G) = xq(G - a_1)$. Otherwise, for any edge (a_1, b) it holds that $q(G) = q(G - a_1) + q(G^{\{a_1,b\}} - b)$. In any case, since all coefficients of all $q(\cdot)$ are non-negative, it follows that the degree of $q(G)$ is not smaller than the degree of $q(G - a_1)$. By subsequently taking a_i that are not contained in a fixed maximum independent set, we get that the degree of $q(G)$ is an upper bound for the independence number.

From Prop. 5 it follows that the degree of $q(G)$ is a lower bound for the independence number, if G is a supremum gain graph from some G_u. Hence, we get the following.

Proposition 6 (Degree = independence number). *The degree of $p(G_u)$ is equal to the independence number of G_u.*

The example $p(K^3) = 4x$ shows that the pivoting polynomial, in contrast to the independence polynomial [8], may list an independent set more than once.

8   Conclusions

We defined a new graph polynomial for undirected simple graphs, the *pivoting polynomial*. It was derived from the interlace polynomial by extending it to gain graphs over arbitrary fields. Provided that the underlying field is infinite, gain graphs with identical vertex and edge sets were then shown to yield the same extended interlace polynomial, as long as they fulfill a non-singularity condition formulated in terms of their adjacency matrices. Finally, the pivoting polynomial was defined by means of gain graphs that fulfill the non-singularity condition. The recursive computation of the pivoting polynomial can be done such that each end of the recursion corresponds to an independent set, one of which is maximum. We plan to exploit the theoretical results presented in this paper in graph matching and related problems.

References

1. R. Arratia, B. Bollobás, and G. B. Sorkin. The Interlace Polynomial: A New Graph Polynomial. Technical Report RC 21813 (98165), IBM T.J. Watson Reserch Center, Yorktown Heights NY, July 2000.
2. R. Arratia, B. Bollobás, and G. B. Sorkin. The Interlace Polynomial: A New Graph Polynomial. In *Proceeding of the Eleventh Annual ACM-SIAM Symposium on Discrete Algorithms*, pages 237–245, San Francisco, CA, January 2000.
3. I. M. Bomze, M. Budinich, P. M. Pardalos, and M. Pelillo. The maximum clique problem. In D.-Z. Du and P. M. Pardalos, editors, *Handbook of Combinatorial Optimization (Suppl. Vol. A)*, pages 1–74. Kluwer, Boston, MA, 1999.
4. R. W. Cottle, J.-S. Pang, and R. E. Stone. *The Linear Complementarity Problem.* Accademic Press, Boston, MA, 1992.
5. R. Glantz and M. Pelillo. Graph polynomials from principal pivoting. Technical Report CS-2003-2, Dipartimento di Informatica, Università Ca' Foscari di Venezia, 30172 Venezia Mestre (VE), Italy, 2003.
6. A. Massaro and M. Pelillo. Matching graphs by pivoting. *Pattern Recognition Letters*, 24:1099–1106, 2003.
7. A. Massaro, M. Pelillo, and I. M. Bomze. A complementary pivoting approach to the maximum weight clique problem. *SIAM Journal on Optimization*, 12(4):928–948, 2002.
8. R. Stanley. *Enumerative combinatorics, Volume 1.* Wadsworth and Brooks-Cole, Monterey, CA, 1986.
9. M. J. Tsatsomeros. Principal pivot transforms: Properties and applications. *Linear Algebra and Its Applications*, 307:151–165, 2000.

Graph Partition for Matching

Huaijun Qiu and Edwin R. Hancock

Department of Computer Science, University of York
Heslington, York, YO10 5DD, UK

Abstract. Although inexact graph-matching is a problem of potentially exponential complexity, the problem may be simplified by decomposing the graphs to be matched into smaller subgraphs. If this is done, then the process may cast into a hierarchical framework or cast in a way which is amenable to parallel computation. In this paper we demonstrate how the Fiedler-vector can be used to partition graphs for the purposes of decomposition. We show how the resulting subgraphs can be matched using a variety of algorithms. We demonstrate the utility of the resulting graph-matching method on both real work and synthetic data. Here it proves to provide results which are comparable with a number of state-of-the-art graph matching algorithms.

1 Introduction

Graph partitioning is concerned with grouping the vertices of a connected graph into subsets so as to minimize the total cut weight [7]. The process is of central importance in electronic circuit design, map coloring and scheduling [18]. However, in this paper we are interested in the process since it provides a means by which the inexact graph-matching problem may be decomposed into a series of simpler subgraph matching problems. There are a great deal of algorithms of reducing graphs into hierarchical representatives [1]. As demonstrated by Messmer and Bunke [11], error-tolerant graph matching can be simplified using decomposition methods and reduced to a problem of subgraph indexing. Our aim in this paper is to explore whether spectral methods can be used to partition graphs in a stable manner for the purposes of matching by decomposition.

Recently, there has been increased interest in the use of spectral graph theory for characterising the global structural properties of graphs. Spectral graph theory aims to summarise the structural properties of graphs using the eigenvectors of the adjacency matrix or the Laplacian matrix [2]. There are several examples of the application of spectral matching methods for grouping and matching in the computer vision literature. For instance, Umeyama has shown how graphs of the same size can be matched by performing singular decomposition on the adjacency matrices [16]. By developing an indexing mechanism that maps the topological structure of a tree into a low-dimensional vector space, Shokoufandeh, Dickinson, Siddiqi and Zucker [15] have shown how graphs can be retrieved efficiently from a database. This method is further extended to encode topological structure by exploiting the interleaving property of eigenvalues. At York,

E. Hancock and M. Vento (Eds.): IAPR Workshop GbRPR 2003, LNCS 2726, pp. 178–189, 2003.

we have recently embarked on a program of activity aimed at developing robust graph-spectral methods for inexact graph-matching. For instance, Luo and Hancock [10] have extended Umeyama's algorithm to the case of matching graphs in different size using the apparatus of the EM algorithm. Luo, Wilson and Hancock [9] have shown how spectral feature vectors can be use to embed sets of graphs into a pattern-space.

One of the most important spectral attributes of a graph is the Fiedler vector, i.e. the eigenvector associated with the second smallest eigenvalue of the Laplacian matrix. In a useful review, Mohar [12] has summarized some important applications of Laplace eigenvalues such as the max-cut problem, semidefinite programming and steady state random walks on Markov chains. More recently, Haemers [6] has explored the use of interlacing properties for the eigenvalues and has shown how these relate to the chromatic number, the diameter and the bandwidth of graphs. In the computer vision literature, Shi and Malik [14] have used the Fiedler vector to develop a recursive partition scheme and have applied this to image grouping and segmentation. The Fiedler vector may also be used for the Minimum Linear Arrangement problem(MinLA) which involves placing the nodes of a graph in a serial order which is suitable for the purposes of visualisation [3].

The aim in this paper is to consider whether the partitions delivered by the Fiedler vector can be used to simplify the graph-matching problem. In particular the aim is to use the Fiedler vector to decompose graphs by partitioning them into structural units. Our aim is to explore whether the partitions are stable under structural error, and in particular whether they can be used for the purposes of graph-matching. In particular, we explore their use in conjunction with the recently reported discrete relaxation method of Wilson and Hancock [17] and in particular the efficient realisation of it based on Levenshtein distance [13].

The outline of this paper is as follows: In section 2, we introduce some basic concepts about spectral graph theory. Section 3 describes how we use the Fielder vector to partition graphs. In section 4, we provide some comparative experimental evaluation, which includes both a sensitivity study and some real-world examples. Finally, section 5 is the conclusion.

2   Laplacian Matrix and Fiedler Vector

Let the weighted graph G to be the quadruple (V, E, Ω, ω), where V is the set of nodes, E is the set of arcs, $\Omega = \{W_i, \forall i \in V\}$ is a set of weights associated with the nodes and $\omega = \{w_{i,j}, \forall (i,j) \in E\}$ is a set of weights associated with the edges. Further suppose that $D(G) = diag(deg_v; v \in V(G))$ be the diagonal degree matrix with $deg_{i,i} = \sum_{j=1}^{n} a_{i,j}$, where $a_{i,j}$ is the element of the *Adjacency matrix*. The elements of the Laplacian matrix are given by

$$L_{ij}(G) = \begin{cases} \sum_{\langle i,k \rangle \in E} w_{i,k} & \text{if } i = j \\ -w_{i,j} & \text{if } i \neq j \text{ and } (i,j) \in E \\ 0 & \text{otherwise} \end{cases} \qquad (1)$$

The Laplacian matrix has a number of important properties. It is symmetric and positive semidefinite. The eigenvector $(1, 1, \ldots, 1)^T$ corresponds to the trivial eigenvalue zero. If the graph is connected then all other eigenvalues are positive and the smallest eigenvalue is a simple one, which means the number of connected components of the graph is equal to the multiplicity of the smallest eigenvalue. If we arrange all the eigenvalues from the smallest to the largest $0 \leq \lambda_1 \leq \lambda_2 \ldots \leq \lambda_n$, the most important are the largest eigenvalue $\lambda_{max}(G)$ and the second smallest eigenvalue $\lambda_2(G)$, whose corresponding eigenvector is referred to as the *Fiedler Vector*.

Our aim is to decompose the graph by partitioning the nodes using a path-based heuristic. The Fiedler vector may be used to order the nodes in a graph in a path order which minimizes the change in edge weight between successive nodes. Let the permutation vector $\pi(i)$ indicate the location of vertex i in the path. The goal is to find the path π that minimizes the cost

$$LA_\pi(G) = \sum_{<i,j> \in E} w_{ij} \cdot |\pi(i) - \pi(j)| \tag{2}$$

Unfortunately, minimizing LA_π leads to multiple solutions since the minimum is invariant under translations in the permutation order. This problem can be overcome by demanding that the permutation π is independent of all-ones eigenvector $e = (1, 1, 1, ..)^T$, i.e. by demanding that $\pi^T e = 0$. Hence LA_π is minimal when all the vertices are placed at the same location, which is not informative. This can be avoided by adding a normalization constraint on the π vector $\pi^T \pi = 1$, and the problem can be represented by

$$LA_\pi(G) = \min_{\pi^T e = 0, \pi^T \pi = 1} \pi^T L \pi \tag{3}$$

where L is the Laplacian matrix. The first solution is the vector $e = (1, 1, 1...)^T$ which is excluded by the first constraint. So we take the second solution namely, the Fiedler vector.

3   Graph Partition

In our previous work on graph matching, we worked with structural units that consisted of a centre-node and all nodes connected to the centre node by an edge. However, these structural units were overlapping, and as a result they do not constitute a partition of the nodes in the graph. The aim in this paper is to use the Fiedler vector to partition graphs into non-overlapping structural units and to use the structural units generated by this decomposition as input to our existing graph-matching methods. The reasons for wanting to do this are twofold. First, if the units are non-overlapping then they may be matched in parallel, and this offers the scope for efficiency gains. Second, it suggests a hierarchical approach in which the structural units are matched subject to constraints on their relational arrangement.

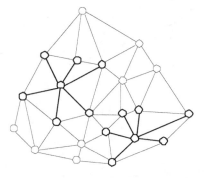

Fig. 1. Supercliques

3.1 Superclique

Generally speaking, graph partition aims to decompose a graph into non-overlapping sets of nodes so as to minimize a cost function. What we are concerned with here is to partition of the graph into *supercliques*. The superclique of the node i consists of its center node, together with its immediate neighbors connected by edges in the graph, i.e., $C_i = \{i\} \cup \{u; (i, u) \in E\}$. An illustration is provided in Figure 1, which shows a graph with two of its supercliques highlighted. Hence, each superclique consists of a *center node* and *immediate neighbors* of the center node, i.e. $N_i = C_i \setminus \{i\}$.

3.2 Graph Partition

The problem addressed here is how to partition the graph into a set of non-overlapping supercliques using the node order defined by the Fiedler vector. Our idea is to assign to each node a measure of significance as the centre of a superclique. We then traverse the path defined by the Fiedler vector selecting the centre-nodes on the basis of this measure.

We commence by assigning weights to the nodes on the basis of their rank-order in the path defined by the Fiedler vector. Studies of the MinLA have revealed that the Fiedler vector can assign the similar labels to the vertices which are located close to each other. We will use this property to assign the weights to the nodes. We first sort the nodes in descending order according to the magnitude of the corresponding component of the Fiedler vector. The order is given by $\Pi^* = \{\pi_1, \pi_2, \pi_3, \ldots, \pi_n\}^T$ with $\pi_i > \pi_{i+1}$. Having the vector sorted, we assign weights on the nodes based on their rank order $W_i = Rank(\pi_i)$. With this weighted graph in hand, we can gauge the significance of each node using the following score function:

$$S_i = \alpha \left(Deg(i) + |N_i \cap P| \right) + \frac{\beta}{W_i} \tag{4}$$

where P is the set of nodes on the perimeter of the graph. The first term depends on the degree of the node and its proximity to the perimeter. Hence, it will

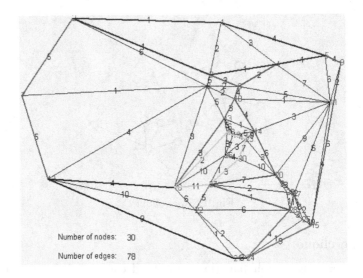

Fig. 2. Graph Partition

sort nodes according to their distance from the perimeter. This will allow us to partition nodes from the outer layer first and then work inwards. The second term ensures that the first ranked nodes in the Fielder vector are visited first.

We use the Fiedler vector of the associated Laplacian matrix $L(G)$ to locate the non-overlapping supercliques of the graph G. Let $\Gamma = < j_1, j_2, j_3, ... >$ be the rank-order of the nodes as defined by the Fiedler vector, i.e. $\pi(j_1) < \pi(j_2) < \pi(j_3)......$ We traverse this list until we find a node k_1 which is neither in the perimeter, i.e. $k_1 \notin P$ and whose score exceeds those of its neighbours, i.e. $S_{k_1} = \arg\max_{i \in k_1 \cup N_{k_1}} S_i$. When this condition is satisfied, then the node k_1 together with its neighbours N_{k_1} represent the first superclique. The set of nodes $C_{k_1} = k_1 \cup N_{k_1}$ are appended to a list A that tracks the set of nodes assigned to supercliques. This process is repeated for all the nodes which have not yet been assigned to supercliques i.e. $R = \Gamma - A$. The procedure terminates when all the nodes of the graph have been assigned to non-overlapping supercliques. An example is shown at Figure 2.

We have used the non-overlapping super-cliques detected in this way as the structural units in the Wilson and Hancock discrete relaxation algorithm [17] and the modification developed by Myers, Wilson and Hancock [13] which effects more efficient computation using Levenshtein distance.

4   Experiments

In this section, we perform two kinds of experiments. First, we commence with a sensitivity study using synthetic data. The aim under here is to measure the sensitivity of our new method to the structural corruption and to compare it with alternatives. The second part evaluates the method on real-world data.

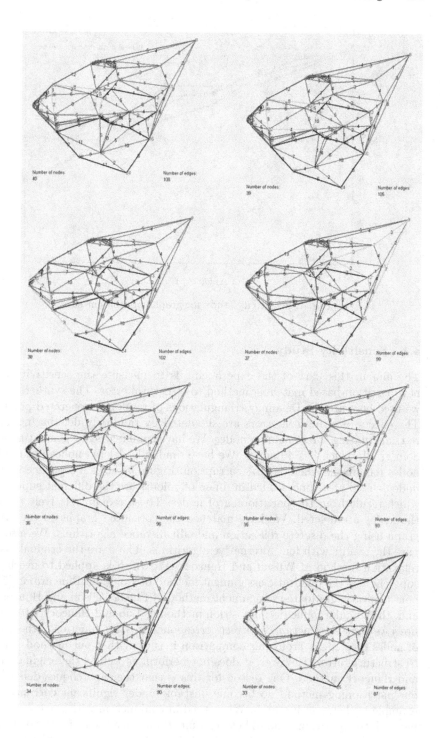

Fig. 3. A sequence of synthetic graphs

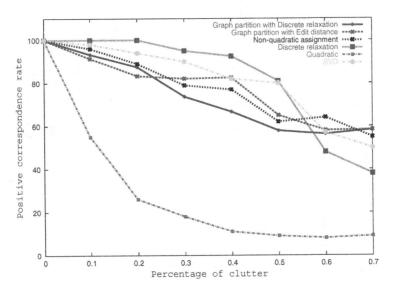

Fig. 4. Sensitivity study for graphs of different size

4.1 Sensitivity Study

The aim in this part of the experiments is to measure the sensitivity of our new partition-based matching method to structural error. The synthetic graphs we used here are the Delaunay triangulations of randomly generated point-sets. The effects of structural errors are simulated by randomly deleting nodes and re-triangulating the remaining nodes. We have commenced by generating a random graph which has 40 nodes. We have randomly deleted nodes until only 12 nodes remain, i.e. there is 70% corruption. Figure 3 shows a sequence with one node deleted at a time. Coded in different colours are the different supercliques which result from the partitioning of nodes. These remain relatively stable as the nodes are deleted. We have matched the corrupted graphs to the original graph using the discrete relaxation and edit distance algorithms. We also compare the results with four alternative algorithms. These are the original discrete relaxation method of Wilson and Hancock [17] which is applied to overlapping supercliques, the quadratic assignment method of Gold and Rangarajan [5], the non-quadratic graduated assignment method of Finch, Wilson and Hancock [4], and, the singular value decomposition method of Luo and Hancock [10]. In Figure 4 we show the fraction of correct correspondences as a function of the fraction of nodes deleted. & From this comparison it is clear that our method is robust to structural error. However, it does not perform as well as the original Wilson and Hancock method. One reason for this is that the supercliques delivered by our partitioning method do become unstable under significant corruption. An example is shown in Figure 5. When used in conjunction with the edit-distance method, the partitions lead to better results than when used with the dictionary-based discrete relaxation method. This is important since the former method is

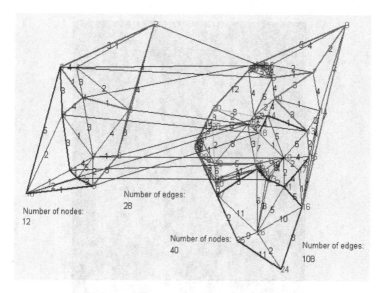

Number of edges:
28

Number of nodes:
12

Number of nodes:
40

Number of edges:
108

Fig. 5. An example in inexact graph matching

more computationally efficient than the latter, since the overheads associated with dictionary construction can grow exponentially if dummy nodes need to be inserted.

4.2   Real-World Data

The real-world data used is taken from the CMU model-house sequence. This sequence is made up of a series of images which have been captured from different viewpoints. In order to convert the images into abstract graphs for matching, we extract point features using corner detector by Luo, Cross and Hancock [8]. Our graphs are the Delaunay triangulations of the corner-features. The supercliques obtained by graph partition are shown in different colors in Figure 6. We have matched the first image to each of the subsequent images in the sequence by using discrete relaxation and edit distance. The results of those two methods are compared with those obtained using the method of Luo, Cross and Hancock [8] in Table 1. This table contains the number of detected corners to be matched, the number of correct correspondence, the number of missed corners and the number of miss-matched corners. Figure 7 shows us the correct correspondence rate as a function of relative position deviation for those three methods according to the data in the Table 1.

From the results, it is clear that our new method degrades gradually and out performs the Luo and Hancock's EM method when the difference in viewing angle is large. Figures 8 to 11 show the results of each pair of graph matching. There are clearly significant structural differences in the graphs including rota-

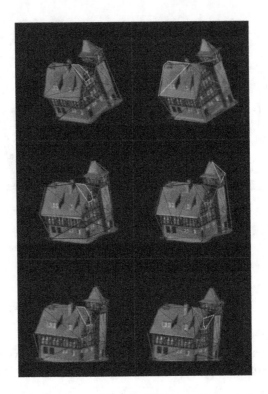

Fig. 6. Graph partition on Delaunay triangulations

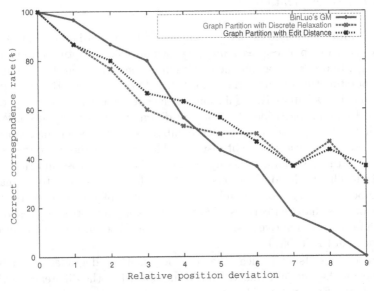

Fig. 7. Results from CMU compared with BinLuo's

Table 1. Correspondence allocation results for three methods

Method	House index	0	1	2	3	4	5	6	7	8	9
	Corners	30	32	32	30	30	32	30	30	30	31
EM	Correct	-	29	26	24	17	13	11	5	3	0
	False	-	0	2	3	8	11	12	15	19	24
	Missed	-	1	2	3	5	6	7	10	8	6
Discrete	Correct	-	26	23	18	16	15	15	11	14	9
Relaxation	False	-	4	6	9	12	14	13	17	16	20
	Missed	-	0	1	3	2	1	2	2	0	1
Edit	Correct	-	26	24	20	19	17	14	11	13	11
Distance	False	-	3	5	8	11	12	16	15	17	19
	Missed	-	1	1	2	0	1	0	4	0	0

tion, scaling and perspective distortion. But even in the worst case, our method has a correct correspondence 36%.

5   Conclusions

In this paper, we have used Fiedler vector of the Laplacian matrix to partition the nodes of a graph into structural units for the purposes of matching. This allows us to decompose the problem of matching the graphs into that of matching the structural units. We investigate the matching of the structural units using a dictionary-based relaxation algorithm and an edit-distance based method. The partitioning method is sufficiently stable under structural error that accuracy of match is not sacrificed.

Our motivation in undertaking this study is to use the partitions to develop a hierarchical matching method, with scope for parallel implementation. The aim is to construct a graph that represents the arrangement of the partitions. By first matching the partition arrangement graphs, we provide constraints on the matching of the individual partitions. Moreover, the matching of different partitions may be conducted in parallel.

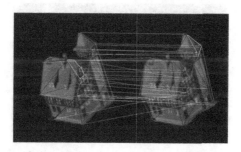

Fig. 8. Correspondences between the first and the third images

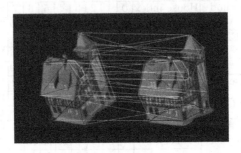

Fig. 9. Correspondences between the first and the fifth images

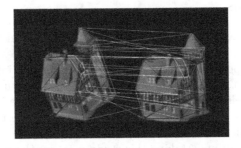

Fig. 10. Correspondences between the first and the seventh images

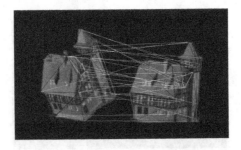

Fig. 11. Correspondences between the first and the tenth images

References

1. Luc Brun and Walter G. Kropatsch. Irregular pyramids with combinatorial maps. *SSPR/SPR*, 1451:256–265, 2000.
2. F.R.K. Chung. *Spectral Graph Theory*. CBMS series 92. American Mathmatical Society Ed., 1997.
3. J. Diaz, J. Petit, and M. Serna. A survey on graph layout problems. Technical report LSI-00-61-R, Universitat Politècnica de Catalunya, Departament de Llenguatges i Sistemes Informàtics, 2000.
4. A.M. Finch, R.C. Wilson, and E.R. Hancock. An energy function and continuous edit process for graph matching. *Neural Computation*, 10(7):1873–1894, 1998.
5. S. Gold and A. Rangarajan. A graduated assignment algorithm for graph matching. *IEEE PAMI*, 18(4):377–388, 1996.
6. W.H. Haemers. Interlacing eigenvalues and graphs. *Linear Algebra and its Applications*, (226-228):593–616, 1995.
7. B.W. Kernighan and S. Lin. An efficient heuristic procedure for partitioning graphs. *The Bell System Technical Journal*, pages 291–307, 1970.
8. B. Luo, A. D. Cross, and E. R. Hancock. Corner detection via topographic analysis of vector potential. *Proceedings of the 9 th British Machine Vision Conference*, 1998.
9. B. Luo, R.C. Wilson, and E.R. Hancock. Spectral embedding of graphs. 2002 Winter Workshop on Computer Vision.
10. Bin Luo and Edwin R. Hancock. Structural graph matching using the em algorithm and singular value decomposition. *IEEE PAMI*, 23(10):1120–1136, 2001.
11. B.T. Messmer and H. Bunke. A new algorithm for error-tolerant subgraph isomorphism detection. *IEEE PAMI*, 20:493–504, 1998.
12. B. Mohar. Some applications of laplace eigenvalues of graphs. *Graph Symmetry: Algebraic Methods and Applications*, 497 NATO ASI Series C:227–275, 1997.
13. Richard Myers, Richard C. Wilson, and Edwin R. Hancock. Bayesian graph edit distance. *IEEE PAMI*, 22(6):628–635, 2000.
14. Jianbo Shi and Jitendra Malik. Normalized cuts and image segmentation. *IEEE PAMI*, 22(8):888–905, 2000.
15. A. Shokoufandeh, S.J. Dickinson, K. Siddiqi, and S.W. Zucker. Indexing using a spectral encoding of topological structure. In *Proc. of the IEEE Conf. on Computer Vision and Pattern Recognition*, pages 491–497, 1999.
16. S. Umeyama. An eigendecomposition approach to weighted graph matching problems. *IEEE PAMI*, 10:695–703, 1988.
17. Richard C. Wilson and Edwin R. Hancock. Structural matching by discrete relaxation. *IEEE PAMI*, 19(6):634–648, 1997.
18. R.J. Wilson and John J. Watkins. *Graphs: an introductory approach: a first course in discrete mathematics*. Wiley international edition. New York, etc., Wiley, 1990.

Spectral Clustering of Graphs

Bin Luo, Richard C. Wilson, and Edwin R. Hancock

Department of Computer Science, Univerisity of York, York Y01 5DD, UK.

Abstract. In this paper we explore how to use spectral methods for embedding and clustering unweighted graphs. We use the leading eigenvectors of the graph adjacency matrix to define eigenmodes of the adjacency matrix. For each eigenmode, we compute vectors of spectral properties. These include the eigenmode perimeter, eigenmode volume, Cheeger number, inter-mode adjacency matrices and intermode edge-distance. We embed these vectors in a pattern-space using two contrasting approaches. The first of these involves performing principal or independent components analysis on the covariance matrix for the spectral pattern vectors. The second approach involves performing multidimensional scaling on the L2 norm for pairs of pattern vectors. We illustrate the utility of the embedding methods on neighbourhood graphs representing the arrangement of corner features in 2D images of 3D polyhedral objects.

1 Introduction

Many of the foundations of current work on graph-structures for vision and pattern recognition were laid by the pioneers of structural pattern recognition in the early 1980's [2,8]. However, despite recent progress in accurately measuring the similarity of graphs and performing robust inexact graph-matching, one of challenges that ramains is how to cluster graphs. Graph-clustering allows the unsupervised learning of the class-structure from sets of graphs. The problem is of practical importance in the organisation of large structural data-bases [9] or the discovery of the view-structure of objects [1].

One of the problems that hinders the graph-clustering that graphs are neither vectorial in nature nor easily transformed into vectors. The reasons for this are twofold. First, there is no canonical ordering of the nodes or edges of a graph. Hence, there is no natural way to map the nodes or edges to the components of a vector. Second, most graph-matching or graph manipulation problems are inexact in nature. That is to say that the graphs are noisy in nature and hence contain different numbers of nodes or edges. Hence, even if an ordering can be established then there needs to be a means of dealing with pattern-vectors of different length. Since they are not easily vectorised, it is not straightforward to characterise the mean and variance of a set of graphs. Hence, standard pattern recognition methods can not be used to analyse or cluster sets of graphs. One way around this problem is to adopt a pairwise clustering approach [3]. This involves measuring the pairwise similarity of the graphs and clustering them by searching for sets of graphs which exhibit a strong mutual affinity to one-another.

E. Hancock and M. Vento (Eds.): IAPR Workshop GbRPR 2003, LNCS 2726, pp. 190–201, 2003.

We have recently attempted to overcome this problem by adopting a spectral representation of graphs [6] which allows the structure of a graph onto a vector of fixed length. We work with the spectral decomposition (or eigendecomposition) of the adjacency matrix. Each component of the vector is taken to represent a different spectral mode of the original graph adjacency matrix. The order of the components of the vector is the magnitude order of the eigenvalues of the adjacency matrix. For each spectral mode, we use the components of the associated eigenvectors to compute spectral attributes. In this way we solve the problem of finding correspondences between nodes and vector-components. Empirical results showed that the feature-vectors resulted in well structured pattern spaces, where siliar graphs were cloese together. The aim in this paper is to investigate in more detail how the spectral feature-vectors can be used to the purposes of clustering. The aim is to investigate a numnber of different strategies including principal components analysis, independant components analysis and multidimensional scaling.

2 Graph Spectra

In this paper we are concerned with the set of graphs $G_1, G_2, .., G_k, ..., G_N$. The kth graph is denoted by $G_k = (V_k, E_k)$, where V_k is the set of nodes and $E_k \subseteq V_k \times V_k$ is the edge-set. Our approach in this paper is a graph-spectral one. For each graph G_k we compute the adjacency matrix A_k. This is a $|V_k| \times |V_k|$ matrix whose element with row index i and column index j is

$$A_k(i,j) = \begin{cases} 1 & \text{if } (i,j) \in E_k \\ 0 & \text{otherwise} \end{cases}. \tag{1}$$

From the adjacency matrices $A_k, k = 1...N$ at hand, we can calculate the eigenvalues λ_k by solving the equation $|A_k - \lambda_k I| = 0$ and the associated eigenvectors ϕ_k^ω by solving the system of equations $A_k \phi_k^\omega = \lambda_k^\omega \phi_k^\omega$, where ω is the eigenmode index. We order the eigenvectors according to the decreasing magnitude of the eigenvalues, i.e. $|\lambda_k^1| > |\lambda_k^2| > \ldots |\lambda_k^{|V_k|}|$. The eigenvectors are stacked in order to construct the modal matrix $\Phi_k = (\phi_k^1 | \phi_k^2 | \ldots | \phi_k^{|V_k|})$.

With the eigenvalues and eigenvectors of the adjacency matrix to hand, the spectral decomposition for the adjacency matrix of the graph indexed k is

$$A_k = \sum_{\omega=1}^{|V_k|} \lambda_k^\omega \phi_k^\omega (\phi_k^\omega)^T. \tag{2}$$

If $\Lambda_k = diag(\lambda_k^1, ..., \lambda_k^{|V_k|})$ is the diagonal matrix with the eigenvalues of A_k as diagonal elements, then the spectral decomposition of the adjacency matrix can be written as $A_k = \Phi_k \Lambda_k \Phi_k^T$.. Associated with the eigenmode with index ω is the mode adjacency matrix $S_k^\omega = \phi_k^\omega (\phi_k^\omega)^T$.. The aim in this paper is to explore whether the properties of these matrices can be used to construct feature vectors for the graphs under study. We explore two different approaches. The first of these involves computing features for individual mode adjacency matrices. The second involves the use of relational features which describe the arrangement of mode adjacency matrices.

For each graph, we use only the first n eigenmodes of the adjacency matrix. The truncated modal matrix is $.\Phi_k = (\phi_k^1|\phi_k^2|\ldots|\phi_k^n)..$

3   Spectral Features

Our aim is to use spectral features computed from the eigenmodes of the adjacency matrices for graphs under study to construct feature-vectors. To overcome the correspondence problem, we use the order of the eigenvalues to establish the order of the components of the feature-vectors. We study a number of features suggested by spectral graph theory.

3.1   Unary Features

We commence by considering unary features for the eigenmodes of the adjacency matrix. The features studied are listed below:

Leading Eigenvalues: Our first vector of spectral features is constructed from the ordered eigenvalues of the adjacency matrix. For the graph indexed k, the vector is $B_k = (\lambda_k^1, \lambda_k^2, ..., \lambda_k^n)^T..$ This vector represents the spectrum of the graph G_k.

Eigenmode Volume: The volume $Vol(S)$ of a subgraph S of a graph G_k is defined to be the sum of the degrees of the nodes belonging to the subgraph, i.e

$$Vol_k(S) = \sum_{i \in S} D_k(i), \tag{3}$$

where $D_k(i) = \sum_{j \in V_k} E_{i,j}.$ is the degree of node i in the graph G_k. If D_k is the degree vector for the graph G_k, then the vector of volumes for the eigenmodes is found using the projection $Vol_k = \Phi_k^T D_k$. In other words, the volume associated with the eigenmode indexed ω in the graph-indexed k is

$$Vol_k(\omega) = \sum_{i \in V_k} \Phi_k(i, \omega) D_k(i) \tag{4}$$

The eigenmodes volume feature-vector for the graph-indexed k is $B_k = (Vol_k(1), Vol_k(2)), ., Vol_k(n))^T$.

Eigenmode Perimeter: For a subgraph S the set of perimeter nodes is $\Delta(S) = \{(u,v)|(u,v) \in E \wedge u \in S \wedge v \notin S\}$. The perimeter length of the subgraph is defined to be the number of edges in the perimeter set, i.e. $\Gamma(S) = |\Delta(S)|$. Again, by analogy, the perimeter length of the adjacency matrix for the eigenmode indexed ω is

$$\Gamma_k(\omega) = \sum_{\nu \neq \omega} \sum_{i \in V_k} \sum_{j \in V_k} \Phi_k(i, \omega) \Phi_k(j, \nu) A_k(i, j) \tag{5}$$

The perimeter values are ordered according to the modal index to form the graph feature vector $B_k = (\Gamma_k^1, \Gamma_k^2, ..., \Gamma_k^n)^T$.

Cheeger Constant: The Cheeger constant for the subgraph S is defined as

$$H(S) = \frac{|\Delta(S)|}{min[Vol(S), Vol(\hat{S})]}. \tag{6}$$

The analogue of the Cheeger constants for the eigenmodes of the adjacency matrix is

$$H_k(\omega) = \frac{\Gamma_k(\omega)}{min[Vol_k(\omega), Vol_k(\hat{\omega})]}, \tag{7}$$

where

$$Vol_k(\hat{\omega}) = \sum_{\omega=1}^{n} \sum_{i \in V_k} \Phi_k(i, \omega) D_k(i) - Vol_k(\omega) \tag{8}$$

is the volume of the complement of the eigenmode indexed ω. Again, the eigenmode Cheeger numbers are ordered to form a spectral feature-vector $B_k = (H_k(1), H_k(2), ..., H_k(n))^T$.

3.2   Binary Features

In addition to the unary features, we have studied pairwise attributes for the eigenmodes.

Mode Association Matrix: Our first pairwise representation is found by projecting the adjacency matrix onto the basis spanned by the eigenvector to compute a matrix of mode associations. The projection or inter-mode adjacency matrix is $U_k = \Phi_k^T A_k \Phi_k$. The element of the matrix with row index u and column index v is

$$U_k(u, v) = \sum_{i \in V_k} \sum_{j \in V_k} \Phi_k(i, u) \Phi_k(j, v) A_k(i, j) \tag{9}$$

This is infact the association between clusters u and v. These matrices are converted into long vectors. This is done by stacking the columns of the matrix U_k in eigenvalue order. The resulting vector is $B_k = (U_k(1,1), U_k(1,2), ..., U_k(1,n), U_k(2,1)....., U_k(2,n,), ...U_k(n,n))^T$. Each entry in the long-vector corresponds to a different pair of spectral eigenmodes.

Inter-mode distances: The between mode distance is defined as the path length, i.e. the minimum number of edges, between the most significant nodes associated with each eigenmode of the adjacency matrix. The most significant node associated a particular eigenmode of the adjacency matrix is the one having the largest co-efficient in the associated eigenvector. For the eigenmode indexed u in the graph indexed k, the most significant node is

$$i_u^k = \arg\max_i \Phi_k(i, u) \tag{10}$$

To compute the distance, we note that if we multiply the adjacency matrix A_k by itself l times, then the matrix $(A_k)^l$ represents the distribution of paths of length l in the graph G_k. In particular, the element $(A_k)^l(i,j)$ is the number of paths of length l edges between the nodes i and j. Hence the minimum distance between the most significant nodes of the eigenmode indexed u and v is

$$d_{u,v} = \arg\min_l (A_k)^l(i_u^k, i_v^k). \tag{11}$$

If we only use the first n leading eigenvectors to describe the graphs, the between mode distances for each graph can be written as a n by n matrix which can be converted to a $n \times n$ long-vector $B_k = (d_{1,1}, d_{1,2}, ...d_{1,n}, d_{2,1}.....d_{n,n})^T$.

4   Embedding the Spectral Vectors in a Pattern Space

In this section we describe three methods for embedding graphs in eigenspaces. The first of these involves performing principal components analysis on the covariance matrices for the spectral pattern-vectors. The second involves independent component analysis. The third method involves performing multidimensional scaling on a set of pairwise distance between vectors.

4.1   Pattern Space Embedding by PCA

Our first method makes use principal components analysis and follows the parametric eigenspace idea of Murase and Nayar [7]. The graphs extracted from each image are vectorised in the way outlined in Section 3. The N different image vectors are arranged in view order as the columns of the matrix $T = [B_1|B_2|\ldots|B_k|\ldots|B_N]..$ Next, we compute the covariance matrix for the elements in the different rows of the matrix T. This is found by taking the matrix product $C = TT^T..$ We extract the principal components directions for the relational data by performing an eigendecomposition on the covariance matrix C. The eigenvalues λ_i are found by solving the eigenvalue equation $|C - \lambda I| = 0$, and the corresponding eigenvalues e_i are found by solving the eigenvector equation $Ce_i = \lambda_i e_i..$

We use the first 3 leading eigenvectors to represent the graphs extracted from the images. The co-ordinate system of the eigenspace is spanned by the three orthogonal vectors by $e = (e_1, e_2, e_3)$. The individual graphs represented by the long vectors $B_k, k = 1, 2, \ldots, N$ can be projected onto this eigenspace using the formula $x_k = e^T B_k..$ Hence each graph G_k is represented by a 3-component vector x_k in the eigenspace.

4.2   Pattern Space Embedding by ICA

Our second approach uses Independent Components Analysis(ICA) to embed the graphs in a pattern space. We explore how to decompose a set of graphs into significantly different independent components. These can then be used for graph clustering by projecting the original graphs into the pattern space spanned by the independent components.

The ICA algorithm used in this paper is Cardoso and Soulourniac's JADE algorithm[3]. JADE is a statistically based algorithm. The main features of the algorithm are as follows. As with other ICA algorithms, the first step is data whitening or sphering. The aim is to eliminate correlations from the data. This can be achieved by removing the mean of the data and using PCA on the data covariance matrix. As a result, the whitened vector set is $Z = WB$, where W is the estimated whitening matrix. The second step of JADE is estimate the 4th-order cumulants Q_z. In the noiseless case, Q_z can be calculated as follows,

$$Q_Z(I_n) = E\{|Z|^2 ZZ^T\} - (n+1)I_n, \tag{12}$$

where I_n is the n-order identity matrix and $E(.)$ is the expectation operator. Next, a joint diagonalisation is performed to find a matrix $\hat{V}$ to minimise the non-diagonal entries of the cumulants matrices,

$$\hat{V} = \arg\min \Sigma_i Off(V^T Q_Z V). \tag{13}$$

Again we use the first 3 most significant independent components to represent the graphs extracted from the images. The co-ordinate system of the pattern-space is spanned by the three independent components by $\hat{e} = (\hat{V}_l, \hat{V}_2, \hat{V}_3)$. The individual graphs represented by the long vectors Z_k, $k = 1, 2, ..., N$ can be projected onto this pattern space using the formula $x_k = e^T Z_k$. Hence each graph G_k is represented by a 3-component vector x_k in the pattern space.

4.3   Multidimensional Scaling

Multidimensional scaling(MDS) is a procedure which allows data specified in terms of a matrix of pairwise distances to be embedded in a Euclidean space. The classical multidimensional scaling method was proposed by Torgenson[10] and Gower[5]. Shepard and Kruskal developed a different scaling technique called ordinal scaling[4]. Here we intend to use the method to embed the graphs extracted from different viewpoints in a low-dimensional space.

To commence we require pairwise distances between graphs. We do this by computing the L2 norms between the spectral pattern vectors for the graphs. For the graphs indexed $i1$ and $i2$, the distance is

$$d_{i1,i2} = \sum_{\alpha=1}^{K} \left[B_{i1}(\alpha) - B_{i2}(\alpha) \right]^2. \tag{14}$$

The pairwise similarities $d_{i1,i2}$ are used as the elements of an $N \times N$ dissimilarity matrix D, whose elements are defined as follows

$$D_{i1,i2} = \begin{cases} d_{i1,i2} & \text{if } i_1 \neq i2 \\ 0 & \text{if } i1 = i2 \end{cases}. \tag{15}$$

In this paper, we use the classical multidimensional scaling method to embed the view-graphs in a Euclidean space using the matrix of pairwise dissimilarities

D. The first step of MDS is to calculate a matrix T whose element with row r and column c is given by

$$T_{rc} = -\frac{1}{2}[d_{rc}^2 - \hat{d}_{r.}^2 - \hat{d}_{.c}^2 + \hat{d}_{..}^2],$$

(16)

where

$$\hat{d}_{r.} = \frac{1}{N} \sum_{c=1}^{N} d_{rc}$$

(17)

is the average dissimilarity value over the rth row, $\hat{d}_{.c}$ is the similarly defined average value over the cth column and

$$\hat{d}_{..} = \frac{1}{N^2} \sum_{r=1}^{N} \sum_{c=1}^{N} d_{r,c}$$

(18)

is the average similarity value over all rows and columns of the similarity matrix T.

We subject the matrix T to an eigenvector analysis to obtain a matrix of embedding co-ordinates X. If the rank of T is $k, k \leq N$, then we will have k non-zero eigenvalues. We arrange these k non-zero eigenvalues in descending order, i.e. $\lambda_1 \geq \lambda_2 \geq \ldots \geq \lambda_k > 0$. The corresponding ordered eigenvectors are denoted by e_i where λ_i is the ith eigenvalue. The embedding co-ordinate system for the graphs obtained from different views is $X = [f_1, f_2, \ldots, f_k]$, where $f_i = \sqrt{\lambda_i} e_i$ are the scaled eigenvectors. For the graph indexed i, the embedded vector of co-ordinates is $x_i = (X_{i,1}, X_{i,2}, X_{i,3})^T$..

5 Experiments

Our experimental vehicle is provided by object recognition from 2D views of 3D objects. We have collected sequences of views for a number of objects. For the different objects the image sequences are obtained under slowly varying changes in viewer angle. From each image in each view sequence, we extract corner features. We use the extracted corner points to construct Delaunay graphs. In our experiments we use three different sequences. Each sequence contains images with equally spaced viewing directions. The sequences contain different numbers of frames, but for each we have selected 10 uniformly spaced images. In Figure 1 we show example images and their associated graphs from the three sequences. From left-to-right the images belong to the CMU, MOVI and chalet sequences. There is considerable variability in the number of feature-points detected in the different sequences. For the CMU sequence the minumum number of points is 30 and the maximum is 35. For the MOVI sequence the minimum is 130 and the maximum is 140, For the chalet sequence the minimum is 40 and the maximum 113; this is the most variable sequence in terms of view structure.

We commence by investigating which combination of spectral feature-vector and embedding strategy gives the best set of graph-clusters. In other words,

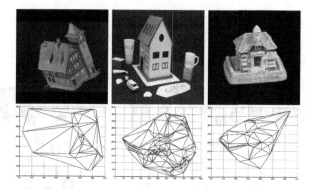

Fig. 1. Example images from the three sequences and their corner-feature Delaunay graphs.

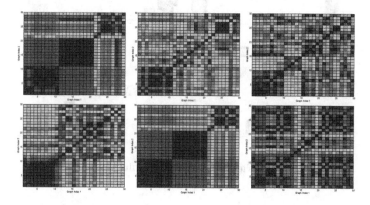

Fig. 2. Distance maps using the spectral features of binary adjacency graph spectra, volumes, perimeters, Cheeger constants, inter-mode adjacency matrix and inter-mode distances.

we aim to explore which method gives the best definition of clusters for the different objects. In Figure 2 we compare the results obtained with the different spectral feature vectors. In the figure we show the matrix of pairwise Euclidean distances between the feature-vectors for the different graphs (this is best viewed in colour). The matrix has 30 rows and columns (i.e. one for each of the images in the three sequences with the three sequences concatenated), and the images are ordered according to the position in the sequence. From left-to-right and top-to-bottom, the different panels show the results obtained when the feature-vectors are constructed using the eigenvalues of the adjacency matrix, the volumes, the perimeters, the Cheeger constants, the inter-mode adjacency matrix and the inter-mode distance. From the pattern of pairwise distances, it is clear that the eigenvalues and the inter-mode adjacency matrix give the best block structure in the matrix. Hence these two attributes may be expected to result in the best clusters.

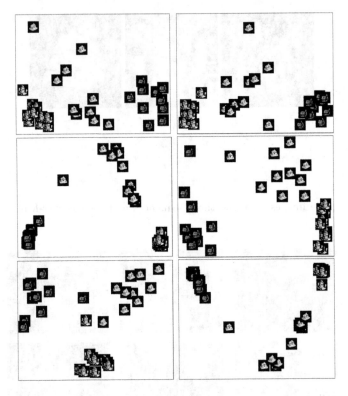

Fig. 3. Clustering Results.

Next, we show the clustering results for the two well behaved features. i.e. the vector of leading eigenvalues and the inter-mode adjacency matrix. In Figure 3, we show the clustering results obtained with vectors of different spectral attributes. For clarity, the results are displayed by positioning image "thumbnails" at the position associated with the corresponding spectral feature vector in relevant eigenspace. In each case we visualise the thumbnails in the space spanned by the leading two eigenvectors generated by the embedding method (PCA, ICA or MDS). The left-hand column shows the results obtained using the vector of leading eigenvalues of the adjacency matrix. The top row shows the result obtained using PCA, the second row that obtained using ICA and the bottom row obtained using MDS. There are a number of conclusions that can be drawn from these plots. First, the cluster associated with the Swiss chalet is always the least compact. Second, the best clusters are produces using MDS. That is to say the clusters are most compact and best separated. This is perhaps surprising, since this method uses only the set of pairwise distances between graphs, and overlooks the finer information residing in the vectors of spectral features. Although the clusters delivered by PCA are poorer, the results can be improved by using ICA. However, in none of the three cases is there any overlap between clusters.

More importantly, they can be separated by straight-lines, i.e. they are linearly separable.

In the right-hand column of Figure 3 we repeat this sequence of experiments for the inter-mode adjacency matrix. This is a pairwise relational attribute. In each case the clusters are slightly improved. This is most marked when PCA is used to perform the embedding.

Finally, we provide some comparison of the different spectral representations. To do this we plot the normalised squared eigenvalues

$$\hat{\lambda}_i^2 = \frac{\lambda_i^2}{\sum_{i=1}^n \lambda_i^2}$$

against the eigenvalue magnitude order i. In the case of the parametric eigenspace, these represent the fraction of the total data variance residing in the direction of the relevant eigenvector. In the case of multidimensional scaling, the normalised squared eigenvalues represent the variance of the inter-graph distances in the directions of the eigenvectors of the similarity matrix.

In Figure 4 we show the eigenvalues of the spectral-vector covariance matrix. The plots in the left-hand column are for the CMU sequence, the plots in the middle column are for the MOVI sequence, and the plots in the right-hand column are for the chalet sequence. The top row is for the unary features while the bottom row is for the binary features. The main feature to note is that the vector of adjacency matrix eigenvalues has the fastest rate of decay, i.e. the eigenspace has a lower latent dimensionality, while the vector of Cheeger constants has the slowest rate of decay, i.e. the eigenspace has greater dimensionality. In the case of the binary attributes, the inter-mode adjacency matrix results in the eigenspace of lower dimensionality. In Figure 5 we repeat the sequence of plots for the three house data-sets, but merge the curves for the unary and binary attributes into a single plot by using MDS space embedding. Again the vector of adjacency matrix eigenvalues gives the space of lower dimensionality, while the vector of inter-mode distances gives the space of greatest dimensionality.

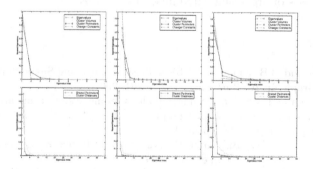

Fig. 4. Comparison of graph spectral features in eigenspaces.

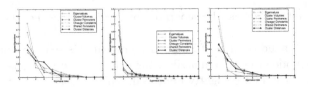

Fig. 5. Comparison of graph spectral features in MDS spaces.

6   Conclusions

In this paper we have investigated how vectors of graph-spectral attributes can be used for the purposes of graph clustering. To do this we must first select attributes from which to construct feature vectors, and then select a means by which to embed the feature-vectors in a pattern space. The attributes studied are the leading eigenvalues, the volumes, perimeters, Cheeger numbers, inter-mode adjacency matrices and inter-mode edge-distance for the eigenmodes of the adjacency matrix. The embedding strategies are PCA, ICA and MDS. Our empirical study is based on corner adjacency graphs extracted from 2D views of 3D objects. We investigate two problems. The first of these is that of clustering the different views of the three objects together. The second problem is that of organising different views of the same object into well structured trajectories, in which subsequent views are adjacent to one another and there are distinct clusters associated with different views. In both cases the best results are obtained when we apply MDS to the vectors of leading eigenvalues.

Hence, we have shown how to cluster purely symbolic graphs using simple spectral attributes. The graphs studied in our analysis are of different size, and we do not need to locate correspondences. Our future plans involve studying in more detail the structure of the pattern-spaces resulting from our spectral features. Here we intend to study how support vector machines and the EM algorithm can be used to learn the structure of the pattern spaces. Finally, we intend to investigate whether the spectral attributes studied here can be used for the purposes of organising large image data-bases.

References

1. C.M. Cyr and B.B. Kimia. 3D Object Recognition Using Shape Similarity-Based Aspect Graph. In *ICCV01*, pages I: 254–261, 2001.
2. M.A. Eshera and K.S. Fu. An image understanding system using attributed symbolic representation and inexact graph-matching. *Journal of the Association for Computing Machinery*, 8(5):604–618, 1986.
3. T. Hofmann and J.M. Buhmann. Pairwise data clustering by deterministic annealing. *PAMI*, 19(2):192–192, February 1997.
4. Kruskal J.B. Nonmetric multidimensional scaling: A numerical method. *Psychometrika*, 29:115–129, 1964.
5. Gower J.C. Some distance properties of latent root and vector methods used in multivariate analysis. *Biometrika*, 53:325–328, 1966.

6. Luo N, Wilson R.C and Hancock E.R. Spectral Feture Vectors for Graph Clustering multivariate analysis. *Proc. SSPR02*, LNCS 2396, pp. 83–93, 2002.
7. H. Murase and S.K. Nayar. Illumination planning for object recognition using parametric eigenspaces. *IEEE Transactions on Pattern Analysis and Machine Intelligence*, 16(12):1219–1227, 1994.
8. A. Sanfeliu and K.S. Fu. A distance measure between attributed relational graphs for pattern recognition. *IEEE Transactions Systems, Man and Cybernetics*, 13(3):353–362, May 1983.
9. K. Sengupta and K.L. Boyer. Organizing large structural modelbases. *PAMI*, 17(4):321–332, April 1995.
10. Torgerson W.S. Multidimensional scaling. i. theory and method. *Psychometrika*, 17:401–419, 1952.

Comparison of Distance Measures for Graph-Based Clustering of Documents

Adam Schenker[1], Mark Last[2], Horst Bunke[3], and Abraham Kandel[1]

[1] University of South Florida, Department of Computer Science and Engineering,
4202 E. Fowler Ave. ENB 118, Tampa, FL, 33620, USA
{aschenke, kandel}@csee.usf.edu
[2] Ben-Gurion University of the Negev,
Department of Information Systems Engineering,
Beer-Sheva 84105, Israel
mlast@bgumail.bgu.ac.il
[3] University of Bern, Department of Computer Science,
Neubrückstrasse 10, CH-3012 Bern, Switzerland
bunke@iam.unibe.ch

Abstract. In this paper we describe work relating to clustering of document collections. We compare the conventional vector-model approach using cosine similarity and Euclidean distance to a novel method we have developed for clustering graph-based data with the standard k-means algorithm. The proposed method is evaluated using five different graph distance measures under three clustering performance indices. The experiments are performed on two separate document collections. The results show the graph-based approach performs as well as vector-based methods or even better when using normalized graph distance measures.

1 Introduction

Clustering of natural language documents is an important research area for two major reasons. First, clustering a document collection into categories enables it to be more easily browsed and used. Second, clustering can improve the performance of search and retrieval on a document collection. Hierarchical clustering methods [8], for example, are used often for this purpose. When representing documents for clustering the vector model is typically used [13]. In this model, each possible term that can appear in a document becomes a feature (dimension). The value assigned to each dimension of a document indicates the number of times that a particular term appears on it.

The vector model is simple and allows the use of traditional clustering methods that deal with numerical feature vectors. However, it discards information such as the order in which the terms appear, where in the document the terms appear, how close the terms are to each other, and so forth. By keeping this kind of structural information we could possibly improve the performance of the clustering. The problem is that traditional clustering methods are often restricted to working on purely numeric feature vectors due to the need to compute distances between data items or to find some representative element of a cluster of items

E. Hancock and M. Vento (Eds.): IAPR Workshop GbRPR 2003, LNCS 2726, pp. 202–213, 2003.

(i.e. a centroid of a cluster), both of which are easily accomplished with numerical feature vectors. Thus either the original data needs to be converted to a vector of numeric values by discarding possibly useful structural information (which is what we are doing when using the vector model to represent documents) or we need to develop new, customized algorithms for a specific representation.

In order to overcome this problem, we have introduced an extension of classical clustering methods that allows us to work with graphs as fundamental data structures instead of being limited to vectors of numeric values [15]. Our approach has two main benefits: (1) it allows us to keep the inherent structure of the original documents by modeling each document as a graph, and (2) we can apply straightforward extensions to use existing clustering algorithms rather than having to create new algorithms from scratch. In this paper we will address comparison of different graph similarity measures in the context of document clustering. We will use the k-means clustering algorithm to cluster two document collections. Each collection will be clustered with seven distance measures in order to compare and evaluate the performance of each measure. We will use the Euclidean distance and cosine similarity measures with the vector model representation as well as five different distance measures based on the concept of maximum common subgraph for the graph-based method. The vector-based methods, which are well established in the information retrieval literature, are used as a baseline for comparison with our approach.

Clustering with graphs is well established in the literature. However, the paradigm in those methods is to treat the entire clustering problem as a graph: nodes represent the items to be clustered and weights on edges connecting two nodes indicate the distance (dissimilarity) between the objects the nodes represent. The usual procedure is to create a minimal spanning tree of the graph and then remove the remaining edges with the largest weight in the MST until the number of desired clustered (connected components) is achieved [8][18]. After applying the algorithm the edges indicate which objects belong to which clusters. Recently there has been some progress with performing clustering directly on graph-based data. For example, an extension of self-organizing maps (SOMs) which allows the procedure to work with graphs has been proposed [6]; graph edit distance and weighted mean of a pair of graphs were introduced to deal with graph-based data under the SOM algorithm. Clustering of shock trees using tree edit distance has also been considered [9]. Both of these methods have in common that they use graph (or tree) edit distance for their graph distance measures. One drawback of this approach is that the edit cost functions must be specified for each application. Sanfeliu *et al.* have investigated clustering of attributed graphs using their own "function-described graphs" as cluster representatives [14]. However, their method is rather complicated and much more involved than our straightforward extension of a classical, simple clustering algorithm.

This paper is organized as follows. In Sect. 2 we briefly describe the extension of the k-means algorithm that allows the use of graphs instead of numerical vectors. In Sect. 3 we give the details of our experimental setup and the methods by which clustering performance is measured. Experimental results for both data sets are given in Sect. 4. Conclusions are presented in Sect. 5.

Inputs:	the set of n data items and a parameter, k, defining the number of clusters to create
Outputs:	the centroids of the clusters and for each data item the cluster (an integer in [1,k]) it belongs to

Step 1.	Assign each data item randomly to a cluster (from 1 to k).
Step 2.	Using the initial assignment, determine the centroids of each cluster.
Step 3.	Given the new centroids, assign each data item to be in the cluster of its closest centroid.
Step 4.	Re-compute the centroids as in Step 2. Repeat Steps 3 and 4 until the centroids do not change.

Fig. 1. The basic k-means clustering algorithm

2   Clustering with k-Means

The k-means clustering algorithm is a simple and straightforward method for clustering data [11]. The basic algorithm is given in Fig. 1. The common paradigm is to represent each data item, which consists of m numeric values, as a vector in the space $\Re^m$. The distance measure used by the algorithm is usually the Euclidean distance, however in vector-based models for information retrieval the cosine similarity measure is often used due to its length invariance property. With the vector-model representation, documents are represented as a vector in $\Re^m$ by having each dimension indicate the number of times a particular term appears in the document [13]. It is common to restrict the number of dimensions by selecting some small set of discriminating or important terms, as the number of possible terms that can occur across a collection of documents can be quite large.

We determined that if methods of computing distance between graphs and constructing a representative of a set of graphs are available it is possible to extend many clustering and classification methods to work directly on graphs [15]. First, any distance calculations between data items to be clustered, which are represented by graphs and not vectors, is accomplished with a graph-theoretical distance measure as we will discuss in Sect. 2.1. Second, since it is necessary to compute the distance between data items and cluster centers, it follows that the cluster centers (centroids) must also be graphs. Therefore, we compute the representative "centroid" of a cluster as the median graph of the set of graphs in that cluster (as we will describe in Sect. 2.2). The graphs we use in this paper are the usual directed graphs with node and edge labels.

2.1   Graph Distance Measures

As we mentioned above, we need a graph-theoretical distance measure in order to use graphs for clustering. We have implemented several distance measures and will compare their clustering performance. For brevity we will refer to the distance measures below as *MCS*, *WGU*, *UGU*, *MMCS*, and *MMCSN*.

The first distance measure (MCS) is the "classic" graph distance measure based on the maximum common subgraph [3]:

$$d_{MCS}(G_1, G_2) = 1 - \frac{|mcs(G_1, G_2)|}{max(|G_1|, |G_2|)} \tag{1}$$

where G_1 and G_2 are the graphs to compare, $mcs(G_1, G_2)$ is their maximum common subgraph, $|\ldots|$ is the size of a graph, and $max(\ldots)$ is the usual maximum operation. Here we define the size of a graph to be the sum of the number of nodes and edges in the graph.

A second distance measure (WGU) which has been proposed by Wallis *et al.* [17], based on the idea of graph union, is:

$$d_{WGU}(G_1, G_2) = 1 - \frac{|mcs(G_1, G_2)|}{|G_1| + |G_2| - |mcs(G_1, G_2)|} \tag{2}$$

By "graph union" we mean that the denominator represents the size of the union of the two graphs in the set theoretic sense; specifically adding the size of each graph ($|G_1|+|G_2|$) then subtracting the size of their intersection ($|mcs(G_1, G_2)|$) leads to the size of the union (the reader may easily verify this using a Venn diagram).

A third distance measure (UGU) which similar to WGU but that is not normalized to the interval [0, 1] is [1]:

$$d_{UGU}(G_1, G_2) = |G_1| + |G_2| - 2|mcs(G_1, G_2)| \tag{3}$$

The fourth distance measure (MMCS), proposed by Fernández and Valiente, is based on both the maximum common subgraph and the minimum common supergraph [7]:

$$d_{MMCS}(G_1, G_2) = |MCS(G_1, G_2)| - |mcs(G_1, G_2)| \tag{4}$$

where $MCS(G_1, G_2)$ is the minimum common supergraph of graphs G_1 and G_2.

The fifth distance measure (MMCSN) is a version of MMCS which is normalized to [0, 1]:

$$d_{MMCSN}(G_1, G_2) = 1 - \frac{|mcs(G_1, G_2)|}{|MCS(G_1, G_2)|} \tag{5}$$

Given any one of these distance measures, we can define the median of a set of graphs, which is explained in the next section.

We will describe our graph representation of documents in detail in Sect. 3.1. However, we wish to mention here an interesting feature our graph representation has on the time complexity of determining the distance using Eqs. (1–5). In the general case the computation of the maximum common subgraph is known to be an NP-Complete problem [10]. However, with our graph representation for documents each node in a graph has a unique label (representing a unique term) that no other node in the graph has. Thus the maximum common subgraph,

G_{mcs}, of a pair of graphs, G_1 and G_2, can be created by the following procedure: (1) find the nodes V_{mcs} by determining the subset of labels that the original graphs have in common with each other and create a node for each common label; (2) find the edges E_{mcs} by examining all pairs of nodes from the previous step and introduce edges that connect pairs of nodes in both of the original graphs. We see that the complexity of this method is $O(|V_1| \cdot |V_2|)$ for the first step and $O(|V_{mcs}|^2)$ for the second step. Thus it is $O(|V_1| \cdot |V_2| + |V_{mcs}|^2) \leq O(|V|^2 + |V_{mcs}|^2) = O(|V|^2)$ overall if we substitute $V = max(|V_1|, |V_2|)$.

2.2   Median of a Set of Graphs

The second ingredient required to apply clustering to graphs is that of a graph-theoretic cluster "centroid" or representative of a set of graphs. For this we have used the concept of graph median, which is the graph which has the minimum average distance to all graphs in the cluster:

Definition 1. The *median of a set of n graphs*, $S = \{G_1, G_2, \ldots, G_n\}$, is a graph G such that G has the smallest average distance to all elements in S [2]:

$$G = \arg \min_{\forall s \in S} \left(\frac{1}{n} \sum_{i=1}^{n} dist(s, G_i) \right) \qquad (6)$$

Here S is the set of n graphs (and thus $|S| = n$) and G is the median. The median is defined to be a graph in set S. Thus the median of a set of graphs is the graph from that set which has the minimum average distance to all the other graphs in the set. The distance $dist(\ldots)$ is computed using one of Eqs. (1–5) above. There also exists the concepts of the generalized median and weighted mean [2], where we don't require that G be a member of S, but we will not consider them here because they are quite expensive to compute. In the case where the median is not unique (i.e. there is more than one graph that has the same minimum average distance) we select one of those graphs at random as the centroid for the k-means algorithm.

3   Experimental Background

Our experiments were performed on two different collections of documents, called the *F-series* and the *J-series*. These data sets are available under these names at: ftp://ftp.cs.umn.edu/dept/users/boley/PDDPdata/. Each collection contains documents in HTML format (i.e. the documents are web pages). The F-series originally contained 98 documents assigned to one or more of 17 sub-categories of four major category areas: *manufacturing*, *labor*, *business & finance*, and *electronic communication & networking*. Because there are multiple sub-category classifications from the same category area for many of these documents, we have reduced the categories to just the four major categories mentioned above in order to simplify the problem. There were five documents that had conflicting classifications (i.e., they were classified to belong to two or more of the four major categories) which we removed, leaving 93 total documents. The J-series contains 185 documents and ten classes: *affirmative action, business capital, information systems, electronic commerce, intellectual property, employee rights, materials processing, personnel management, manufacturing systems,* and *industrial partnership.* We have not modified this data set.

In the case of the vector-model representations there are already several pre-created term–document matrices available for our experiments. We selected the matrices with the smallest number of dimensions. For the F-series documents there are 332 dimensions (terms) used, while the J-series has 474 dimensions. For our graph-based experiments we created graph representations from the original documents as described in Sect. 3.1 below. Clustering performance is measured using three commonly used methods, which we describe in detail in Sect. 3.2. For each approach, whether vector or graph-based, we repeated the same experiment ten times and took the average in order to account for the variance between runs due to the random initialization of the clustering algorithm.

3.1   Graph Representation of Documents

We represent each document as a graph using the following method. First, each term (word) appearing in the document, except for stop words such as "the", "of", and "and" which convey little information, becomes a vertex in the graph representing that document. This is accomplished by labeling each node with the term it represents. Note that we create only a single vertex for each word even if a word appears more than once in the text. Thus each vertex in the graph represents a unique word and is labeled with a unique term not used to label any other node. Second, if word a immediately precedes word b somewhere in a "section" s of the document, then there is a directed edge from the vertex corresponding to a to the vertex corresponding to b with an edge label s. We take into account certain punctuation (such as a period) and do not create an edge when these are present between two words. Sections we have defined are: *title,* which contains the text related to the document's title and any provided keywords (meta-data); *link,* which is text appearing in clickable hyper-links on the document; and *text,* which comprises any of the readable text in the document (this includes link text but not title and keyword text). Next, we perform a simple stemming method and conflate terms to the most frequently occurring form by re-labeling nodes and updating edges as needed. Finally, we remove the most infrequently occurring words on each document, leaving at most m nodes per graph (m being a user provided parameter). This is similar to the dimensionality reduction process for vector representations [13]. For our experiments we varied m in order to examine the effects this parameter has on clustering performance. It is important to note that even the largest graph we use is based on much fewer terms than the vector representations of the same documents (332 and 474 for F-series and J-series, respectively). An example of this type of graph representation is given in Fig. 2. The ovals indicate nodes and their corresponding term labels. The edges are labeled according to title (TI), link

(L), or text (TX). The document represented by the example has the title "YA-HOO NEWS", a link whose text reads "MORE NEWS", and text containing "REUTERS NEWS SERVICE REPORTS". Note there is no restriction on the form of the graph and that cycles are allowed.

Fig. 2. Example of a graph representation of a document

3.2   Clustering Performance Measures

We use the following three clustering performance measures to evaluate the performance of each clustering. The first two indices measure the matching of obtained clusters to the "ground truth" clusters, while the third index measures the quality of clustering in general.

The first index is the *Rand index* [12], which is defined as:

$$R_I = \frac{A}{A + D} \tag{7}$$

where A is the number of "agreements" and D is the number of "disagreements" (described below). We perform a pair-wise comparison of all pairs of objects in the data set after clustering. If two objects are in the same cluster in both the ground truth clustering and the clustering we wish to measure, this counts as an agreement. If two objects are in different clusters in both the ground truth clustering and the clustering we wish to investigate, this is also an agreement. Otherwise, there is a disagreement. Thus the Rand index is a measure of how closely the clustering created by some procedure matches ground truth. It produces a value in the interval $[0, 1]$, with 1 representing a clustering that perfectly matches ground truth.

The second performance measure we use is *mutual information* [4][16], which is defined as:

$$\Lambda^M = \frac{1}{n} \sum_{l=1}^{k} \sum_{h=1}^{g} n_l^{(h)} log_{k \cdot g} \left(\frac{n_l^{(h)} \cdot n}{\sum_{i=1}^{k} n_i^{(h)} \sum_{i=1}^{g} n_l^{(i)}} \right) \tag{8}$$

where n is the number of data items, k is the desired number of clusters, g is the actual number of ground truth categories, and $n_i^{(j)}$ is the number of items in cluster i classified to be category j. Note that k and g may not necessarily be equal, which would indicate we are attempting to create more (or fewer) clusters than those that exist in the ground truth clustering. However, for the experiments described in this paper we will create an identical number of clusters as is present in ground truth. Mutual information represents the overall degree of agreement between the clustering and the categorization provided by the ground truth with a preference for clusters that have high purity (i.e. are homogeneous with respect to the classes of objects clustered).

The third performance measure we use is the *Dunn index* [5], which is defined as:

$$D_I = \frac{d_{min}}{d_{max}} \tag{9}$$

where d_{min} is the minimum distance between any two objects in different clusters and d_{max} is the maximum distance between any two items in the same cluster. The numerator captures the worst-case amount of separation between clusters, while the denominator captures the worst-case compactness of the clusters. Thus the Dunn index is an amalgam of the overall worst-case compactness and separation of a clustering, with higher values being better. It does not, however, measure clustering accuracy compared to ground truth as the other two methods do. Rather it is based on the basic underlying assumption of any clustering technique: items in the same cluster should be similar (i.e. have small distance, thus creating compact clusters) and items in separate clusters should be dissimilar (i.e. have large distance, thus creating clusters that are well separated from each other).

4   Results

The results of our experiments for the F-series documents are given in Fig. 3 for Rand index (top), mutual information (middle) and Dunn index (bottom). The results of the vector-model clusterings are shown by horizontal lines. The results for our graph-based method are given for graph sizes ranging from 10 to 100 nodes per graph. Similarly, the results of our experiments for the J-series documents are given in Fig. 4 for graph sizes of 10 to 60 nodes per graph. The legend for each graph indicates the order of performance of each distance measure for the maximum graph size.

We see that the graph-based methods that use normalized distance measures (MCS, WGU, and MMCSN) performed similarly to each other and as well or better than vector-based methods using cosine similarity or Euclidean distance. Distance measures that were not normalized to the interval $[0, 1]$ (UGU and MMCS) also behaved similarly to each other but performed poorly when the maximum graph size became large. To see why this occurs, we have provided the following example. Let $|G_1| = 50$, $|G_2| = 50$, $|mcs(G_1, G_2)| = 0$, $|G_3| = 80$, $|G_4| = 80$, and $|mcs(G_3, G_4)| = 10$. Clearly graphs G_3 and G_4 are more similar to each other than graphs G_1 and G_2 since G_1 and G_2 have no common subgraph. However, the distances computed for these graphs are $d_{MCS}(G_1, G_2) = 1.0$, $d_{MCS}(G_3, G_4) = 0.875$, $d_{UGU}(G_1, G_2) = 100$ and $d_{UGU}(G_3, G_4) = 140$. Thus the distance for un-normalized graph union (UGU) is actually greater for the pair of graphs that are more similar. This is both counter-intuitive and the opposite of what happens in the cases of the normalized distance measures. Since the size of the graph includes the number of edges, which can grow at a

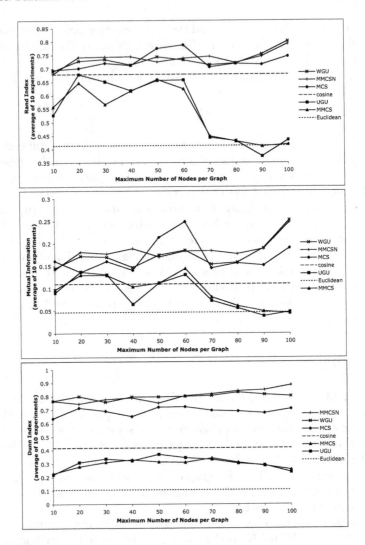

Fig. 3. Results for F-series data set: Rand index (top), mutual information (middle), Dunn index (bottom)

rate of $O(|V|^2)$, we can see that the potential size variance which causes this phenomenon becomes more pronounced as we increase the number of nodes per graph (i.e. as $|V|$ increases).

We see from the experimental results that the quality of clustering as measured by the Dunn index is more or less constant with the size of the graphs, with the distance measure seeming to have more effect than the data representation itself. We note similar behavior for mutual information and Rand index. For these two indices we see performance peaking at certain graph sizes but otherwise the performance is fairly constant. This could be due to the fact that the terms selected for those graph sizes provide particularly good discriminating

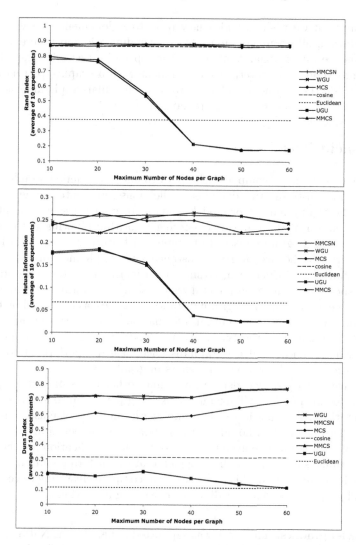

Fig. 4. Results for J-series data set: Rand index (top), mutual information (middle), Dunn index (bottom)

information on the pre-defined set of categories. For larger graphs that indicate a decline in performance, it may be the case that we have polluted the graphs with unimportant terms which hurt the performance of the clustering algorithm. We also see that the results for each distance measure under all three performance indices are consistent across both data sets. The fact that the graph-based methods performed as well or better than the vector-based method, even when the size of the graphs was relatively small, is encouraging and may indicate the quality of the added structural information that is captured in the graphs but not in the vector representations. Further, with the graph model there is also a potential for computational and space savings in terms of the data represented.

For example, if we wish to add a new term to a document in a vector model representation, all documents must have their dimensionality increased by one (and thus the dimensionality of the feature space increases by one). However, if we want to add a new term to a document with a graph representation, we need only add the new node and edges to that particular graph, and the distance calculations between other graphs remains unaffected.

5   Conclusions

In this paper we have compared the performance of different graph similarity measures for clustering two document collections represented by graphs rather than by the traditional vector model. The distance measures were implemented in our extension of the k-means clustering algorithm allowing it to work with data expressed as graphs instead of numerical feature vectors. This is accomplished through the use of graph distance and graph median instead of the usual distance (e.g. Euclidean, cosine similarity) and cluster centroid calculations, respectively. Our experimental results show an improvement in clustering performance with normalized graph distance measures, as measured by the Rand and Dunn indices as well as mutual information, when representing documents as graphs compared to the typical vector model representation approach. On the other hand, un-normalized graph distance measures performed poorly when there was a potential for large differences in graph size.

For future research we wish to investigate other types of graph representations for documents. We will examine clustering performance for those different representations. Automatically determining the number of clusters is another important problem when no information about the number of actual categories is available. An additional avenue to explore is the modification of other clustering algorithms, such as hierarchical agglomerative clustering and fuzzy c-means, for use with graphs. As in this paper, one can compare clustering of vector-based data and graph-based data for different clustering algorithms.

Acknowledgments. This work was supported in part by the National Institute for Systems Test and Productivity at the University of South Florida under U.S. Space and Naval Warfare Systems Command grant number N00039–01–1–2248.

References

1. Bunke, H.: On a relation between graph edit distance and maximum common subgraph. Pattern Recognition Letters 18 (1997) 689–694
2. Bunke, H., Günter, S. and Jiang, X.: Towards bridging the gap between statistical and structural pattern recognition: two new concepts in graph matching. In: Singh, S., Murshed, N., and Kropatsch, W. (eds.): Advances in Pattern Recognition – ICAPR 2001, LNCS 2013. Springer-Verlag (2001) 1–11
3. Bunke, H. and Shearer, K.: A graph distance metric based on the maximal common subgraph. Pattern Recognition Letters 19 (1998) 255–259

4. Cover, T. M. and Thomas, J. A.: Elements of Information Theory. Wiley (1991)
5. Dunn, J.: Well separated clusters and optimal fuzzy partitions. Journal of Cybernetics 4, 95–104.
6. Günter, S. and Bunke, H.: Self-organizing map for clustering in the graph domain. Pattern Recognition Letters 23 (2002) 405–417
7. Fernández, M.-L. and Valiente, G.: A graph distance metric combining maximum common subgraph and minimum common supergraph. Pattern Recognition Letters 22 (2001) 753–758
8. Jain, A. K., Murty, M. N. and Flynn, P. J.: Data clustering: a review. ACM Computing Surveys 31 (1999) 264–323
9. Luo, B., Robles-Kelly, A., Torsello, A., Wilson, R. C. and Hancock, E. R.: Clustering shock trees. 3rd IAPR-TC15 Workshop on Graph-based Representations in Pattern Recognition (2001) 217–228
10. Messmer, B. T. and Bunke, H.: A new algorithm for error-tolerant subgraph isomorphism detection. IEEE Transactions on Pattern Analysis and Machine Intelligence 20 (1998) 493–504
11. Mitchell, T. M.: Machine Learning. McGraw-Hill, Boston (1997)
12. Rand, W. M.: Objective criteria for the evaluation of clustering methods. Journal of the American Statistical Association 66 (1971) 846–850
13. Salton, G.: Automatic Text Processing: the Transformation, Analysis, and Retrieval of Information by Computer. Addison-Wesley, Reading (1989)
14. Sanfeliu, A., Serratosa, F. and Alquézar, R.: Clustering of attributed graphs and unsupervised synthesis of function-described graphs. Proceedings of the 15th International Conference on Pattern Recognition (ICPR'2000) 2 (2000) 1026–1029
15. Schenker, A., Last, M., Bunke, H., and Kandel, A.: Clustering of web documents using a graph model. Web Document Analysis: Challenges and Opportunities, Antonacopoulos, A. and Hu, J. (eds.). To appear
16. Strehl, A., Ghosh, J., and Mooney, R.: Impact of similarity measures on web-page clustering. AAAI-2000: Workshop of Artificial Intelligence for Web Search (2000) 58–64
17. Wallis, W. D., Shoubridge, P., Kraetz, M. and Ray, D.: Graph distances using graph union. Pattern Recognition Letters 22 (2001) 701–704
18. Zahn, C. T.: Graph-theoretical methods for detecting and describing gestalt structures. IEEE Transactions on Computers C-20 (1971) 68–86

Some Experiments on Clustering a Set of Strings

Jean-Michel Jolion

Lyon Research Center for Images and Information Systems
Bât. J. Verne, INSA Lyon
69621 Villeurbanne Cedex
France
jolion@rfv.insa-lyon.fr
http://rfv.insa-lyon.fr/~jolion

Abstract. We introduce in this paper the concept of set deviation as a tool to characterize the deviation of a set of strings around its set median. The set deviation is defined as the set median of the positive edit sequences between any string and the set median. We show how the set deviation can be efficiently used in well known statistical estimation and particularly with the minimum volume ellipsoid estimator. This concept is illustrated on several examples and particularly in clustering a set of shapes coded as strings using the Freeman code.

1 Introduction

During the last IAPR-TC15's workshop on Graph based Representations [10] and also during the last workshop on Visual Form in Capri [1], the discussions focused on the need to better explore the relations between the structural approach of pattern recognition (and especially graphs) and the statistical approach.

As pointed out by H. Bunke in [5], among the various open problems in graph matching, the introduction of concepts that are well established in the area of statistical pattern recognition to the structural domain is a great challenge. For instance, Massaro and Pellilo introduced in [16] a link between classic graph matching and linear optimization. This is indeed of interest, not only because it results in realistic solutions, but also because it links the graph domain with the optimization domain and lets us hope for other exchanges between them.

In a set of recent works, relationships between the graph edit distance and other well known concepts from graph theory were studied (see [6]). This leads to works such as finding the median of a set of graphs [8] which can be considered as a first step (as well as computing the weighted mean of a pair of strings [3] or graphs [4] ...) toward a conceptual exchange between structural and statistical pattern recognition. Many applications of such new concepts are now available such as [14,13,19] in OCR and shape analysis.

Our goal is to contribute to this challenge of extending traditional statistical techniques to the domain of symbolic structures and, in this paper, more particularly to strings.

We will first recall in section 2 some already published results on the computation of the median and the deviation of a set of strings. Then in section 3 we

E. Hancock and M. Vento (Eds.): IAPR Workshop GbRPR 2003, LNCS 2726, pp. 214–224, 2003.

will introduce the MVE estimator and show how it can be generalised to a set of strings. Some examples will be described in section 4 especially when working with shapes and words, and in section 5 we will discuss the generalization of these novel concepts.

2   First and Second Order Moment for a Set of Strings

2.1   First Order Moment

Let us first recall some basic definitions and properties related to symbolic spaces. In the following, we will assume a set S of patterns in a symbolic representation space U. Our goal is to statistically characterize the set S by a list of moments, e.g. the mean, the variance .... In the following, we will only consider sets of strings. We need to assume a distance function $d(a, b)$ to quantify the similarity between any two strings of U. We will assume the so-called edit distance [12,20, 15] based on elementary edit operations which transform a strings A in a strings B, because this distance is the most popular one and is also used in the graph domain.

The first moment, the *generalized median* of U is defined as

$$\bar{s} = arg \min_{p \in U} \sum_{q \in S} d(p, q) \ . \tag{1}$$

If U is not finite or too huge (the computation is NP-hard [7]), it is often of interest to constrain the search to the set S, yielding the *set median* of S as

$$\hat{s} = arg \min_{p \in S} \sum_{q \in S} d(p, q) = sms(S) \ . \tag{2}$$

The complexity of this search for strings, using the edit distance, is $\mathcal{O}(n^2 m^2)$ where n is the size of the set S and m is the size of the strings.

2.2   Second Order Moment

The set median is of interest to summarize the inside information of the set S. When $U \equiv \Re$, it is however obvious that the mean (or the median or the mode) is not enough to efficiently capture the structure of a given set. Higher order statistical moments provide information about the deviation (second order moment or variance), the symmetry (third order moment or skewness coefficient), the shape (fourth order moment or kurtosis).

We introduced in [9], the *set deviation string* as a proposed solution to the second order moment of a set of strings. We also generalized this approach for higher order moments but we will not use them in this paper.

The underlying concept we try to quantize with the variance operator V is the deviation over a set of elements. However, with this operator, the deviation is not of the same nature as the element because V makes use of a square. That is why, in $\Re$, we use the standard deviation.

We can characterize a pair of strings (s_i, s_j) by its distance $d(s_i, s_j)$ and by a sequence of elementary edit operations (substitution, insertion and deletion) e, *i.e.* $d(s_i, s_j) = c(e)$ where $c()$ is the cost function of the Levenstein edit distance. $e \in \mathcal{E}$, the set of all edit sequences between any pair of $U \times U$. We introduce the *positive edit string* between two strings A and B.

Definition 1. *Let e be the edit sequence from the string A to the string B, $e = (A \to B)$. The positive edit sequence $e_+ = (A \to B)_+$ is the restriction of e to edit operations having a non null cost.*

Definition 2. *Let S be a set of strings. The edit string set, $\mathcal{E}(S)$ is the set of all edit sequences between any pair of strings of S.*

Definition 3. *Let S be a set of strings. The positive edit string set, $\mathcal{E}_+(S)$ is the set of all positive edit sequences between any pair of strings of S.*

So any member of $\mathcal{E}_+(S)$ is called a positive edit string. The positive edit string, as well as the edit sequence, is a string, *i.e.* a word over an alphabet built from the initial alphabet and the edit operations. As does the variance, the positive edit string does not belong to the same space, *i.e.* $e_+ \notin U$, $e_+ \in \mathcal{E}_+(U)$.

The definition of the variance is "the expectation of the set of deviations from the expectation". We introduce the *set deviation* as the expectation of the positive edit strings from the set median string.

Definition 4. *Let S be a set of strings, and $\hat{s} = sms(S)$ its set median string. The set deviation string is given by $sds(S) = sms(\{(s_i \to \hat{s})_+, s_i \in S\}$.*

In order to compute the set deviation, one must first define the cost function for the set of edit operations; for instance, the cost of substituting a deletion by a substitution, $c((a_i \to \lambda) \to (a_j \to b_k))$. This cost function depends of course on the application and on the exact nature of the initial set of patterns.

The set deviation, as the set median, is not unique and of course depends on the cost function. A more general definition of the set deviation which try to overcome the uniqueness problem is proposed in [9].

Note that this definition is similar to the median of absolute deviation [18]. Referring to the real numbers, the set deviation is thus more similar to the standard deviation than to the variance even though it is not of the same nature as the input strings ($sds(S) \in \mathcal{E}_+$).

For a more detailed presentation of this second order moment, as well as higher moments, see [9].

3   Using the Set Deviation

The MVE (Minimum Volume Ellipsoid) estimator has been proposed by Rousseeuw [18] in order to extract the best represent in a set of corrupted data. The algorithm is as follows: first randomly select p elements of the data set of size n, say Ji, and compute the estimated mean m_i and variance s_i^2 [1]. In a second step, we can relate all the data set referring to these values by

$$x_j \to y_j^i = \frac{(x_j - m_i)^2}{s_i^2} \qquad (3)$$

and then sort the new data set of y_j^i values. Let z_j^i the ordered list of values. The MVE estimator, say mve, is the mean m_i associated with the minimum value of z_α^i for all the subset J_i, *i.e.*

$$mve = m_k \text{ such that } z_\alpha^k = Min_{J_i} z_\alpha^i . \qquad (4)$$

This search involves a huge amount of subsets J_i. A stochastic approach is thus used and only some subsets (related to the *a priori* percentage of outliers) are used yielding an approximation of the MVE estimator. Rousseeuw proposed to use a *kernel* of half the size of the data set, *i.e.*, $\alpha = 50\%$ which is suppose to be the upper limit of the breakdown point of any estimator. This algorithm has been generalized by Jolion *et al.* [11] and successfully applied in many domains related to pattern recognition and computer vision where real data are always corrupted by noise.

When working with strings, one can use the MVE estimator as we now have a well defined mean and deviation. Indeed, the set deviation can be used for normalization purposes. However, it is not possible to use it to transform a string, y_i will not be a string, but it can be used to weight the computation of any distance between two strings by using an sds(S) based cost function.

Definition 5. *Let c be a cost function defined on an alphabet such that* $c() = 0$ *or* $c() \geq 1$ *and sds(S) the resulting set deviation using this cost function. The sds(S)-based cost function* $c_{sds(S)}$ *is given by*

$$c_{sds(S)}(a_i \to b_j) = \begin{cases} 0 & \text{if } c(a_i \to b_j) = 0 \\ \frac{1}{\max(1, length(sds(S)))} & \text{if } (a_i \to b_j) \in sds(S) \\ c(a_i \to b_j) & \text{otherwise} \end{cases}$$

Any computation using this new cost function is done by applying the classic dynamic programming procedure [12]. Only the individual costs for symbols of the alphabet change as they do not only related to some *a priori* knowledge but also to a data specific knowledge by means of $sds(S)$. If one edit operation belong to $sds(S)$, its cost is decreased relatively to its original *a priori* cost. The mean cost of the procedure is increased by $length(sds(S))$ as any edit operation must be compared to the set deviation string. As a consequence of this new cost function, any string s_i such that $(s_i \to sms(S)) = sds(S)$ has now a unit distance to the set median. A set of strings can thus be ordered relatively to its set median and its set deviation.

---

[1] In a k-dimensional space, one has to compute the mean vector and the $k \times k$ covariance matrix.

Figure 1 presents an example of such a procedure. The set median is
5577007755 and the set deviation is $(7 \to \lambda)(7 \to \lambda)(2 \to 7)(2 \to 7)$.

The initial cost function is as follows $c(a_i \to a_i) = 0$, $c(a_i \to b_j) = 1$ for all
the other cases (including insertion and deletion).

Label	s_i	$d(s_i, sms(S))$	$(s_i \to sms(S))_+$
0	557777002222	6	$(7\lambda)(7\lambda)(27)(27)(25)(25)$
1	117777	6	$(15)(15)(\lambda 0)(\lambda 0)(\lambda 5)(\lambda 5)$
2	00766700	8	$(05)(05)(6\lambda)(6\lambda)(\lambda 7)(\lambda 7)(\lambda 5)(\lambda 5)$
3	007755007755	4	$(05)(05)(5\lambda)(5\lambda)$
4	77707202	6	$(75)(\lambda 5)(\lambda 0)(27)(05)(25)$
5	5577007755	0	
6	557777002255	4	$(7\lambda)(7\lambda)(27)(27)$
7	007667	7	$(\lambda 5)(\lambda 5)(\lambda 7)(\lambda 7)(67)(65)(75)$
8	57887552112	9	$(\lambda 5)(87)(80)(\lambda 0)(\lambda 7)(2\lambda)(1\lambda)(1\lambda)(2\lambda)$
9	55770022777755	4	$(2\lambda)(2\lambda)(7\lambda)(7\lambda)$

Label	s_i	$d_{sds(S)}(s_i, sms(S))$	$(s_i \to sms(S))$
0	557777002222	5	$(7\lambda)(7\lambda)(27)(27)(25)(25)$
1	117777	12	$(15)(15)(\lambda 0)(\lambda 0)(\lambda 5)(\lambda 5)$
2	00766700	16	$(\lambda 5)(05)(07)(\lambda 0)(60)(67)(05)(05)$
3	007755007755	8	$(05)(05)(5\lambda)(5\lambda)$
4	77707202	10.25	$(\lambda 5)(\lambda 5)(70)(27)(05)(25)$
5	5577007755	0	
6	557777002255	1	$(7\lambda)(7\lambda)(27)(27)$
7	007667	14	$(\lambda 5)(\lambda 5)(\lambda 7)(\lambda 7)(67)(65)(75)$
8	57887552112	14.5	$(7\lambda)(85)(87)(50)(50)(27)(17)(15)(25)$
9	55770022777755	1.5	$(27)(27)(7\lambda)(7\lambda)(7\lambda)(7\lambda)$

Fig. 1. Example of ordering of string based on the sms (first table) and on sms and sds
(*second table*). The positive edit strings have been compacted as follows $(a_i \to b_j) \equiv (a_i b_j)$

The first ordering is $5, (3, 6, 9), (0, 1, 4), 7, 2, 8$ and the second ordering is
$5, 6, 9, 0, 3, 4, 1, 7, 8, 2$. On this small example, the ordering remains mainly the
same however we can notice that the strings 3 and 0 permute because the edit
string transforming 0 in 5 (the set median) mainly uses the edit operations contained in the set deviation. So 0 more resembles to 5 than 3. The edit string
associated to 9 also changed during the normalization as their exists a new edit
string, longer than the initial one, but made of edit operations belonging to the
set deviation. The normalized cost is thus 1.5.

4   Applications

Several applications on words processing has been proposed in [9]. We will
present here an illustration on shape clustering. We consider a set of shapes

encoded using the freeman code. The main advantage of this coding scheme is that it is associated to simple cost functions for the general case, *i.e.*

$$c(a_i \rightarrow b_j) = min(|a_i - b_j|, |8 - a_i|, b_j) \qquad (5)$$

where a_i and b_j are two elements of the Freeman code.

More interesting is the fact that the cost function for the second order moment is also easy to define as being partially only distance between vectors.

Our data set is made of three clusters. The first one, say F, is composed of 185 "France" as shown in figure 2. This clusters is quite noisy [2].

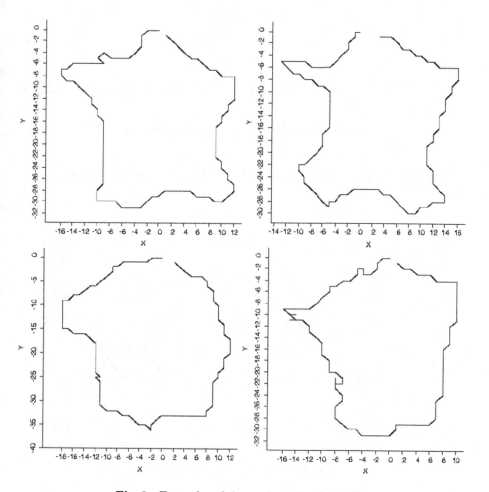

Fig. 2. Examples of shapes from the cluster F

---

[2] These 185 contours has been extract from a dataset of several hundreds of drawings we used in a previous study [17].

The second cluster, C, is made of 113 synthetic squares moderately noisy as shown on Figure 3.

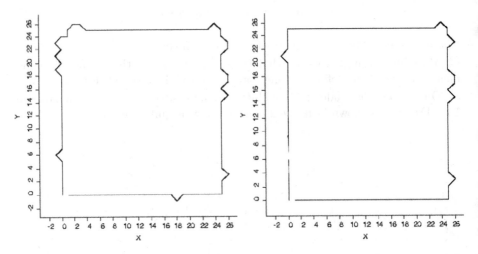

Fig. 3. Examples of shapes from the cluster C

Finally, the third cluster, B, is made of 115 random strings, (see Figure 4).

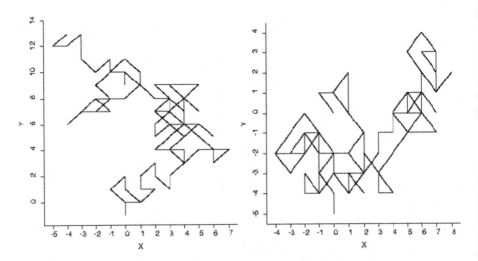

Fig. 4. Examples of random shapes

All the 413 shapes have been digitized using 100 points, *i.e.* the 413 strings are of the same length. In order to focus on the benefit we can get from the set deviation, we implement two version of the MVE estimator. The first one, yielding the estimator mve_1 follows the algorithm described in the previous section. The second algorithm, yielding the estimator mve_2, does not use the set deviation and the strings are ordered using only their distance to the set median. The parameters of this experiments are thus the amount of subsets NJ, the amount of strings in a subset, p, and the percentage α of the main cluster we are focusing on.

Table 1 shows, for some values of the parameters, the cluster the MVE estimator belongs to for both algorithms. We also show the execution time. In order to better appreciate these values, one has to compare with the execution time for the computation of the sms of the entire data set (4′) and for the computation of the sds of the entire data set (9′30″).

Table 1. Results obtained for the two implementations of the MVE estimator, mve_1 using the set deviation and mve_2 without the set deviation

p	NJ	α	mve_1	time	mve_2	time
3	10	50	C	1′00	B	5″
3	10	30	F	-	F	-
3	10	20	C	-	C	-
3	20	50	F	2′35	B	8″
3	20	30	F	-	F	-
3	20	20	C	-	C	-
3	30	50	F	3′56	B	14″
3	30	30	F	-	F	-
3	30	20	C	-	C	
5	10	50	C	1′05	B	5″
5	10	30	F	-	F	-
5	10	20	C	-	F	-
5	20	50	F	2′11	B	16″
5	20	30	F	-	F	-
5	20	20	C	-	C	-
10	10	50	C	1′23	F	5″
10	10	30	C	-	F	-
10	10	20	C	-	C	-

The behavior of the estimator mve_2 only depends on the size of the *kernel*. For $\alpha = 50\%$, there is no cluster of this size in the data set (F is about 45% of the data set and C is about 27%). When looking at the clusters, as the criteria is mainly the distance between the candidate sms and its 207th closest strings, the algorithm does not focus in a dense area of the set. Figure 5 (right) shows the ordered list of distances between the strings and the sms. This distribution does not show any specific breakpoint. This can be compared with the distribution obtained with the estimator mve_1 on the same figure (left). The returned sms belongs to cluster F and the figure clearly shows a nice breakdown close to the size of the cluster F.

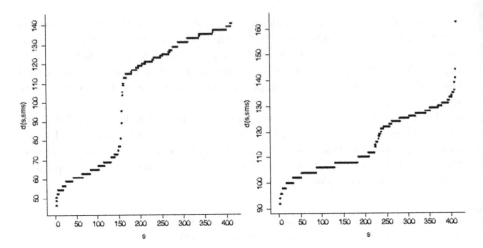

Fig. 5. Ordered list of distances between any string of the data set and the sms obtained with the estimators mve_1 (*left*) and mve_2 (*right*)

For $\alpha = 30\%$, the mve_2 estimator always find a sms inside the cluster F and always inside the cluster C for $\alpha = 20\%$. The last also holds for the mve_1 estimator. However, this one takes advantage of the computation of the sds and never returns any sms in the cluster B, $i.e.$, the noise whatever is α.

Figure 6 shows another difference between the algorithms. In both cases ($p = 5$, $NJ = 20$ and $\alpha = 20\%$), the sms belongs to the clusters C. The estimator mve_2 is associated to a curve (see Figure 6 right) with a large breakpoint which is useful to extract a cluster. This cluster is of size 113 and only made of strings of cluster C. However, the remaining part of the curve does not exhibit any particular breakpoint and no more cluster can be extracted. One has to run the estimator again on the remaining data set (after removing the strings belonging to the first detected cluster). The estimator mve_1 is associated to a curve with many breakpoints of comparative heights. The first detected cluster is of size 108 and only made of strings of cluster C. The second cluster is of size 99 and made of 90 strings of cluster B and 9 of cluster F. The third cluster is the worst one as being made of 5 strings from C, 6 from B and 24 from F. The remaining part of the data set is thus made of 21 strings from B and 152 from F. The mve_1 estimator allows a better discrimination among the actual clusters of the data set.

Of course, the main difference between these two algorithms is the computational complexity. Using the sds is a lot costly (between 8 and 20 times the cost of the algorithm without the computation of sds). However, the stochastic approach including the computation of the set deviation (mve_1 estimator) can be of the same cost of the computation of the sms on the entire set.

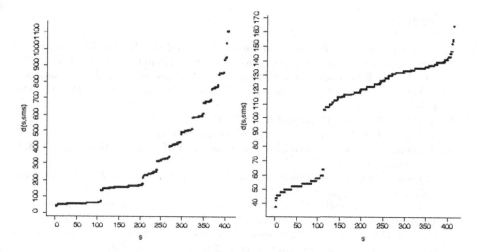

Fig. 6. Ordered list of distances between any string of the data set and the sms obtained with the estimators mve_1 (*left*) and mve_2 (*right*)

5 Discussion

We introduce in this paper a new framework for statistical moments of a set of strings and show one particular use of this framework for robust clustering of set of shapes. Many works still have to be done on a theoretical point of view as well as on applications. First we can easily extend to the definition of the covariance as well as the correlation coefficient. Next, the high order moments, e.g. greater than 2, have to be studied in order to better understand the information and properties of the data set they are related to. Basically, we should have access to the skewness and kurtosis. Our current work also include the search for a complete probabilistic framework of a set of strings.

So far, these new concepts are limited to the set median and it would be of greater interest to extend them to the generalized median which provides a better representation of a set of strings. As the computational cost is very high, e.g. the problem is NP-hard, one must use recent approximate algorithms to overcome this computational limitation. Nothing in our novel framework is "set median" dependent and so the generalized deviation can easily be deduced from the set deviation (up to a high computational cost).

The generalization to graphs is straightforward as the edit distance had also been used to define the median graph, the weighted mean of a pair of graphs... However an alternative approach can be to convert first graphs to strings, by a seriation, using for instance a spectral approach like in [2].

Our hope is that this preliminary work will be considered as one another contribution toward bridging the gap between statistical and structural pattern recognition.

References

1. C. Arcelli, L.P. Cordella and G. Sanniti di Baja, 4th IAPR Int. Workshop on Visual Form, Capri, Italy, may 2001, Springer, Lecture Notes in Computer Science, 2059.
2. J.E. Atkins, E.G. Boman and B. Hendrickson, A spectral algorithm for seriation and the consecutive ones problem, *SIAM Journal on Computing*, 28(1), 297–310, 1998.
3. H. Bunke, X. Jiang, K. Abegglen and A. Kandel, On the weighted mean of a pair of strings, *Pattern Analysis and Applications*, 2003 (to appear).
4. H. Bunke and S. Günter, Weighted mean of a pair of graphs, *Computing*, 67, 209–224, 2001.
5. H. Bunke, Recent Developments in Graph Matching, *Proc. of ICPR'00*, Barcelona, Spain, 2, 117–124, 2000.
6. H. Bunke, On a relation between graph edit distance and maximum common subgraph, *Pattern Recognition Letters*, 18, 689–694, 1997.
7. C. de la Higuera and F. Casacuberta, Topology of strings: Median string is NP-complete, *Theoretical Computer Science*, 230(1/2), 39–48, 2000.
8. X. Jiang, A. Münger and H. Bunke, On Median Graphs: Properties, Algorithms, and Applications, *IEEE trans. on Pattern Analysis and Machine Intelligence*, 23(10), 1144–1151, 2001.
9. J.M. Jolion, On the deviation of a set of strings, to appear in *Pattern Analysis and Application*, 2003.
10. J.M. Jolion, Graph matching: what are we talking about ?, J.M. Jolion, W. Kropatsch, and M. Vento eds, *Proc. of Third IAPR-TC15 Workshop on Graph-based Representations in Pattern Recognition*, may 23-25, 2001, Ischia, Cuen editor, Italy (isbn 88 7146 579-2), 170–175, 2001.
11. J.M. Jolion, P. Meer and S. Bataouche, Robust clustering with applications in computer vision, *IEEE Trans. on Pattern Analysis and Machine Intelligence*, 13(8), 791–802, 1991.
12. A. Levenstein, Binary codes capable of correcting deletions, insertions and reversals, *Sov. Phy. Dohl.*, 10, 707–710, 1966.
13. T. Lewis, R. Owens and A. Baddeley, Averaging feature maps, *Pattern Recognition*, 32(9), 331–339, 1999.
14. D. Lopresti and J. Zhou, Using Consensus Sequence Voting to Correct OCR Errors, *Computer Vision and Image Understanding*, 67(1), 39–47, 1997.
15. A. Marzal and E. Vidal, Computation of Normalized Edit Distance and Applications, *IEEE trans. on Pattern Analysis and Machine Intelligence*, 15(9), 926–932, 1993.
16. A. Massaro and M. Pelillo, A Linear Complementarity Approach to Graph Matching, *Proc. of Third IAPR-TC15 Workshop on Graph-based Representations in Pattern Recognition*, may 23-25, 2001, Ischia, Cuen editor, Italy (isbn 88 7146 579-2), 160–169, 2001.
17. M. Mattenet and J.M. Jolion, Analyse de Formes Déformées (in french), *Mappemonde*, 3, 5–9, 1992.
18. P.J. Rousseeuw and A.M. Leroy, *Robust Regression & Outlier Detection*, Wiley, 1987 (p. 45).
19. G. Sánchez, J. Lladós and K. Tombre, A mean string algorithm to compute the average among a set of 2D shapes, *Pattern Recognition Letters*, 23, 203–213, 2002.
20. R.A. Wagner and M.J. Fisher, The string to string correction problem, *Journal of the ACM*, 21(1), 168–173, 1974.

A New Median Graph Algorithm

Adel Hlaoui and Shengrui Wang

DMI, University of Sherbrooke, Sherbrooke, QC, J1K2R1, Canada
{hlaoui,wang}@dmi.usherb.ca

Abstract. Median graph is an important new concept introduced to represent a set of graphs by a representative graph. Computing the median graph is an NP-Complete problem. In this paper, we propose an approximate algorithm for computing the median graph. Our algorithm performs in two steps. It first carries out a node reduction process using a clustering method to extract a subset of most representative node labels. It then searches for the median graph candidates from the reduced subset of node labels according to a deterministic strategy to explore the candidate space. Comparison with the genetic search based algorithm will be reported. This algorithm can be used to build a graph clustering algorithm.

1 Introduction

In structural pattern recognition, the median graph, introduced by Jiang, Munger and Bunke [1,2,3], is a powerful concept that can be used to represent a set of graphs. It aims to extract the essential information from a set of graphs into a single prototype. Given a set of graphs S, the median is defined as the graph derived from S which has the smallest sum of distances (SOD) to all the graphs in the set. Here the concept of distance should be interpreted as the dissimilarity between graphs. If the condition that the median belongs to the set of graphs is required, the set median is obtained, otherwise the generalized median is obtained.

The median graph can have a variety of applications in areas such as image retrieval, shape matching, hand-written character recognition, and Web-mining. In particular, since it is an extension of the concept of the center of a set of vectors (points in a vector space), it can be used in a direct approach to graph clustering involving application of a traditional clustering algorithm to a set of graphs. While an indirect approach such as that proposed by Luo, Wilson, and Hancock [4,5] using vectors of graph-spectral features is computationally more efficient, a direct approach using the median graph concept allows for the direct use of graph similarity (and dissimilarity) obtained from a graph matching procedure and has a potential to result in a better partition of graphs in terms of within-cluster scatter and between-cluster separation. The direct approach has not yet been widely explored, partly due to the lack of an efficient algorithm for computing the median graph.

Let L_V and L_E denote the set of node and edge labels, respectively. A labeled, undirected graph G is a 4-tuple $G = (V, E, \mu, \xi)$, where

- V is the set of nodes,
- $E \subseteq V \times V$ is the set of edges,

E. Hancock and M. Vento (Eds.): IAPR Workshop GbRPR 2003, LNCS 2726, pp. 225–234, 2003.

- $\mu : V \rightarrow Lv$ is a function assigning labels to the nodes, and
- $\xi : E \rightarrow L_E$ is a function assigning labels to the edges.

Let U be the set of all graphs that can be constructed from a given set $S = \{g_1, g_2, g_3, ..., g_n\} \subset U$, i.e., any graph whose number of nodes is between 1 and the sum of all the nodes of the g_i and whose node and edge labels are taken from L_v and L_E. The generalized median graph and the set median graph are defined respectively by

$$\overline{g} = \arg\min_{g \in U} \sum_{i=1}^{n} d(g, g_i)$$

and

$$\hat{g} = \arg\min_{g \in S} \sum_{i=1}^{n} d(g, g_i)$$

The generalized median graph is a more interesting concept because it has a high potential to capture the essential information. But it is also more difficult to compute, as its time complexity grows exponentially with both the number of graphs and their size. The set median graph, on the other hand, can be computed by a straightforward procedure involving only $O(n^2)$ steps. Note that for each step, we need to compute a distance between two graphs, which is an NP-complete task. Hence, approximate techniques are needed to cope with these high computational complexities. To our knowledge, the only approximate algorithm so far put forward for computing the generalized median graph is the method using the genetic search algorithm proposed by Jiang et al. [6,7,8]. This algorithm simultaneously finds the generalized median graph and the optimal mapping between the median graph and each graph in the set. In Section 3, we will present a comparison between the genetic search algorithm and the algorithm proposed here.

Recently, we proposed an approximate algorithm [9] for computing the distance (dissimilarity) between graphs, which can be directly used to compute the set median graph. In this paper, however, our focus is on computation of the generalized median graph.

The new algorithm we have proposed for computing the generalized median graph carries out the search process in two stages. In the first stage, the potential node labels used to build the generalized median graph are selected and an edit distance is computed. Then the sum of distances between the median graph and each graph in the set S is computed. In the second stage, we assign a label to each edge in the generalized median graph according to the matching results from the first stage. In the rest of this paper, we use labeled, undirected graphs. The paper is organized as follows. Section 2 describes the new median graph algorithm. Section 3 gives the results of the experiments designed to demonstrate the effectiveness of the new algorithm. A summary and directions for future work are presented in Section 4.

2 The New Median Graph Algorithm

In this paper, we propose an approximate algorithm to compute the generalized median graph, based on a process that reduces the space of graphs in which the median is sought. A naive approach would be to compute the SOD for each graph in the search space and choose the optimal one, i.e., the one that yields the smallest SOD. Unfortunately, this solution suffers from very high computational complexity, since the search space of graphs contains all graphs which can be constructed using node and edge labels from S. The idea of our algorithm is based on the fact that the distance between the median graph and a given set of graphs $d(g, g_i)$ includes two parts: a node-to-node distance and an edge-to-edge distance. The formula given above for the SOD thus becomes:

$$SOD = \sum_{i=1}^{n} d_n(g, g_i) + \sum_{i=1}^{n} d_e(g, g_i) = SOD_n + SOD_e$$

The heuristic used in our algorithm (which is what makes it an approximate algorithm) involves dividing the search process into two parts: a node searching process and an edge searching process. In other words, the algorithm first searches for the subset of nodes most likely to be part of the final median graph, by minimizing SOD_n. It then searches for the best subset of edges connecting the selected nodes. We will describe these two processes individually.

2.1 Initialization Steps

As described in [1], it can be proved that the generalized median graph cannot have more nodes than the sum of all nodes of all graphs in S. The first question that needs to be answered concerns the cardinality of the median graph. Since the search process involves computing a node-to-node edit distance, the edit operations of substitution, insertion, and deletion influence the distance according to their costs. The cost of each operation is described as follows:

- Node substitution: $C_{ns}(u, v) = d(u, v)$.
- Node insertion: $C_{ni}(u) = u$.
- Node deletion: $C_{nd}(u) = u$.

Intuitively, using an insertion or deletion operation to match two graphs tends to increase the edit distance. For the node insertion operation, the edit distance is increased by the value of the label for the added node and the labels for all the outgoing edges originating from this node. Similarly, for the deletion operation, the edit distance is increased by the value of the label for the deleted node and all the labels for the incoming edges ending at this node. The substitution operation increases the edit distance only by the substituted node or edge. Using the substitution operation thus helps to reduce the edit distance. In order to favour the use of the substitution operation, a good choice for the size of the median graph is the average number of nodes in set S. In the work described here, we assume the cardinality of the generalized median graph of a given set of n graphs to be the average cardinality of the n graphs.

$$|V| = \frac{1}{n} \sum_{i=1}^{n} |V_i|$$

This hypothesis is partially validated by the first experiment reported in Section 3.
Since we have set the cardinality of the median graph, a naive approach can be used, which iteratively takes $|V|$ nodes, computes the *SOD* and compares this sum with the best one until the stack of combinations is empty. This approach is not appropriate for a larger set L_V, since the number of combinations grows exponentially with the size of the set L_V. The approach adopted in this work involves a process of reducing the set L_V to a set designated NL_V (New L_V). The well-known k-means algorithm is applied to L_V to find clusters among the node labels. As the k-means is an unsupervised clustering algorithm, we need to provide the number of clusters. The greater this number, the more the generalized median graph tends toward the optimal one and the greater the time needed to compute the median. In the work described here, we used a number equal to twice the cardinality of the median graph. The center of each cluster represents a potential node label. If the center does not belong to the set L_V, we assign the closest label to this center.

2.2 Minimization of *SOD*

Following the selection of the subset NL_V, we need to choose $|V|$ labels from the set NL_V to minimize the SOD_n. A straightforward approach would be to check every subset of $|V|$ labels from NL_V by computing its SOD_n and retain the one that yields the minimum SOD_n as the best median graph candidate. This search would require evaluating $C_{|NL_V|}^{|V|}$ subsets of $|V|$ labels from NL_V, a number which could be very high, depending on the difference between $|V|$ and $|NL_V|$. The search method we propose below is particularly useful when $|V| \le (3/4)|NL_V|$.

This search method is based on a strategy of iterative generation and exploration of the subsets of NL_V composed of the labels most similar to the labels from L_V. Those first generated subsets (represented here by g) that minimize the edit distance $d_n(g, g_s)$ from one of the graphs (represented by g_s) in the set S are considered as *promising candidates* (to be placed on a list). When a particular promising candidate has been generated n times in the iterative process, i.e., it has been compared with all graphs in S, it is considered as a *median graph candidate* and is compared, in terms of SOD_n, with the best median graph candidate found so far. The best median graph candidate is updated accordingly and a termination condition is tested for.

This method is implemented in steps 3 to 6 of the algorithm *Median_Node* below. In this algorithm, u and v denote a label from the set NL_V and a label from the set L_V, respectively. The edit distance between u and v is the main search thread leading to the generation of promising candidates and then, of median graph candidates. Each time a new (u, v) mapping is considered (Step 3), a target graph g_s (from the set S) containing the node v is identified. One or several promising candidates containing the (u, v) mapping will be generated if there are enough (i.e., at least $|V|$-1) other mappings from NL_V to the node set of the target graph g_s (Step 4). These candidates are evaluated in Step 5. In particular, in Step 5.2, such a candidate will be considered as a median graph candidate and be compared with the best median graph candidate found so far if it has been matched to every (target) graph in the set S in the previous steps. In Step 6, a sufficiency condition is tested in order to guarantee the best median graph candidate found so far is indeed the optimal median graph candidate in terms of SOD_n. We are still investigating other possible less restrictive conditions.

Algorithm *Median_Node*:

Input: L_V, the set of labels, and S, the set of graphs.

Output: A subset of $|V|$ nodes forming a median graph candidate.

Step 1: Compute the cardinality of the generalized median graph $|V|$. (Section 2.1)

Step 2: Choose a subset NL_V of node labels from the set L_V using the k-means algorithm. (Section 2.1)

Step 3: Choose the (u,v) mapping which has the smallest edit distance. Let g_s be the (target) graph from the set S that has the node with the label v.

Step 4: Generate all possible promising candidates containing the current (u,v) mapping and $|V|$-1 mappings previously selected that have g_s as the target graph.

Step 5: For each of the candidates (represented by g)generated in Step 4:
5.1: Compute the edit distance $d_n(g, g_s)$.
5.2: If the current candidate g has already been matched to all the target graphs in S, compute the SOD_n and update the best median graph accordingly.

Step 6: If the SOD_n is smaller than the edit distance of the current (u,v) mapping, then go to Step 7, otherwise go to Step 3.

Step 7: Output the best median graph candidate.

End.

After finding the best median graph candidate in terms of SOD_n, we now focus on searching for the edge subset that minimizes SOD_e, the second term of SOD. The main idea adopted here is to assign individually, to each edge in the median graph candidate, the label from the set L_E that is closest to the average value of all corresponding edges in the different graphs in S.

Procedure *Median_Graph*:

Input: Median graph candidate g and the set of graphs S.

Output: Generalized median graph.

Step 1: For each pair of nodes (u_1, u_2) in the median graph candidate g
1.1 Initialize Sum = 0;
1.2 For each graph g_s in S
Find the nodes (v_1, v_2) in g_s that have been matched to (u_1, u_2);
Compute Sum += label of the edge(v_1, v_2)

Step 2: Assign the label Sum / $|V|$ to the edge (u_1, u_2).

End.

2.3 Example

We now present a simple example to illustrate how the proposed algorithm works. We will show how the algorithm finds promising candidates. Suppose that we have three graphs g_1, g_2 and g_3, as shown in 0. The initial set of labels L_v (0) contains 12 labels. The first reduction will be performed by k-means clustering on the set L_v (of course, for the small set here the use of k-means clustering is not absolutely necessary). From this reduction procedure, 8 labels will be retained to form the set NL_v (see 0).

Now, from the set NL_v, only 4 labels will be selected to form the median graph candidate. We select a (u, v) mapping, generate a promising candidate, evaluate its edit distance from the target graph in the set S, and finally, if it has previously been compared to the other graphs in the set S, evaluate its SOD_n. The first (u, v) mapping to be selected is (0.5,0.5) and the target graph g_s is 1, since the label $v(0.5)$ belongs to the first graph. At this stage there is no possible promising candidate since there are no other previously selected mappings. The first promising candidate arises when the (1,1) mapping is selected. This candidate is composed of the following labels {0.5, 0.2, 0.6, 1}, which belongs to the set NL_v, and mapped to the following labels taken from the set L_v: {0.5, 0.2, 0.6, 1}. Its edit distance is null. At this stage, the current candidate is compared only with the first graph. Thus it is not yet considered as a median graph candidate. This example also shows that the promising candidate {0.6, 0.5, 0.4, 1} generated at the steps 42, 52 and 78 has the SOD_n value of 1. In the subsequent iterations, no other median graph candidate obtained has a smaller SOD_n value. The details of the search procedure are given in 0.

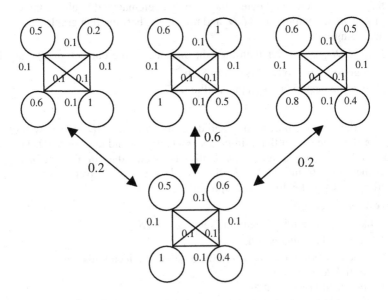

Fig. 1. Three input graphs and their median graph

Table 1. The set L_v

0.5	0.2	0.6	1	0.6	1	0.5	1	0.6	0.5	0.8	0.4

Table 2. The set NL_v. Eight labels selected from L_v by the K-means algorithm.

0.5	0.2	0.6	1	0.4	1	0.5	0.8

Table 3. Description of the interesting iterations of the search procedure.

Index	(u, v)	G_s	Promising candidates	$d_n(g, g_s)$	SOD_n
1	0.5,0.5	1	√	√	√
2	0.5,0.5	2	√	√	√
3	0.5,0.5	3	√	√	√
4	0.2,0.2	1	√	√	√
5	0.6,0.6	1	√	√	√
6	0.6,0.6	2	√	√	√
7	0.6,0.6	3	√	√	√
8	1,1	1	0.5 * 0.2 * 0.6 * 1	0	√
9	1,1	2	√	√	√
10	1,1	2	√	√	√
11	0.4,0.4	3	√	√	√
12	1,1	1	0.5 * 0.2 * 0.6 * 1	0	√
13	1,1	2	0.5 * 0.6 * 1 * 1	0	√
42	0.4,0.2	1	0.5 * 0.6 * 1 * 0.4	0.2	√
52	1,0.8	3	0.5 * 0.6 * 1 * 0.4	0.2	√
78	0.5,1	2	0.5 * 0.6 * 1 * 0.4	0.6	1

3 Experimental Results

The aim of this section is to demonstrate the performance of the algorithm and show how the reduction process for the original set labels L_v can be a good solution to the problem of finding an approximate generalized median graph. We conducted three experiments. The first of these concerns the average cardinality assumption adopted in this work for the size of the generalized median graph. The second concerns the performance of the algorithm in terms of computation time. The third experiment is a comparative study of the new algorithm and the genetic search based algorithm proposed by Jiang [6,7,8]. The comparison between the two algorithms is performed in terms of run time and the sums of distances (*SOD*) computed using the generalized median graphs yielded by the two algorithms.

For the purpose of testing the best cardinality to use, 10 sets of graphs of different sizes with random labels at each node and edge were generated. We gradually increased the number of graphs in each set, starting with 5 and terminating at 50. We assigned a random number of nodes, varying between 3 and 9, to each graph. Note that all of the graphs are labeled and undirected. We assigned a random real label, taken from {0.1, 0.2, ...,0.9}, to each node and edge.

In each test, we varied the size of the generalized median graph from three to eight nodes. We computed the sum of distances for each test and for each cardinality of the median. For each test set, we also computed the average number of nodes per graph. 0 shows the minimum SOD between the optimal generalized median graph of a particular size and the graphs in each test set. From this table, we observe that the best cardinality to use for the median graph in the majority of cases is the average number of nodes for the corresponding test set.

Table 4. SOD between the optimal generalized median graph of a particular size and the graphs in a test set. The best cardinality found for a set of graphs is close to the average size of the graphs in that set.

Cardi-nality of the median	Number of graphs used in the set S (Average number of nodes per graph)									
	5 (4.8)	10(4)	15(7)	20(6)	25(6)	30(6)	35(4.2)	40(4.3)	45(4.1)	50(5)
3	7.4	12.5	28.4	32.1	39	51.5	36.3	48.7	49.7	58.2
4	6.4	12	25.3	26.3	36.2	51.2	33.1	48.8	49.1	55.8
5	6.2	12.4	19.8	25.2	34.1	47.3	38.2	56.7	55.3	63.1
6	7.2	15	21.2	24.1	31.2	42.1	42.3	77	64	77.7
7	9.4	19	30.3	29	34.4	56.1	52.3	98	72	111.1
8	9.7	19.3	29.7	33.3	38.7	57.2	66.3	126.3	98.7	134.1

The second experiment concerns the algorithm's computation time. In this experiment, we generated six sets of graphs of varying sizes. We gradually increased the number of graphs in each set from 10 to 50. For each graph, we randomly assigned the following parameters: the number of nodes (from three to eight), and a real label taken from {0.1, 0.2, ...,0.9}. In each test, we varied the cardinality of the median graph from three to eight nodes and set the number of clusters for the k-means algorithm at $2 \times |V|$. It can be observed that the time needed to compute the generalized median graph increases linearly with respect to the number of graphs in S, but remains exponential w.r.t. the size chosen for the generalized median graph. It is interesting to note that when the size is chosen to be the average size for the test set (as is the case with the proposed algorithm), the time required is significantly lower than the time required for a larger size. Further experiments are required in order to better understand the time-size relation.

Table 5. Time needed to compute the generalized median graph (in seconds)

Cardinality Used	Number of graphs used in the set S					
	10	15	20	30	40	50
3	0.063	0.109	0.125	0.156	0.219	0.390
4	0.172	0.219	0.250	0.375	1.078	1.453
5	2.344	3.891	5.532	8.703	11.750	14.750
6	11.953	18.250	28.547	40.360	54.594	72.422
7	49.313	76.281	104.891	149.766	221.156	272.313
8	292.843	441.578	616.266	896.255	1226	1542

In the last experiment, we conducted a comparison with the genetic search based algorithm proposed by Jiang et al [6,7,8]. The main idea of the genetic solution involves using a chromosome to represent the transformations from a candidate graph G to each graph in the set S. This representation is followed by a node labelling process and the specification of edges and their labelling. We conducted four tests, in each of which we varied the number of graphs in the set S from 5 to 20. For this experiment, we set the number of nodes in each graph at 5 and the number of clusters in the K-means algorithm at 10 clusters. A label randomly selected from {0.1, 0.2, ...,0.9} was assigned to each node and each edge. For the genetic solution, we set the population size, the crossover probability, the mutation probability and the number of generations at 50, 0.9, 0.1 and 100, respectively. Table 6 and Table 7 show comparisons in terms of SOD and computation time. We observe that the new algorithm proposed here is able to compute a better generalized median graph and is also faster. Extensive tests are needed, however, to clarify the specific advantages of each algorithm. Nevertheless, we believe that the deterministic nature of the algorithm proposed here is one of its main advantages.

Table 6. SOD of the four tests

		Number of graphs in each set			
		5	**10**	**15**	**20**
New algorithm	SOD_n	2.9	7.4	10.2	12.4
	SOD_e	10.3	22.8	35.1	47
	SOD	13.2	30.2	45.3	59.4
Genetic solution	SOD_n	4.9	12.6	17.8	23.8
	SOD_e	11.3	23	32	47.8
	SOD	16.2	35.6	49.8	71.6

Table 7. Computation time in seconds

	Number of graphs in each set			
	5	**10**	**15**	**20**
New algorithm	3.350	7.310	10.110	13.840
Genetic solution	3.070	8.120	14.660	22.680

4 Concluding Remarks

We have developed an approximate algorithm for the generalized median graph. Our algorithm guarantees that a good solution will be found. The efficiency of the new algorithm rests on reduction of the set of node labels and on a search strategy aiming at exploring the most likely median graph candidates first. We have also shown that the assumption adopted as to the cardinality of the median graph yields a good approximation of the true generalized median graph and there is no need to use the whole set of labels; only a subset of most representative labels (obtained from a clustering process) is sufficient to generate a median graph close to the optimal one.

Currently, we are using this algorithm to design a graph clustering algorithm for content-based image retrieval.

Acknowledgement. This work has been supported by a Strategic Research Grant from Natural Sciences and Engineering Research Council of Canada (NSERC) to the team composed of Dr. F. Dubeau, Dr. J. Vaillancourt, Dr. S. Wang and Dr. D. Ziou. Dr. S. Wang is also supported by NSERC via an individual research grant.

References

1. X. Jiang, A. Munger and H. Bunke. On median graphs: properties, algorithms, and applications. IEEE Trans. on PAMI, vol. 23, no. 10, October 2001.
2. X. Jiang and H. Bunke. Optimal lower bound for generalized median problems in metric space. IAPR International Workshop on Structural, Syntactic and Statistical Pattern Recognition (S+SSPR), Windsor, Canada, LNCS 2396, pp. 143–151, 2002.
3. H. Bunke. Recent advances in structural pattern recognition with applications to visual form analysis. C. Arcelli, L. Cordella, G. Sanniti di Baja (eds.): Visual Form 2001, Springer Verlag, LNCS 2059, 2001, 11–23.
4. B. Luo, R. C. Wilson and E. R. Hancock. Spectral feature vectors for graph clustering. IAPR International Workshop on Structural, Syntactic and Statistical Pattern Recognition (S+SSPR), Windsor, Canada, LNCS 2396, pp. 83–93, 2002.
5. B. Luo, R. C. Wilson and E. R. Hancock. The independent and principal component of graph spectra. Proc. 16th International Conference on Pattern Recognition, Quebec, Canada, Volume-2, 2002.
6. X. Jiang, L. Schiffmann and H. Bunke. Computation of median shapes. Proc. 4th Asian Conference on Computer Vision, Taipei, Taiwan, 300–305, 2000.
7. X. Jiang, A. Munger and H. Bunke. Synthesis of representation graphical symbols by computing generalized median graph. Proc. of 3rd IAPR Int. Workshop on Graphics Recognition, Jaipur, India, 187–194, 1999.
8. X. Jiang, A. Munger and H. Bunke. Computing the generalized median of a set of graphs. Proc. of 2nd ICPAR Workshop on Graph-based Representations, Haindorf, Austria, 115–124, 1999.
9. A. Hlaoui and S. Wang. A new algorithm for inexact graph matching. In 16th International Conference on Pattern Recognition, Quebec, Canada, 2002.

Graph Clustering Using the Weighted Minimum Common Supergraph

Horst Bunke[1], P. Foggia[2], C. Guidobaldi[2], and M. Vento[3]

[1] Institut für Informatik und angwandte Mathematik,
Universität Bern, Neubrückstrasse 10, CH-3012 Bern, Switzerland
bunke@iam.unibe.ch
[2] Dipartimento di Informatica e Sistemistica,
Università di Napoli "Federico II", Via Claudio, 21 I-80125 Napoli (Italy)
cguidoba@unina.it
[3] Dipartimento di Ingegneria dell'Informazione ed Ingegneria Elettrica,
Università di Salerno, via P.te Don Melillo, I-84084 Fisciano (SA), Italy
mvento@unisa.it

Abstract. Graphs are a powerful and versatile tool useful for representing patterns in various subfields of science and engineering. In many applications, for example, in pattern recognition and computer vision, it is required to measure the similarity of objects for clustering similar patterns. In this paper a new structural method, the Weighted Minimum Common Supergraph (*WMCS*), for representing a cluster of patterns is proposed. Using this method it becomes easy to extract the common information shared in the patterns of a cluster and separate this information from noise and distortions that usually affect graphs representing real objects. Moreover, experimental results show that *WMCS* is suitable for performing graph clustering.

1 Introduction

Clustering has become a mature discipline, that is important to a number of areas, including pattern recognition, machine learning, computer vision and related fields. There are applications in these fields in which a structural representation of patterns is particularly suitable. In those cases the representation of patterns by attributed graphs (simply graphs, in the following) is extremely convenient and currently very investigated. In unsupervised learning or clustering there is no explicit teacher, and the system forms clusters or *natural groupings* of input patterns, where *natural* is always defined explicitly or implicitly in the clustering system itself [4].

If graphs are used for the representation of structured objects, then grouping similar objects becomes equivalent to find those graphs that are similar to each other in a set of graphs. Even if a number of graph clustering methods are known from the literature [5,7,8,10,11,13,14,15], the investigation on clustering of graphs has only recently started and it is still widely unexplored.

E. Hancock and M. Vento (Eds.): IAPR Workshop GbRPR 2003, LNCS 2726, pp. 235–246, 2003.

In many pattern recognition and artificial vision applications, a cluster is composed of many distorted representations of the same object. In these cases the information contained in the cluster is the representation of the original object itself, while the differences that occur in the distorted objects can be considered as a noise.

When graph clustering is performed and a grouping is achieved, a very interesting open question is how to represent the structural information contained in the cluster. In case the patterns are represented by vectors and hence there is no structural information, the cluster representation problem is very well established and many possibilities have been exploited [4]. Instead relatively new methods for representing a cluster of graphs have been proposed in [8,11,15].

In this paper a new method for representing a cluster of graphs is presented. A cluster of graphs will be described by a graph that is, in general, different from each graph of the given cluster. This new graph, that we call the Weighted Minimum Common Supergraph (WMCS) of the cluster, is a graph summarizing all (and only) the properties of all the graphs belonging to the cluster. Moreover weights indicating the occurence of each property in the cluster are added during the construction process, so obtaining a weighted graph. The main advantage of using WMCS for representing a cluster of graphs is that the structural information contained in the cluster is preserved and that it is also easy to separate this information from the noise.

The main contribution of this paper is a formal definition of WMCS and an approximate procedure for its computation. Moreover two experiments have been conducted. The aim of the first experiment is to evaluate the quality of WMCS approximation for representing a cluster of graphs. A second experiment has been conducted to investigate if WMCS is suitable to perform graph clustering tasks.

The remainder of the paper is organized as follows. In Section 2, basic terminology and concepts are introduced. Next, in Section 3 a new approach for representing a cluster of graphs is proposed. In Section 4 an approximated method is derived from the proposed approach, while experimental results are reported in Section 5. Finally future work is discussed and some conclusions are drawn in Section 6.

2 Basic Definitions

A *graph isomorphism* is a bijective mapping between the nodes of two graphs that have the same number of nodes, identical labels and identical edge structure. Similarly, a *subgraph isomorphism* between two graphs g_1 and g_2 is an isomorphism between g_1 and a subgraph of g_2.

The *maximum common subgraph* of two graphs g_1 and g_2, $mcs(g_1,g_2)$, is a subgraph of both g_1 and g_2 and has, among all those subgraphs, the maximum number of nodes. The *Minimum Common Supergraph* of two graphs g_1 and g_2, $MCS(g_1,g_2)$, is a supergraph of both g_1 and g_2 and has, among all those supergraphs, the minimum number of nodes. Notice that, both $mcs(g_1,g_2)$ and $MCS(g_1,g_2)$ are not necessarily unique for two given graphs. The *difference* between g_2 and a subgraph g_1, $g_2 - g_1$ is obtained by removing g_1 from g_2, including edges connecting g_1 with the rest of the graph. These edges are called the *embedding* of g_1 in g_2, $E = emb(g_1,g_2)$. The *union* of g_1 and g_2 including E, $g_1 \cup_E g_2$ is a graph composed from the graphs g_1 and g_2 jointed by edges E. In [3] more details are given on these definitions and a formal proof of the following theorem has been given.

Theorem 2.1: Let g_1 and g_2 be graphs. Then

$$MCS(g_1,g_2) = mcs(g_1,g_2) \cup_{E1} (g_1 - mcs(g_1,g_2)) \cup_{E2} (g_2 - mcs(g_1,g_2)) \qquad (1)$$

where
$E_1 = emb(mcs(g_1,g_2),g_1)$ and $E_2 = emb(mcs\ (g_1,g_2),g_2)$

The computation of *mcs* is a NP-compete problem. Several algorithms for *mcs* are known from literature. Theorem 2.1 shows a way how *MCS* can be actually computed for two given graphs g_1 and g_2.

3   Optimal Representation of a Cluster of Graphs

Let G be a set of n graphs. Let us suppose that the graphs of G represent the same object with different distortions, due, for instance, to a not optimal system used to obtain the graph representations from real objects. The graphs of this set are different instances of the same pattern and we can call this set a cluster. The easiest way to represent a cluster is to choose a criterion to decide which element of the cluster is the most representative. Of course the representation of a cluster by means of one of the patterns can cause a loss of information and the inclusion of noise. Thus we want to represent a cluster of graphs using a graph that is in general different from each graph of the cluster. For our purpose, firstly we need to define a *weighted graph*, i.e. a graph with a label on each node (edge) for storing the occurrences of nodes (edges) of the graphs of the cluster.

Def. 3.1: A *weighted graph* is a 6-tuple $g = (V, \lambda, E, \varepsilon, \alpha, \beta)$, where
- V is the finite set of vertices (also called nodes)
- $\lambda: V \rightarrow N^+$ is a function assigning positive weights (labels) to the nodes
- $E \subseteq V \times V$ is the set of edges
- $\varepsilon: E \rightarrow N^+$ is a function assigning positive weights (labels) to the edges
- $\alpha: V \rightarrow L$ is a function assigning attributes to the vertices
- $\beta: E \rightarrow L$ is a function assigning attributes to the edges

In the following, a graph g and a weighted graph g' in which $\lambda(v)=1$ $\forall v \in V$ and $\varepsilon(e)=1$ $\forall e \in E$, will be considered equivalent.
On the basis of Def 3.1, it is possible to define the Weighted Maximum Common Subgraph of the cluster G, *WMCS(G)*. The *WMCS(G)* is a suitable instrument for representing a cluster of graphs. It is the smallest graph containing as subgraph each

graph of the given cluster. As a consequence it is the smallest graph representing each property appearing in the cluster. The *WMCS(G)* is formally defined in Def.3.2.

Def. 3.2: Let $G = \{g_1,...,g_n\}$ be a set of graphs; a *Weighted Common Supergraph* of G, *WCS(G)*, is a weighted graph $g = (V, \lambda, E, \varepsilon, \alpha, \beta)$ such there exist subgraph isomorphisms from g_i to g $\forall i \in \{1,...,n\}$. We call g a *Weighted Minimum Common Supergraph* of the set of graphs G, *WMCS(G)*, if there exists no other weighted common supergraph of G that has fewer nodes than g.
Let $f_i : V_i \rightarrow V$ be an isomorphism between g_i and a subgraph of g. The weight of each node v of g is k if there exist exactly k distinct isomorphisms f_j such that $f_j(v_i) = v$.
Let (u,v) be a pair of nodes of V. The weight of each edge $e = (u,v)$ is h if there exist exactly h distinct isomorphisms $f_j(e_{il}) = (f_j(u_i), f_j(v_l))$ such that $f_j(e_{il}) = e$.

It is worth notice that in general there exists more than an isomorphism $f_i : V_i \rightarrow V$ $\forall i \in \{1,...,n\}$. Thus the *WMCS(G)* is not uniquely defined. In [16], weights on nodes of attributed graphs are introduced. If the graphs are patterns of a cluster, then these weights represent the occurrence of a node (edge) in the cluster. According to Def.3.2 the weight of a node (edge) of *WMCS(G)* is k iff that node (edge) is contained exactly in k different graphs of the cluster; thus k is always included between 1 and n.

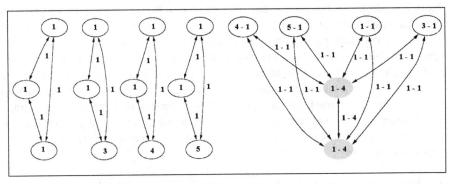

Fig. 3.1. A set of graphs G and a *WMCS(G)*. On each node (edge) of *WMCS(G)* there is a second label for the weight. This weight is the occurrence of the node (edge) in the cluster.

A very interesting problem is to extract from the graphs of the cluster the properties belonging to the original object and to distinguish them from noise introduced by the system. A property of a pattern represented by a graph can be represented by a subgraph of that graph. In the rest of this paragraph *subgraph* and *property* will be used as synonymous. A possibility to select the more representative properties in a cluster of graphs G, i.e. properties appearing more than p times, is to select nodes (edges) of *WMCS(G)* having a weight greater than p. This is formally obtained in Def .3.3.

Def. 3.3: Let g be a *WMCS(G)*. A *WMCS$_p$(G)*, is a subgraph of g containing all the nodes of g whose weight is at least p and all the edges of g having weight at least p, that connect those nodes in V having weight at least p.

The $WMCS(G)$ is a suitable instrument for finding the common properties of a cluster of graphs for several reasons:

- Nodes (edges) of those subgraphs that are present in many different graphs of the cluster, i.e. the information of the cluster, will correspond to nodes and edges of the $WMCS(G)$ having large weights. Conversely nodes (edges) of subgraphs that are present in a few graphs of the cluster, i.e. the noise of the cluster, will correspond to nodes and edges of the $WMCS(G)$ having small weights.

- A possibility to distinguish the information from the noise is the computation of the $WMCS_p(G)$, where p is a threshold of noise-rejection. The selection of an optimal threshold p is useful to restore the real information contained in the cluster.

- The construction of the $WMCS_p(G)$ can be viewed as a process of generalization. Indeed only those subgraphs present at least in p graphs of the given cluster, are also present in $WMCS_p(G)$. The $WMCS_p(G)$ summarizes all and only those properties appearing at least in p different graphs of the cluster. $WMCS_p(G)$ with high value of p may be considered as some kind of mean graph; conversely, if we select all the vertices and edges which have a value lower than a given threshold can such a graph be interpreted as the variance of the cluster.

- The $WMCS_p(G)$ can be used to build an expert system. Indeed a $WMCS_p(G)$ not only represents the common properties found in a cluster of graphs, but these properties can be easily displayed and interpreted by an expert of the underlying domain. In other words, the knowledge of the system is not hidden, but is easily interpretable by an expert.

4 WMCS Approximation

The proposed definition of $WMCS$ is expensive to compute. Indeed the computation of a $WMCS$ on a pair of graph is an NP-complete problem, that is exponential in the number of nodes of one of the two given graphs. Moreover the complexity of computing a $WMCS$ of a set of graphs is exponential in the number of graphs. Consequently we propose a procedure to approximately compute a $WMCS$ of a set of graphs G.

Firstly, it is necessary to extend the definition of the $WMCS$ to introduce the $WMCS$ of a pair of weighted graphs. Similarly to the MCS of a pair of graphs, that can be derived from the mcs, also the $WMCS$ of a pair of weighed graphs can be derived from the $wmcs$, i.e. the *weighted maximum common subgraph*. Both $wmcs$ and $WMCS$ of a pair of weighted graphs are defined below.

Def. 4.1: Let $g_1 = (V_1, \lambda_1, E_1, \varepsilon_1, \alpha_1, \beta_1)$ and $g_2 = (V_2, \lambda_2, E_2, \varepsilon_2, \alpha_2, \beta_2)$ be two weighted graphs. A *weighted common subgraph* of g_1 and g_2, $wcs\ (g_1, g_2)$, is a weighted graph $g=(V,\lambda,E,\varepsilon,\alpha,\beta)$ such there exist subgraph isomorphisms from g to g_1 and from g to g_2. We call g a *weighted maximum common subgraph* of g_1 and g_2, $wmcs(g_1, g_2)$, if there exists no other common subgraph of g_1 and g_2 that has more nodes than g. Let m be the size of $wcs(g_1,g_2)$, and let $V=\{v_1,...,v_m\}$ the set of nodes of g. Furthermore let $V_i =$

$\{v'_{i1},...,v'_{im}\} \subseteq V$, $i \in \{1,2\}$ be the subsets of nodes corresponding to $wcs(g_1,g_2)$. There exist subgraphs isomorphisms $f_i(v_j):V\rightarrow V'_i$, $\forall i \in \{1,2\}, \forall j \in \{1,...,m\}$. The weight of each node v_j is $\lambda(v_j) = \lambda_1(f_1(v_j)) + \lambda_2(f_2(v_j))$. Let $f_i(e_{jk}) = (f_i(v_j), f_i(v_k))$, $i \in \{1,2\}$ and $\forall j, \forall k \in \{1,...,m\}$. The weight of each edge e_{jk} is $\varepsilon(e_{jk})=\varepsilon_1(f_1(e_{jk}))+ \varepsilon_2(f_2(e_{jk}))$.

The computation of $wmcs(g_1,g_2)$, is based on for the computation of the *mcs* of two graphs. For this aim a version of McGregor algorithm [12] has been used.

Def. 4.2: Let g_1 and g_2 be weighted graphs. A *Weighted Common Supergraph* of g_1 and g_2, $WCS(g_1,g_2)$, is a weighted graph g such that there exist subgraph isomorphisms from g_1 to g and from g_2 to g. We call g a *Weighted Minimum Common Supergraph* of g_1 and g_2, $WMCS(g_1,g_2)$, if there exists no other weighted common supergraph of g_1 and g_2 that has less nodes than g.

Theorem 2.1 can be also extended in case of weighted graphs, obtaining a procedure for constructing $WMCS(g_1,g_2)$. The weights of $WMCS(g_1,g_2)$ are defined according to the construction procedure of Fig 4.1.

```
input: g₁,g₂; output: WMCS(g₁,g₂);
procedure WMCS(g₁,g₂)
begin
    E₁ = emb(wmcs(g₁,g₂),g₁); E₂ = emb(wmcs(g₁,g₂),g₂)
    WMCS(g₁,g₂) =
        wmcs(g₁,g₂)∪E₁(g₁-wmcs(g₁,g₂))∪E₂(g₂-wmcs(g₁,g₂));
end procedure
```

Fig. 4.1. Construction procedure for the weights of $WMCS(g_1,g_2)$.

Firstly, each graph of the set G is transformed in a weighted graph, assigning to each node (edge) the weight 1. Then an ordering θ is chosen for the n graphs of the set G: $\theta = ord(G)$. After this step the procedure $WMCS(g_1,g_2)$ is iterated n-1 times for computing an approximated $WMCS(G)$. This procedure is shown in Fig 4.2.

```
procedure WMCS(θ)
input: g₁,…,gₙ,θ output: WMCS(θ);
begin
    assign to each node and edge the weight 1;
    order the graphs according to θ;
    W = g₁ ;
    for i = 2 to n
    begin
        W = WMCS(W,gᵢ);
    end for
    WMCS(θ) = W;
end procedure
```

Fig. 4.2. Linear approximation of $WMCS(G)$

The main advantage of this approximate procedure is that the complexity of the computation becomes linear in the number of graphs. Using this procedure, the built graph is surely a common supergraph, but the main disadvantage is that in most cases it is not minimal, i.e. it will usually include some extra nodes. It is also possible that the weight of some node has not anymore the proper value but it becomes smaller. Indeed, it can happen that a node already present in the graph under construction is not recognized and thus it is inserted once more. The weight is updated, for each iteration of the procedure, on only one of the two copies of the node, thus it is not guaranteed that at the end of the procedure the weight will be correct. More in details, it is only ensured that the sum of the weights of all the copies of a node will be equal to the correct value. In general some nodes could have a weight smaller than the real occurrence of the property in the cluster. As a consequence, the capability of the approximated WMCS to learn from examples is reduced, since some more node is introduced, and the occurence of the some real property is not always correctly represented, but could be underestimated. This phenomenon can be interpreted as a kind of hypergeneralization.

A possibility to reduce the error generated by the proposed procedure, is to perform a random search algorithm. The first step is to realize a number m of different randomly generated orderings $\theta_1,...,\theta_m$ of the graphs of the cluster (*shuffles*). For each shuffle θ_j, the $WMCS(\theta_j)$ have to be evaluated and then the best approximation is chosen.

The quality of the approximation is measured through an entropy function $E(WMCS(\theta_j))$:

$$E = -\sum_{i=1}^{t} \frac{w_i}{n} \cdot \log\left(\frac{w_i}{n}\right) - \sum_{i=1}^{t} \sum_{j=1}^{t} \frac{w_{ij}}{n} \cdot \log\left(\frac{w_{ij}}{n}\right) \qquad (2)$$

where n is the number of graphs of the cluster, t is the number of nodes of the graph $WMCS(\theta_j)$, w_i is the weight of the node i and w_{ij} is the weight of the edge (i,j). Each weight is associated to the occurrences of the node (edge) in the graph, thus the ratio w_i/n (w_{ij}/n) can be interpreted as a probability. The minimum of E is obtained if the cluster is composed of all isomorphic graphs, while the maximum is obtained if there is no overlap between any pair of graphs of the cluster. The $WMCS(\theta_j)$ minimizing the entropy will be assumed as the best $WMCS(\theta_j)$.

5 Experimental Results

The aim of the first experiment, is the evaluation of the quality of the WMCS approximation for representing the cluster G. For this aim, let us suppose that the graphs of a cluster G represent the same pattern with different distortions, due, for instance, to a not optimal system used to obtain the graph representations from real objects.

Firstly an attributed graph that we call *seed* is generated. This graph is a *mesh* [2,9] in which the number of nodes is chosen and the size of the alphabet of attributes is fixed. The method for generating this graph is explicated in details in [1]. A degree of distortion is added to the seed, using a graph distortion algorithm, obtaining a second graph. This operation is repeated k times, thus a cluster of k distorted graphs is obtained. These k graphs are the elements of the cluster G. The seed is not included in the cluster, but it is only used in the generation process. In our experiments a distortion is defined as the substitution of an attribute of a node (edge) with another attribute randomly chosen from the same alphabet, the considered alphabet has $M = 50$ different symbols. For each node (edge) of the seed a probability p of distortion is defined. For instance, if $p = 0.3$, this means that the probability that the attribute of each node (edge) of the seed is changed is 0.3. Experiments are repeated with 1, 10 and 100 shuffles.

In Fig.5.1 the size of the approximated WMCS is shown when the distortion increase. Obviously, when the graph distortion increases, also the WMCS size increases, due the smaller overlap of the graphs in the cluster. It is worth noting that there is no significant improvement when the number of shuffles of the graphs is increased. Moreover, it is interesting to notice, in Fig.5.2, that the maximum weight of the approximated WMCS, is increasing when the number of shuffles increases form 1 to 10, thus a better approximation is obtained, but only a small further improvement is achieved when the number of shuffles becomes 100. From our experiments, it follows that a good approximation of the WMCS can be obtained with a small number of shuffles, and further shuffles do not cause very significant variations. We assume that using shuffles, the algorithm can avoid most of the local maxima; the fact that increasing the number of shuffles there is only a small improvement in the approximation, can be an indication that the approximate WMCS is not too far from the exact one. Finally it should be pointed out that, when there is a significant overlap in the graphs of the cluster, the WMCS is very compact.

A second experiment has been conducted to investigate if WMCS is suitable to perform graph clustering tasks. In this experiment, patterns are graphs extracted from the pictures obtained using the plex grammar tool [5] (see Fig.5.3).

Each component of the picture is represented using a node. If a picture represents, for instance, a man, then there will be a node representing the hat, another node representing the head and so on. On each node one attribute is used: the area of the bounding box of the component. The area is normalized between 0 and 99 and two nodes are assumed to have the same area, if the difference between the two areas is smaller than 5. Nodes are connected through edges. Each edge has two attributes: sine and cosine of the angle between the segment connecting the centers of the two components connected by the edge and the horizontal line. Both the two attributes are normalized between 0 and 19 and two attributes are assumed to have the same value if their difference is smaller than 2. It is interesting to notice that only a part of the information of a picture is stored in the graph: for instance no information on the colors and on the shapes of the components are taken into account. The classification is obtained only using areas of the components and their relative positions.

Three different classes of patterns are considered: men, ships and houses. For each class, 10 patterns are considered and the WMCS is computed, using 100 shuffles.

What we expect is that the entropy of the WMCS has a small variation if a new pattern, similar to the patterns of the class, is added to the cluster. Indeed, in this case, weights of those nodes (edges) with larger frequencies will be incremented. Conversely if a pattern very dissimilar from the patterns of the class is added to the cluster, we expect a larger entropy variation, because nodes (edges) with small frequencies will be added to the *WMCS*. In Tab.5.1 results are summarized. Each column represents the entropy variation ΔE of a *WMCS* when a new pattern (i.e. different from the 10 patterns used to built the *WMCS*) is added to a cluster. It is noteworthy that, for each row, the minimum ΔE is obtained when a new pattern of the same class of the *WMCS* is added to the cluster. This result confirms that *WMCS* can be used to distinguish patterns of a given class from patterns of other classes. Results shown in Tab.5.1 are in average: the insertion of a new pattern in a cluster has been repeated 100 times for each category.

In Tab.5.2 the classification results (in percentage) are shown. Each row summarizes the classification of patterns belonging to a class, i.e. the second row represents the classification results for the houses. Results are obtained using a nearest-neighbour criterion, i.e. firstly a pattern is assigned to each *WMCS*, then the entropy variation ΔE is evaluated and finally the pattern is assigned to the class minimizing ΔE. Non-diagonal elements represent misclassifications.

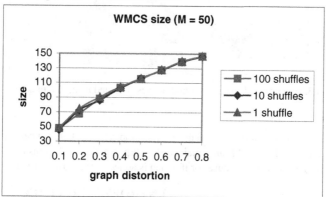

Fig. 5.1. The size of the approximated *WMCS* increase for an increasing distortion of the graphs. It is worth noting that a very small improvement can be obtained raising the number of shuffles of the graphs. Results are in average, each computation has been repeated 100 times.

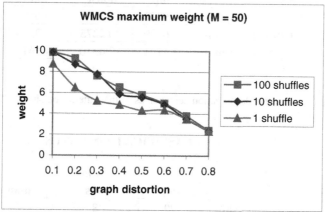

Fig. 5.2. When the distortion of the graph increases, then the maximum weight of the approximated *WMCS* decrease, due a smaller overlap of the graphs. Moreover, if more then 10 shuffles are produced, than only a small improvement can be obtained in the approximation. Results are in average, each computation has been repeated 100 times.

Fig. 5.3. Examples of men, ships and houses obtained using the plex grammar tool.

Table 5.1. Each column represents the entropy variation ΔE of a *WMCS* when a pattern is added to a cluster. Lower values of ΔE are obtained if a ship is added to the cluster of ships, or if an house is added to the cluster of houses or if a man is added to the cluster of men.

ENTROPY VARIATION

		Element		
		ship	house	man
	ship	**0.9922**	1.913	3.6501
Cluster	house	3.6129	**0.2275**	4.3503
	man	2.4141	2.4310	**1.1613**

Table 5.2. In each row the classification percentages of the patterns are shown. Bold values are the correct classifications.

CLASSIFICATION RATES

		Class		
		ship	house	man
	ship	**90**	8	2
Element	house	0	**100**	0
	man	0	0	**100**

It is worth notice that for *houses* and *men* there is no misclassification, even if no information is used for describing shapes and colours in the classification system. The worst case is for ships: 8% of ships are classified as houses.

6 Conclusions and Perspectives

Clustering has become a mature discipline that is important to a number of areas, including pattern recognition and related fields. But the clustering of graphs is still widely unexplored. In this paper a new method for representing a cluster of graph, the *WMCS*, has been described. The main advantage of using *WMCS* for representing a cluster is that the structural information contained into the cluster is preserved and that is also easy to separate this information from the noise. The main disadvantage is the high computational complexity of the method, thus an approximation is proposed. The accuracy of the proposed approximation is dependent on the chosen number of orderings of the elements of the cluster. Experiments have been realized to describe the behavior of *WMCS* when the number of orderings is changed and to describe the behavior of *WMCS* when the noise of the cluster is increasing. Preliminary test shows that a good *WMCS* approximation can be obtained with a small number of orderings of the set of graphs. Moreover, an entropy function $E(WMCS)$ has been defined and the variation ΔE has been used to perform graph clustering on a three classes problem. Preliminary tests show that *WMCS* is a suitable approach for performing graph clustering even if the representation is rough and doesn't contain all the information of the original patterns.

References

[1]   H. Bunke, P. Foggia, C. Guidobaldi, C. Sansone, M. Vento: A Comparison of Algorithms for Maximum Common Subgraph on Randomly Connected Graphs SSPR/SPR2002, pp 123–132. 2002.

[2]   H. Bunke, M. Vento, Benchmarking of Graph Matching Algorithms, Proc. 2nd Workshop on Graph-based Representations, Haindorf, pp. 109–114, 1999.

[3]   H. Bunke, X. Jiang, A. Kandel, On the Minimum Common Supergraph of two Graphs, Computing 65, No. 1, pp. 13–25, 2000.

[4]   R.O. Duda, P.E. Hart, D.G. Stork. Pattern Classification, II Ed., Wiley-Interscience Publication. 2000.

[5]   M. Hagenbuchner, A.C. Tsoi, The Traffic Policeman Benchmark. European Symposium on Artificial Neural Networks, Michel Verleysen, D-Facto, pp. 63–68, April, 1999.

[6]   C. De Mauro, M. Diligenti, M. Gori, M. Maggini, Similarity Learning for Graph-Based Image Representations. Proceedings of the 3rd IAPR-TC15 Workshop on Graph-based Representations in Pattern Recognition, pp. 250–259. 2001.

[7]   R. Englert, R. Glantz, Towards the Clustering of Graphs. Proceedings of the 2nd IAPR-TC15 Workshop on Graph-based Representations in Pattern Recognition, pp. 125–133. 1999.

[8]   S. Günter, H. Bunke, Self-organizing Map for Clustering in the Graph Domain, Pattern Recognition Letters 23, pp. 401–417. Accepted for Publication.

[9]   P. Foggia, C. Sansone, M. Vento, A Database of Graphs for Isomorphism and Sub-Graph Isomorphism Benchmarking, Proc. of the 3rd IAPR TC-15 International Workshop on Graph-based Representations, Italy, pp. 176–187, 2001.

[10] B. Luo, A. Robles-Kelly, A. Torsello, R.C. Wilson, E. R. Hancock, Clustering Shock Trees. Proceedings of the 3rd IAPR-TC15 Workshop on Graph-based Representations in Pattern Recognition, pp. 217–228. 2001.

[11] B. Luo, R.C. Wilson, E. R. Hancock, Spectral Feature Vectors for Graph Clustering. Proceeding of the Joint IAPR International Workshops SSPR and SPR, pp. 83–93. 2002.

[12] J.J. McGregor, Backtrack Search Algorithms and the Maximal Common Subgraph Problem, Software Practice and Experience, Vol. 12, pp. 23–34, 1982.

[13] D. Seong, H. S. Kim, K.H. Park, Incremental clustering of attributed graphs. IEEE Transactions on Systems, Man and Cybernetics. Vol. 23 - 5, pp. 1399–1411. 1993.

[14] A. Sanfeliu, F. Serratosa, R. Alquezar, Clustering of attributed graphs and unsupervised synthesis of function-described graphs. Proceedings of 15th International Conference on Pattern Recognition, pp. 1022–1025. 2000.

[15] A. Sanfeliu, F. Serratosa, R. Alquezar, Synthesis of Function-Described Graphs and Clustering of Attributed Graphs. International Journal of Pattern Recognition and Artificial-Intelligence, Vol.16, No.6, pp. 621–655. 2002.

[16] A.K.C. Wong, J. Constant, M.L. You, Random Graphs, in Syntactic and Structural Pattern Recognition - Theory and Applications 4:197–234, H. Bunke and A. Sanfeliu (Eds.), World Sci. Pub. Co., 1990.

ACM Attributed Graph Clustering for Learning Classes of Images

Miguel Angel Lozano and Francisco Escolano

Robot Vision Group
Departamento de Ciencia de la Computación e Inteligencia Artificial
Universidad de Alicante, Spain
{malozano,sco}@dccia.ua.es
http://rvg.ua.es

Abstract. In a previous work we have adapted the Asymmetric Clustering Model (ACM) to the domain of non-attributed graphs. We use our Comb algorithm for graph matching, a population-based method which performs multi-point explorations of the discrete space of feasible solutions. Given this algorithm we define an incremental method to obtain a prototypical graph by fusing the elements of the ensemble weighted by their prior probabilities of belonging to the class. Graph-matching and incremental fusion are integrated in a EM clustering algorithm. In this paper, we adapt the latter ACM clustering model to deal with attributed graphs, where these attributes are probability density functions associated to nodes and edges. In order to do so, we modify the incremental method for obtaining a prototypical graph to update these pdf's provided that they are statistically compatible with those of the corresponding nodes and edges. This graph-clustering approach is successfully tested in the domain of Mondrian images (images built on rectangular patches of colored textures) because our final purpose is to the unsupervised learning of image classes.

1 Introduction

Our interest in methods for clustering ensembles of graphs is partially motivated by the development of structural methods for learning classes of images. In [9], we proposed a method for integrating image segmentation (texture and color) and graph matching for recognizing indoor images. Image regions were inferred with an adaptive version of the ACM clustering algorithm developed Hoffman and Puzicha [4] [11]. After segmentation, adjacency graphs were built and matched with our Comb algorithm, which basically optimizes the Gold and Rangarajan's cost function [3] with attributes by embodying multi-point local search inside a global random algorithm that explores the discrete space of feasible solutions. The Comb strategy (Common structure of best local maxima) was originally applied to solve MRF-labeling problems [8].

Our preliminary recognition experiments were very promising in terms of defining a distance between images relying both structural and appearance dissimilarities. Then, we addressed the problem of using such a distance in order

E. Hancock and M. Vento (Eds.): IAPR Workshop GbRPR 2003, LNCS 2726, pp. 247–258, 2003.

to effectively cluster the images. In order to do so, we adapted to the domain of graphs the extended ACM clustering algorithm used for segmentation [10]: In the E-step it is assumed that class prototypes are known and the distances between each graph in the ensemble and each prototype are computed on the basis of the Comb-matching algorithm. Such computations yield the re-estimation of the class-membership variables, and in turn those variables are used in the M-step to re-estimate the prototypes. The M-step relies on an incremental algorithm that builds the prototype of each class by considering the current probabilities of belonging to each class. Finally, adaptation is implemented by dividing the computation in epochs. Starting from a high number of classes, as in [2], in each epoch we try to merge the two classes with closer prototypes.

However, in the latter EM algorithm we did not consider attributed graphs, because we were interested in evaluating our method when only pure structural information were used. However, in the present work we incorporate node and edge attributes because these attributes, specially those referred to image regions, were key to image clustering. Although our model is related to the random graph approach for building prototypes [14][15][16] or to its first-order variants [13][6] [12][1], our notion of graph prototype is closer to the median-graph model [5]. From our point of view, when input graphs are not attributed, a graph prototype registers the sub-structures with higher frequencies in the ensemble, and latter discards those with low probability. For instance, if many instances of chairs graphs are three-legged and only few are four legged, the prototype will assume that the chair class is three-legged. Extending this approach to the attributed case, we assume that nodes and edges attributes are probability-density functions (pdf's) and we consider that a sub-structure is repeated if their pdf's are statistically compatible with those of another isomorphism sub-structure. If compatibility arised these pdf's are updated conveniently. Finally, only nodes and edges whose pdf's are repeatedly compatible are retained in the prototype.

Translating this approach to the problem of learning image classes, we retain as belonging to the prototype those regions which are very frequent, not only in terms of their appearance, but in terms of their local structure. Consequently, spurious noisy regions are removed from the prototype.

The paper is organized as follows: In Section 2 we will describe a simple incremental method for building prototypes by considering nodes and pdf's as attributes, and we illustrate how it works with synthetic Mondrian images [7]. We consider different patterns for providing inter-class variability. In order to introduce intra-class variability we generate structural variations (by randomly inserting regions) and appearance variations (by introducing Gaussian noise) of each basic pattern. In Section 3 we will present how our ACM clustering algorithm for graphs works, and we apply it to Mondrian images after mapping to their regions a limited set of coloured textures. We follow the same method for introducing variability. Finally, in Section 4 we will summarize our clustering results and we will outline our future work.

2 Graph Prototypes

2.1 Incremental Fusion

Given N input attributed graphs $G_i = (V_i, E_i)$, whose nodes $a \in V_i$ and edges $(a, b) \in E_i$ are random variables defined respectively over D_V and D_E, we register their probability density functions (pdf's)

$$\Gamma_i(a) : D_V \to [0, 1] \text{ and } \Delta_i(a, b) : D_E \to [0, 1], \tag{1}$$

and we want to obtain a representative prototype $\bar{G} = (\bar{V}, \bar{E})$ of them, including the pdf's of all its nodes and egdes. In order to do so, we define such a prototype in terms of the following operators

$$\bar{G} = \bigoplus_{i=1}^{N} (\pi_i \odot G_i) \text{ where } \sum_{i=1}^{N} \pi_i = 1. \tag{2}$$

Firstly, the $\odot$ operator implements the weighting of each graph G_i by π_i which we will interpret in terms of the probability of belonging to the class defined by the prototype. The definition of resulting graph $\tilde{G}_i = \pi_i \odot G_i$, with $\tilde{V}_i = V_i$ and $\tilde{E}_i = E_i$ is extended in order include a node-weighting function $\tilde{\mu}_i : \tilde{V}_i \to [0, 1]$, addressed to register the relative frequency of each node, and an edge-weighting function $\tilde{\delta}_i : \tilde{E}_i \to [0, 1]$ which must register the relative frequency of each edge. Then, weighting consists of assigning the following priors:

$$\tilde{\mu}_i(a) = \pi_i \ \forall a \in \tilde{V}_i \text{, and } \tilde{\delta}_i(a, b) = \pi_i \ \forall (a, b) \in \tilde{E}_i, \tag{3}$$

and also of properly weighting the pdf's of all nodes and edges

$$\tilde{\Gamma}_i(a) = \pi_i \Gamma_i(a) \ \forall a \in \tilde{V}_i \text{, and } \tilde{\Delta}_i(a, b) = \pi_i \Delta_i(a, b) \ \forall (a, b) \in \tilde{E}_i. \tag{4}$$

Secondly, given a set of weighted graphs, the $\bigoplus$ operator builds the prototype $\bar{G}$ by fusing all of them. Ideally, the result should not depend on the order of presentation of the weighted graphs. However, such a solution would be too computational demanding because we need to find a high number of isomorphisms, that is, to solve a high number of NP-complete problems. In this work we propose an incremental approach. Initially we select randomly a weighted graph $\tilde{G}_X = \pi_X \odot G_X$, as a temporary prototypical graph $\bar{G}$. Then we select, also randomly a different graph $\tilde{G}_Y = \pi_Y \odot G_Y$. The new temporary prototype results from fusing these graphs:

$$\bar{G} = \tilde{G}_X = \pi_X \odot G_X$$
$$\bar{G} = \tilde{G}_X \oplus \tilde{G}_Y = (\pi_X \odot G_X) \oplus (\pi_Y \odot G_Y)$$
$$\cdots$$

An individual fusion $\tilde{G}_X \oplus \tilde{G}_Y$ relies of finding the matching M maximizing

$$\tilde{F}(\tilde{G}_X, \tilde{G}_Y; M) = \sum_{a=1}^{|\tilde{V}_X|} \sum_{i=1}^{|\tilde{V}_Y|} w_{ai} M_{ai} \sum_{b=1}^{|\tilde{V}_X|} \sum_{j=1}^{|\tilde{V}_Y|} M_{bj} \tilde{X}_{ab} \tilde{Y}_{ij} \tilde{P}_{ai} \tilde{Q}_{abij} . \qquad (5)$$

Firstly, $\tilde{X} = X$, $\tilde{Y} = Y$, are the respective adjacency matrices and M is a matching matrix which encodes a global solution ($M_{ai} = 1$ if a and i match and 0 otherwise) and must satisfy all the row and column constraints

$$\sum_{i=1}^{|\tilde{V}_X|} M_{ai} \leq 1, \forall a \text{ and } \sum_{a=1}^{|\tilde{V}_Y|} M_{ai} \leq 1, \forall i . \qquad (6)$$

Secondly, $\tilde{P}_{ai}$ and $\tilde{Q}_{abij}$ are look-up tables registering the degree of compatibility between the pdfs of pairs of nodes and edges, respectively. For nodes,

$$\tilde{P}_{ai} = \begin{cases} \exp\{-KL(\tilde{\Gamma}_X(a), \tilde{\Gamma}_Y(i))\} & \text{if } KS_\alpha(\tilde{\Gamma}_X(a), \tilde{\Gamma}_Y(i)) \text{ fails} \\ -\infty & \text{otherwise} \end{cases}$$

where $KL(.,.)$ is the symmetric Kullback-Leibler divergence between the pdf's figuring as arguments, and $KS_\alpha(.,.)$ is the result of a Kolmogorov-Smirnov test with $\alpha = 0.05$ (if both pdf's are statistically compatible the test fails). On the other hand, edge-compatibility $\tilde{Q}_{abij}$ is defined in a similar way, simply by replacing the node's pdf's by the edges' ones.

Finally, $w_{ai} = (1 + \epsilon(\tilde{\mu}_X(a) + \tilde{\mu}_Y(i))$, being $\epsilon > 0$ a small positive constant closer to zero (in our implementation $\epsilon = 0.0001$) is a weight addressed to deal with ambiguity: the $\tilde{F}(\tilde{G}_X, \tilde{G}_Y; M)$ cost is an extension of the Gold-Rangarajan one

$$F(\tilde{G}_X, \tilde{G}_Y; M) = \sum_{a=1}^{|\tilde{V}_X|} \sum_{i=1}^{|\tilde{V}_Y|} M_{ai} \sum_{b=1}^{|\tilde{V}_X|} \sum_{j=1}^{|\tilde{V}_Y|} M_{bj} \tilde{X}_{ab} \tilde{Y}_{ij} \tilde{P}_{ai} \tilde{Q}_{abij} , \qquad (7)$$

and if there are many matchings equal in terms of $F(\tilde{G}_X, \tilde{G}_Y; M)$, the alternative in which the nodes with the higher weights so far are matched is preferred.

Following the latter definitions, the fused graph, that is, the new prototype $\bar{G} = \tilde{G}_X \oplus \tilde{G}_Y$ registers the weights of its nodes and edges through its weighting functions $\bar{\mu} : \bar{V} \to [0, 1]$ and $\bar{\delta} : \bar{E} \to [0, 1]$. Such a registration relies on the optimal match

$$M^* = \arg \max_M \tilde{F}(\tilde{G}_X, \tilde{G}_Y; M) , \qquad (8)$$

subject to the row and column constraints. We obtain such a optimal match through our Comb-matching algorithm. Thereafter, the weights of the nodes $m \in \bar{V}$ are updated as follows:

$$\bar{\mu}(m) = \begin{cases} \tilde{\mu}_X(a) + \tilde{\mu}_Y(i) & \text{if } (M^*_{ai} = 1) \text{ and } (m = a = i) \\ \tilde{\mu}_X(a) & \text{if } (M^*_{ai} = 0 \ \forall i) \text{ and } (m = a \neq i) \\ \tilde{\mu}_Y(i) & \text{if } (M^*_{ai} = 0 \ \forall a) \text{ and } (m = i \neq a), \end{cases} \qquad (9)$$

Similarly, the weighs of the edges $(a, b) \in \bar{E}$ are updated by

$$\bar{\delta}(m, n) = \begin{cases} \tilde{\delta}_X(a, b) + \tilde{\delta}_Y(i, j) & \text{if } (M_{ai}^* = M_{bj}^* = 1) \\ & \text{and } ((m, n) = (a, b) = (i, j)) \\ \tilde{\delta}_X(a, b) & \text{if } ((m, n) = (a, b) \neq (i, j)) \\ \tilde{\delta}_Y(i, j) & \text{if } ((m, n) = (i, j) \neq (a, b)). \end{cases} \quad (10)$$

Consequently, weights are updated by considering that: (i) two matched nodes correspond to the same node in the prototype; (ii) edges connecting matched nodes are also the same edge; and (iii) nodes and edges existing only in one of the graphs must be also integrated in the prototype. In cases (i) and (ii) frequencies are added, whereas in case (iii) we retain the original frequencies.

Finally, the pdf's of the nodes in the temporary prototype are given by

$$\bar{\Gamma}(m) = \begin{cases} \pi_X \tilde{\Gamma}_X(a) + \pi_Y \tilde{\Gamma}_Y(i) & \text{if } (M_{ai}^* = 1) \text{ and } (m = a = i) \\ \pi_X \tilde{\Gamma}_X(a) & \text{if } (M_{ai}^* = 0 \ \forall i) \text{ and } (m = a \neq i) \\ \pi_Y \tilde{\Gamma}_Y(i) & \text{if } (M_{ai}^* = 0 \ \forall a) \text{ and } (m = i \neq a), \end{cases} \quad (11)$$

and the pdf's of the edges are updated in a similar way

$$\bar{\Delta}(m, n) = \begin{cases} \pi_X \tilde{\Delta}_X(a, b) + \pi_Y \tilde{\Delta}_Y(i, j) & \text{if } (M_{ai}^* = M_{bj}^* = 1) \\ & \text{and } ((m, n) = (a, b) = (i, j)) \\ \pi_X \tilde{\Delta}_X(a, b) & \text{if } ((m, n) = (a, b) \neq (i, j)) \\ \pi_Y \tilde{\Delta}_Y(i, j) & \text{if } ((m, n) = (i, j) \neq (a, b)). \end{cases} \quad (12)$$

As it may occur that the resulting pdf's do no add up to the unit (for instance when a given node or edge is discarded in one or more isomorphisms) we proceed to re-normalize the resulting pdf's once all graphs in the set are fused.

After re-normalization, the resulting prototype $\bar{G}$ must be pruned in order to retain only those nodes $a \in \bar{V}$ with $\bar{\mu}(a) \geq 0.5$ and those edges $(a, b) \in \bar{E}$ also with $\bar{\delta}(a, b) \geq 0.5$. This results in simpler median graphs containing nodes and edges with significant frequencies.

2.2   Fusion Results

In Figure 1, we illustrate how an attributed graph prototype is built. In the left column we show the input Mondrian images, assumed to belong to the same class (all of them with unitary weights). These images encode the Hue component of pure colors and they are corrupted with a Gaussian noise with $\sigma = 0.1$ being the intensity range $[0, 1]$. In the central column we show their corresponding attributed graphs, displaying their pdfs (only node pdf's are considered in this paper). In all cases the Kolmogorov-Smirnov test considers an error of $\alpha = 0.05$ and histograms of $n = 16$ bins. In the right column we show the temporary prototypes by fusing the graph corresponding to a given row with the temporary prototype so far. Nodes and edges with low final probabilities will be deleted. In Figure 2 we show the prototypical graph and its corresponding prototypical image.

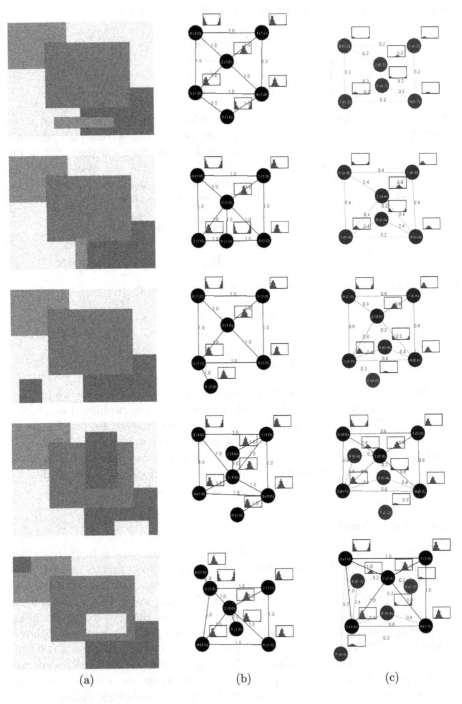

Fig. 1. Incremental fusion (upside-down). (a) Input images; (b) Attributed graphs; (c) Building a prototype step-by-step.

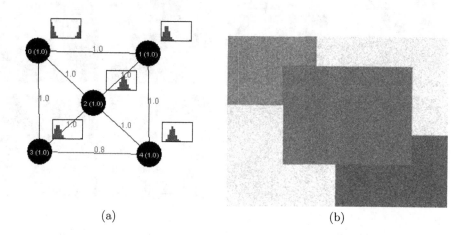

(a)   (b)

Fig. 2. Results of incremental fusion. (a) Prototype after removing low probability nodes and edges; (b) Resulting prototypical image.

3   Graph Clustering

3.1   ACM for Graphs

Given N input graphs $G_i = (V_i, E_i)$ the Asymmetric Clustering Model (ACM) for graphs finds the K graph prototypes $\bar{G}_\alpha = (\bar{V}_\alpha, \bar{E}_\alpha)$ and the class-membership variables $I_{i\alpha} \in \{0, 1\}$ maximizing the following cost function

$$L(\bar{G}, I) = -\sum_{i=1}^{N} \sum_{\alpha=1}^{K} I_{i\alpha}(1 - F_{i\alpha}) , \qquad (13)$$

where $F_{i\alpha}$ are the values of a symmetric and normalized similarity measure between individual graphs G_i and prototypes $\bar{G}_\alpha$. Such a measure is defined by

$$F_{i\alpha} = \frac{\max_M [F(G_i, \bar{G}_\alpha; M)]}{\max[F_{ii}, F_{\alpha\alpha}]} , \qquad (14)$$

where $F_{ii} = F(G_i, G_i; I_{|V_i|})$, $F_{\alpha\alpha} = F(\bar{G}_\alpha, \bar{G}_\alpha; I_{|\bar{V}_\alpha|})$, being $I_{|V_i|}$ and $I_{|\bar{V}_\alpha|}$ the identity matrices defining self-matchings. This result in $F_{i\alpha} = F_{\alpha i} \in [0, 1]$.

Prototypical graphs are built on all individual graphs assigned to each class, but such an assignment depends on the membership variables. Here we adapt to the domain of graphs the EM-approach proposed by Hoffman and Puzicha. As in such approach the class-memberships are hidden or unobserved variables, we start by providing good initial estimations of both the prototypes and the memberships, feeding with them an iterative process in which we alternate the estimation of expected memberships with the re-estimation of the prototypes.

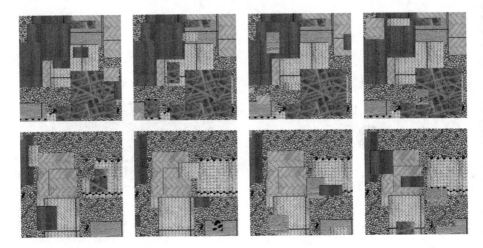

Fig. 3. Mondrian textured images corresponding to two different classes.

Initialization. Initial prototypes are selected by a greedy procedure: First prototype is assumed to be a graph selected randomly, and the following ones are the most dissimilar graphs from any of the yet selected prototypes. Initial memberships $I_{i\alpha}^0$ are then given by:

$$I_{i\alpha}^0 = \begin{cases} 1 & \text{if } \alpha = arg\min_\beta[1 - F_{i\beta}] \\ 0 & \text{otherwise} \end{cases}$$

E-step. Consists of estimating the expected membership variables $I_{i\alpha} \in [0,1]$ given the current estimation of each prototypical graph $\bar{G}_\alpha$:

$$I_{i\alpha}^{t+1} = \frac{\rho_\alpha^t \exp[(1 - F_{i\alpha})/T]}{\sum_{\beta=1}^{K} \rho_\beta^t \exp[(1 - F_{i\beta})/T]} \text{ , being } \rho_\alpha^t = \frac{1}{N} \sum_{i=1}^{N} I_{i\alpha}^t \text{ ,} \tag{15}$$

that is, these variables encode the probability of assigning any graph G_i to class c_α at iteration t, and T is the temperature, a control parameter which is reduced at each iteration (we are using the deterministic annealing version of the E-step, because it is less prone to local maxima than the un-annealed one).

As in this step we need to compute the similarities $F_{i\alpha}$, we need to solve $N \times K$ NP-complete problems.

M-step. Given the expected membership variables $I_{i\alpha}^{t+1}$, the prototypical graphs are re-estimated as follows:

$$\bar{G}_\alpha^{t+1} = \bigoplus_{i=1}^{N} (\pi_{i\alpha} \odot G_i) \text{ , where } \pi_{i\alpha} = \frac{I_{i\alpha}^{t+1}}{\sum_{k=1}^{N} I_{k\alpha}^{t+1}} \text{ ,} \tag{16}$$

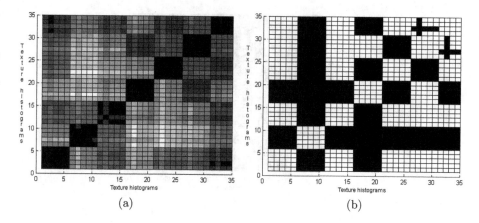

Fig. 4. Similarities between pair of textures. (a) Symmetric Kullback-Leibler divergences. (b) Results of the Kolmogorov-Smirnov.

where variables $\pi_{i\alpha}$ are the current probabilities of belonging to each class c_α. This step is completed after re-estimating the K prototypes, that is, after solving $(N-1) \times K$ NP-complete problems. Moreover, the un-weighted median graph $\bar{G}_\alpha^{t+1}$ resulting from the fusion will be used in the E-step to re-estimate the similarities $F_{i\alpha}$ through maximizing $F(G_i, \bar{G}_\alpha; M)$ with Comb-matching.

3.2 Clustering Results

First of all, in Figure 3 we show some images of our colored-textured Mondrian database (5 classes, each one with 10 examples). Each example is built by mapping one of the 35 available patterns (7 different colored textures corrupted with 5 different distortions including Gaussian noise ($\sigma = 0.05$ in an intensity range of $[0, 1]$), slight horizontal and vertical waving, and Gaussian smoothing ($\sigma = 1.0$). Given these textures, we compute both color features (Hue) and texture features (4 Gabor filters with $\sigma = 2.0$ and orientations $0°, 45°, 90°, 135°$). We have built histograms with 16 components for each feature, and thus we have global histograms of 80 components, which are the pdf's of our regions and nodes.

In Figure 4a, we show the symmetric Kullback-Leibler divergences between pairs of textures (35×35 comparisons): The darker the color the closer the divergences. Basically, textures belonging to the same class are closer (see the blocks in the diagonal) than textures belonging to different classes. In Figure 4b we show the results of the Kolmogorov-Smirnov tests ($\alpha = 0.05$) between pairs of textures: white blocks indicate that the tests fail, whereas black blocks (forbidden assignements, that is, $-\infty$ compatibility, in graph matching) indicate that the tests succeed. Consequently, our experimental set is designed in order to test the behavior of the clustering algorithm when appearance is quite ambiguous. We expect good results in these conditions because in [10] we obtained good results with purely structural graphs.

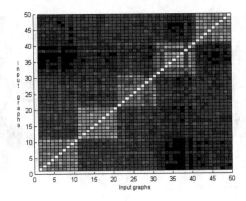

Fig. 5. Optimal similarities F_{ij} between pairs of graphs.

We generated randomly 5 Mondrian patterns with 20 rectangular patches whose weight and height are between the 25% and the 50% of the image size. We randomly mapped one of the 7 uncorrupted textures to each patch. Then, for each of patterns we generated 10 variations by inserting smaller rectangular patches whose weight and height are between the 5% and the 25% of the image size. We randomly mapped one of the 5 corrupted variants of the corresponding uncorrupted texture to each patch.

Given the latter texture-based Mondrian images we use them to test our Comb-matching algorithm. In Figure 5 we show the optimal similarities F_{ij} (see Equation 14) between pairs of graphs (50×50 comparisons of graphs with roughly 25 nodes and 40 edges on average). In this case, high similarities are showed in white, and small similarities are appear in dark grey or black. Graphs belonging to the same class have higher similarities with respect to graphs belonging to different classes.

Finally, in Figure 6 we show the similarities $F_{i\alpha}$ between input graphs and prototype. Higher similarities appear in white whereas lower ones appear in dark grey or black. In Figure 6a we have performed the classification by hand, whereas in Figure 6b we show how the clustering algorithm finds the clusters of images.

4   Conclusions

In this paper we have extended the ACM clustering model for graphs in order to incorporate node and edge attributes, defined by probability density functions. Such an extension consists of introducing the update of these pdf's in the incremental method for building the prototypical graph given the prior probabilities of belonging to the class. We have successfully tested the clustering method in the domain of Mondrian images, obtaining good classification results.

Future work includes the testing the method with bigger graphs which are typically associated to natural images, and this requires the improvement of the

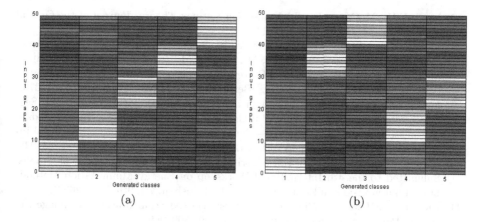

Fig. 6. Similarities $F_{i\alpha}$ between input graphs and prototypes. (a) Classes selected by hand. (b) Classes selected by the clustering algorithm.

Comb-matching. On the other hand we are designing good feature extractor which allow good segmentation results. Finally, we are beginning to apply this graph learning framework to other pattern recognition problems like comparing molecular surfaces in the domain of protein-to-protein docking.

Acknowledgements. This work was partially supported by grant $TIC2002 - 02792$ funded by *Ministerio de Ciencia y Tecnología* and *FEDER*, and grant $CTBPRA/2002/25$ funded by the *Oficina de Ciencia y Tecnología de la Generalitat Valenciana*.

References

1. Bagdanov, A.D., Worring, M.: First Order Gaussian Graphs for Efficient Structure Classification. Pattern Recognition (2002) (to appear).
2. Figueiredo, M.A.T, Leitao, J.M.N, Jain, A.K.: On Fitting Mixture Models. In: Hancock, E.R., Pelillo, M. (eds.): Energy Minimization Methods in Computer Vision and Pattern Recognition. Lecture Notes in Computer Science, Vol. 1654. Springer-Verlag, Berlin Heidelberg New York (1999) 54–69.
3. Gold, S., Rangarajan, A.: A Graduated Assignement Algorithm for Graph Matching. IEEE Trans. on Pattern Analysis and Machine Intelligence, Vol. 18, No. 4 (1996) 377–388.
4. Hofmann, T., Puzicha, J.: Statistical Models for Co-occurrence Data. MIT AI-Memo 1625 Cambridge, MA (1998)
5. Jiang, X., Münger, A., Bunke, H.: On Median Graphs: Properties, Algorithms, and Applications. IEEE Trans. on Pattern Analysis and Machine Intelligence, Vol. 23, No. 10 (2001) 1144–1151
6. Kim, H.Y., Kim, J.H.: Hierarchical Random Graph Representation of Handwritten Characters and its Application to Hangul Recognition. Pattern Recognition 34 (2001) 187–201

7. Land, E.H., McCann, J.J.: Lightness and Retinex Theory. Journal of the Optical Society of America, 61, (1971) 1–11.
8. Li, S.Z.: Toward Global Solution to MAP Image Estimation: Using Common Structure of Local Solutions. In: Pelillo, M., Hancock, E.R.(eds.): Energy Minimization Methods in Computer Vision and Pattern Recognition. Lecture Notes in Computer Science, Vol. 1223. Springer-Verlag, Berlin Heidelberg New York (1997) 361–374.
9. Lozano, M.A., Escolano, F.: Recognizing Indoor Images with Unsupervised Segmentation and Graph Matching. In: Garijo, F.J., Riquelme, J.C., Toro, M.(eds.): Advances in Artificial Intelligence - Iberamia 2002. Lecture Notes on Artificial Intelligence, Vol. 2527. Springer-Verlag, Berlin Heidelberg New York (2002) 933–942.
10. Lozano, M.A., Escolano, F.: EM Algorithm for Clustering an Ensemble of Graphs. Submitted to the 4th International Workshop on Energy Minimization Methods in Computer Vision and Pattern Recognition. Lisbon, 7th-9th July (2003).
11. Puzicha, J.: Histogram Clustering for Unsupervised Segmentation and Image Retrieval. Pattern Recognition Letters, 20, (1999) 899–909.
12. Serratosa, F., Alquézar, R.,Sanfeliu, A.: Synthesis of function-described graphs and clustering of attributed graphs, International Journal of Pattern Recognition and Artificial Intelligence, Vol. 16, No. 6 (2002) 621–655.
13. Serratosa, F., Alquézar, R., Sanfeliu, A.: Function-described graphs for modelling objects represented by sets of attributed graphs, Pattern Recognition, Vol. 23, No. 3 (2003) 781–798
14. Wong, A.K.C, Ghahraman, D.E.: Random Graphs: Structural-Contextual Dichotomy. IEEE Trans. on Pattern Analysis and Machine Intelligence, Vol. 2, No. 4 (1980) 341–348.
15. Wong, A.K.C., You, M.: Entropy and Distance of Random Graphs with Application to Structural Pattern Recognition. IEEE Trans. on Pattern Analysis and Machine Intelligence, Vol. 7, No. 5 (1985) 599–609.
16. Wong, A.K.C., Constant, J., You, M.:Random Graphs. In: Synthactic and Structural Pattern Recognition: Theory and Applications, World-Scientific, Singapore, (1990) 197–234

A Competitive Winner-Takes-All Architecture for Classification and Pattern Recognition of Structures

Brijnesh J. Jain and Fritz Wysotzki

Dept. of Electrical Engineering and Computer Science
Technical University Berlin, Germany

Abstract. We propose a winner-takes-all (WTA) classifier for structures represented by graphs. WTA classification follows the principle *elimination of competition*. The input structure is assigned to the class corresponding to the winner of the competition. In experiments we investigate the performance of the WTA classifier and compare it with the canonical maximum similarity (MS) classifier.

1 Introduction

Let $\mathcal{Y} = \{Y_1, \ldots, Y_N\}$ be a set of N models, each representing a class or category. A canonical form of a classifier is in terms of a set of discriminant functions $\delta_1, \ldots, \delta_N$. Given an input data X, the classifier first computes the N discriminants $\delta_1(X), \ldots, \delta_N(X)$. Then a *comparative maximum selector* assigns X to the category corresponding to the largest discriminant

$$\delta_{i^*}(X) = \max\{\delta_j(X) \; : \; 1 \le j \le N\}.$$

In the domain of graphs the most commonly used discriminant functions correspond to a problem dependent similarity measure σ^1. Thus for a maximum similarity (MS) classifier we have $\delta_i(X) = \sigma(X, Y_i)$ for all $1 \le i \le N$.

MS classifiers for graphs have been applied in various classification, pattern recognition, and clustering tasks by using similarity measures based on the common maximal structural overlap of the graphs under consideration [9], [8], [17], [21], [23]. The problem of computing a similarity between graphs by means of structural overlaps belongs to the class of *graph matching problems*. Since the graph matching problem is well known to be NP complete [6], exact algorithms, which guarantee an optimal solution, are useless in a practical setting for all but the smallest graphs. Therefore, due to its applicability in various fields, many heuristics have been devised to approximately solve the graph matching problem within a *reasonable* amount of time [4], [5], [7], [18], [22], [26]. As well as various other heuristics, artificial neural networks and related methods have been successfully employed to graph matching problems [19], [21], [24], [25], [27]. Nevertheless, in contrast to computing similarities of real valued feature vectors, computation of approximate similarities between graphs remains still a computational

---

[1] For convenience of representation we consider similarity-based classification rather than distance-based classification. This is no severe restriction, since each distance function can be transformed to a similarity function in several ways.

E. Hancock and M. Vento (Eds.): IAPR Workshop GbRPR 2003, LNCS 2726, pp. 259–270, 2003.

intensive procedure. Since decision making of a MS classifier relies on a maximum selector, the computational effort is linearly dependent on the number N of models. Therefore classifying a large number of unseen input data by a MS classification procedure may be computational intractable.

In this paper we propose a two layer *winner-takes-all* (WTA) classifier for graphs. To circumvent the high computational effort of an 1:N matching task, possibly at the expense of classification accuracy, we replace the comparative maximum selector by an inhibitory winner-takes-all network for maximum selection. The basic procedure of the WTA classifier simultaneously matches the input data X with the N models Y_i by using N Hopfield networks $\mathcal{H}_i$, called *subnets*. During evolution the subnets $\mathcal{H}_i$ participate in a winner-takes-all competition. Following the principle *elimination of competition* the WTA classifier focuses on promising subnets and disables unfavorable subnets until one subnet wins the competition and all other subnets are disabled. The input X is assigned to the category corresponding to the winner. This procedure aims to select a winner without completing the computation of a single subnet.

This paper is structured as follows: Section 2 reviews basic components, which will be assembled to a novel WTA classifier for structures. Section 3 briefly summarizes the MD classifier algorithm and introduces the WTA classifier. Section 4 presents and discusses experiments. Finally, Section 5 concludes this contribution.

2   Foundations

This section provides the prerequisites for describing the WTA classification algorithm.

2.1   The Graph Matching Problem

Let V be a set. With $V^{\{2\}}$ we denote the set of all 2-element subsets $\{i,j\} \subseteq V$.

A *weighted graph* is a pair $X = (V, \mu)$ consisting of a finite set $V \neq \emptyset$ of *vertices* and a mapping $\mu : V^{\{2\}} \to \mathbb{R}_+$ assigning each pair $\{i,j\} \in V^2$ a non-negative real valued weight. The elements $\{i,j\} \in V^{\{2\}}$ with positive weight $\mu(\{i,j\}) > 0$ are the *edges* of X. The vertex set of a graph X is referred to as $V(X)$, its edge set as $E(X)$, and its weight mapping as μ_X. A *binary graph* is a weighted graph $X = (V, \mu)$ with $\mu(V^{\{2\}}) \subseteq \{0, 1\}$. A binary graph assigns the weight 1 for its edges and the weight 0 for non-edges. The *adjacency matrix* of a graph X is a matrix $A(X) = (x_{ij})$ with entries $x_{ij} = \mu_X(\{i,j\})$. The number $|X| = |V(X)|$ of vertices of X is called the order of X.

Let X be a graph. A graph Z is an *induced subgraph* of X ($Z \subseteq X$), if $V(Z) \subseteq V(X)$ and $E(Z) = E(X) \cap V(Z)^{\{2\}}$. A subset C_m of $V(X)$ with m vertices is called *clique* of X if $C_m^{\{2\}} \subseteq E(X)$. Thus the graph $X[C_m]$ induced by the vertex set $C_m \subseteq V(X)$ is a fully connected induced subgraph of X. A *maximal clique* is a clique which is not properly contained in any larger clique. A *maximum clique* is a clique with maximum number of vertices.

Let X and Y be graphs. An *isomorphism* between X and Y is a bijective mapping

$$\phi : V(X) \to V(Y), \quad i \mapsto i^\phi$$

with $\{i, j\} \in E(X) \Leftrightarrow \{i^{\phi}, j^{\phi}\} \in E(Y)$ for all $i, j \in V(X)$. If there is an isomorphism between X and Y then X and Y are said to be *isomorphic*. In this case we also write $\phi : X \to Y$ instead of $\phi : V(X) \to V(Y)$. An *overlap* of X and Y is an isomorphism $\phi : X' \to Y'$ between induced subgraphs $X' \subseteq X$ and $Y' \subseteq Y$. By $|\phi| = |X'|$ we denote the *order* of ϕ. A *maximal overlap* is an overlap $\phi : X' \to Y'$, which can not be extended to an overlap $\pi : X'' \to Y''$ with $X' \subset X''$. A *maximum overlap* of X and Y is an overlap $\phi : X' \to Y'$ of maximum order $|\phi|$. By $\mathcal{O}(X, Y)$ we denote the set of all overlaps of X and Y.

The problem of determining a maximum overlap of X and Y is more generally referred to as a *graph matching problem*. Solving that graph matching problem is equivalent to compute a similarity measure

$$\sigma(X, Y) = \max \left\{ f(|\phi|) \ : \ \phi \in \mathcal{O}(X, Y) \right\}. \tag{1}$$

where f is a monotonously increasing function of $|\phi|$. Examples for f are

$$f_1(|\phi|) = \frac{|\phi|}{n} \tag{2}$$

$$f_2(|\phi|) = \frac{|\phi|^2 - |\phi|}{n^2 - n} \tag{3}$$

where n is a scaling factor.

2.2 Graph Matching as Maximum Clique Search

Let X and Y be graphs with adjacency matrices $A(X) = (x_{ij})$ and $A(Y) = (y_{ij})$, respectively. The detection of a maximum overlap of X and Y can be transformed to the problem of finding a maximum clique in an association graph $X \diamond Y$. This concept apparently has been first suggested by Ambler [1], Barrow and Burstall [2], and Levi [15]. Since then it has been applied in several graph matching problems [18], [21], [20]. An association graph $X \diamond Y$ of X and Y is a binary graph with

$$V(X \diamond Y) = \left\{ (i, j) \in V(X) \times V(Y) \ : \ x_{ii} = y_{jj} \right\}$$

$$E(X \diamond Y) = \left\{ \{(i, k), (j, l)\} \in V(X \diamond Y)^{\{2\}} \ : \ x_{ij} = y_{kl}, i \neq j, k \neq l \right\}.$$

Theorem 1 justifies to cast graph matching problems to the problem of finding a maximum clique in an association graph.

Theorem 1. *Let $X \diamond Y$ be the association graph of X and Y. Then the maximum (maximal) cliques of $X \diamond Y$ are in 1-1 correspondence to the maximum (maximal) overlaps of X and Y.*

Proof. We first construct a bijective mapping $\Phi : \mathcal{C}(X \diamond Y) \to \mathcal{O}(X, Y)$ between the set $\mathcal{C}(X \diamond Y)$ of cliques in $X \diamond Y$ and the set $\mathcal{O}(X, Y)$ of overlaps of X and Y.

Let $C_m = \{(i_1, j_1), \ldots, (i_m, j_m)\} \in \mathcal{C}(X \diamond Y)$ be a clique and $X' \subseteq X, Y' \subseteq Y$ the subgraphs induced by $\{i_1, \ldots, i_m\} \subseteq V(X)$ and $\{j_1, \ldots, j_m\} \subseteq V(Y)$, respectively.

We define $\Phi(C_m)$ to be the bijective mapping $\phi : V(X') \to V(Y')$ with $i_k^\phi = j_k$ for all $1 \leq k \leq m$. By definition of $X \diamond Y$ the mapping ϕ is an isomorphism between X' and Y' and therefore an overlap of X and Y. By construction Φ is well defined.

In order to show that Φ is a bijection, we construct a mapping $\Phi' : \mathcal{O}(X, Y) \to \mathcal{C}(X)$ as follows: Let $\phi : X' \to Y'$ be an isomorphism between induced subgraphs $X' \subseteq X$ and $Y' \subseteq Y$. Define $\Phi'(\phi) = \{(i, i^\phi) : i \in V(X)\} \subseteq V(X \diamond Y)$. By definition of $X \diamond Y$ the set $\Phi'(\phi)$ forms a clique in $X \diamond Y$. Then Φ is a bijection, since $\Phi \circ \Phi' = id_{\mathcal{O}(X,Y)}$ and $\Phi' \circ \Phi = id_{\mathcal{C}(X)}$.

From the definitions of Φ and Φ' directly follows the assertion, that the restriction of Φ to the subset of all maximum (maximal) cliques in $X \diamond Y$ is a bijection onto the subset of all maximum (maximal) overlaps. $\qquad\square$

In order to cope with noise in the case of weighted graphs we slightly modify the definition of an association graph according to an approach proposed in [13]. Let $\varepsilon \geq 0$. Then $(i, j) \in V(X \diamond Y)$ if $|x_{ii} - y_{jj}| \leq \varepsilon$. Similarly, $\{(i, k), (j, l)\} \in E(X \diamond Y)$ if $|x_{ij} - y_{kl}| \leq \varepsilon$. Searching a maximum clique in $X \diamond Y$ corresponds to searching a *noisy* maximum overlap of X and Y where the weights of matched vertices and edge differs at most by ε.

2.3   Solving Maximum Clique with Hopfield-Clique Networks

Several different neural network approaches and related techniques have been proposed to solve the maximum clique problem [3], [11]. Following the seminal paper of Hopfield and Tank [10], the general approach to solve combinatorial optimization problems by using neural networks maps the objective function of the optimization problem onto an energy function of the network. The constraints of the problem are included in the energy function as penalty terms, such that the global minima of the energy function correspond to the solutions of the combinatorial optimization problem. Here the combinatorial optimization problem is the problem of finding a maximum clique in an association graph. As in [11] we refer to the term *Hopfield-Clique Network* (HCN) as a generic term for any variant of the Hopfield model for solving the maximum clique problem. As a representative and reference model we consider the HCN described in [12].

Let X be a graph with $|X| = n$. A HCN $\mathcal{H}_X$ associated with X is composed of n fully interconnected units. The synaptic weight w_{ij} between distinct units i and j is $w_E > 0$, if the corresponding vertices are connected by an edge $\{i, j\} \in E$ and $w_I \leq 0$, if $\{i, j\} \notin E$. Connections with weight w_E (w_I) are said to be *excitatory* (*inhibitory*). Thus a graph X uniquely determines the topology of $\mathcal{H}_X$ (up to isomorphism). For this reason we identify units and excitatory connections of $\mathcal{H}_X$ with vertices and edges of X. The dynamical rule of $\mathcal{H}_X$ is given by

$$x_i(t + 1) = x_i(t) + \sum_{j \neq i} w_{ij} o_j(t) \qquad (4)$$

where $x_i(t)$ denotes the activation of unit i at time step t. The output $o_i(t)$ of unit i is a piecewise linear limiter transfer function of the form

$$o_i(t) = \begin{cases} 1 & : \quad x_i(t) \geq \tau_t \\ 0 & : \quad x_i(t) \leq 0 \\ x_i(t)/\tau_t & : \quad \text{otherwise} \end{cases} \tag{5}$$

where $\tau_t > 0$ is a control parameter called *pseudo-temperature* or *gain*. Starting with a sufficient large initial value τ_0 the pseudo-temperature is decreased according to an *annealing-schedule* to a final value τ_f.

Provided an appropriate parameter setting is given, the following Theorem is a direct consequence of the results proven in [12]. It states that the dynamical rule (4) performs a gradient descent with respect to the energy function

$$E(t) = -\frac{1}{2} \sum_i \sum_{j \neq i} w_{ij} o_i(t) o_j(t) \tag{6}$$

where the minima of E correspond to the maximal cliques of X. To simplify the formulation of Theorem 2 we introduce some technical terms: Let $\deg_E(i)$ be the number of excitatory connections incident to unit i. We call

$$\deg_E = \max\{\deg_E(i) \mid 1 \leq i \leq n\}$$
$$\deg_I = \max\{n - \deg_E(i) \mid 1 \leq i \leq n\}$$

the *excitatory* and *inhibitory degree* of $\mathcal{H}_X$, respectively.

Theorem 2. *Let $\mathcal{H}_X$ be a* HCN *associated with a graph X. Suppose that $\deg_E > 0$ and $\deg_I > 0$. If*

$$w_E < \frac{2\tau_t}{\deg_I(n - \deg_E)} \tag{7}$$
$$w_I > \deg_E w_E \tag{8}$$

then the dynamical system (4) performs a gradient descent of the energy function $E(t)$ where the global (local) minima of $E(t)$ are in a 1-1 correspondence with maximum (maximal) cliques of X.

Given a parameter setting satisfying the bounds of Theorem 2 the HCN operates as follows: An initial activation is imposed on the network. Finding a maximum clique then proceeds in accordance with the dynamical rule (4) until the system converges to a stable state $o(t) \in \{0, 1\}^n$. During evolution of the network any unit is excited by all active units with which it can form a clique and inhibits all other units. After convergence the stable state corresponds to a maximal clique of X. The size of a maximal clique can be read out by counting the units with output $o_i(t) = 1$.

2.4   Winner-Takes All Networks for Maximum Selection

This section describes *winner-takes-all* (WTA) networks as a neural network implementation of a maximum selector.

Let $\mathcal{V} = \{v_1, \ldots, v_N\}$ be a set consisting of distinct values $v_i \in \mathbb{R}$. One way to select the maximum from $\mathcal{V}$ within a connectionist framework are *winner-takes-all* (WTA) networks. The WTA net consists of N mutually inhibitory connected units. The activation $z_i(t)$ of unit c_i is initialized with v_i. After initialization the WTA network updates its state according to

$$z_i(t+1) = z_i(t) - w \cdot \sum_{j \neq i} [z_j(t)]_0 + I_i(t) \tag{9}$$

where $-w < 0$ represents the synaptic weight of the inhibitory connections and $I_i(t)$ is the external input of unit c_i. For $z \in \mathbb{R}$ the function

$$[z]_0 := \begin{cases} z & : \quad z > 0 \\ 0 & : \quad z \leq 0 \end{cases}$$

denotes the linear threshold function with lower saturation 0.

The next result is due to [16], [14]. It shows, that under some assumptions, the WTA network converges to a feasible solution within finite time.

Theorem 3. *Let $I(t) = 0 \in \mathbb{R}^N$ and $z(0) \in \mathbb{R}^N$ be the initial activation of the dynamical system (9). Suppose that $z(0)$ has a unique maximum $z_{i^*}(0) > z_j(0)$ for all $j \neq i$. If*

$$0 < w < \frac{1}{N-1}$$

then $z(t)$ converges after a finite number of update steps to a state z with $z_{i^*} > 0$ and $z_j \leq 0$ for all $j \neq i^*$.

For selecting the maximum of a given set $\mathcal{V}$ of input values it does not matter how the winner-takes-all character is implemented. In a practical setting one prefers a simple search algorithm for finding the maximum. However, a neural network implementation of the winner-takes-all selection principle is a fundamental prerequisite to accelerate decision making in classification and pattern recognition tasks of structures.

3   The MS and WTA Classifier for Graphs

Let $X \in \mathcal{X}$ be a data graph, $\mathcal{Y} = \{Y_1, \ldots, Y_N\}$ be the set of model graphs, and σ the similarity measure given in (1).

3.1   The MS Classifier for Graphs

Here we consider a MS classifier based on a structural similarity measure as given in (1), which is directly related to the WTA classifier proposed in Section 3.2.

The MS classifier first computes the similarities $\sigma(X, Y_i)$ by means of HCNs applied on $X \diamond Y_i$ for all $i \in \{1, \ldots, N\}$. Then a comparative maximum selector assigns X to the category corresponding to the the largest similarity

$$\sigma(X, Y_{i^*}) = \max\{\sigma(X, Y_j) \ : \ 1 \leq j \leq N\}.$$

3.2   The WTA Classifier for Graphs

A WTA classifier is a neural network consisting of two layers of neural networks, a matching layer and an output layer. The matching layer has N subnets $\mathcal{H}_i = \mathcal{H}_{X \diamond Y_i}$, each of which is a HCN for matching an input X with model Y_i. The output layer is a competitive WTA network for maximum selection consisting of N inhibitory connected units c_i. Each subnet $\mathcal{H}_i$ is connected to output c_i. Notice that the term WTA classifier refers to the overall system and the term WTA network solely to the inhibitory network in the output layer.

The WTA classifier proceeds first by initializing the inhibition w of the WTA network and the subnets $\mathcal{H}_i$ with respect to the graphs X and Y_i of consideration. Once the classifier has been initialized, the WTA classifier repeatedly executes three basic steps, *similarity matching, competition*, and *adaption* until one output unit wins the competition and all other output units are inhibited:

1. *Similarity Matching.* The classifier simultaneously updates the current state vector $x_i(t)$ of each subnet $\mathcal{H}_i$ according to the dynamical rule (4) giving $x_i(t+1)$ as the new state of $\mathcal{H}_i$. Subsequently the classifier computes *interim values* $\sigma_i(t+1)$ of the target similarities $\sigma(X, Y_i)$ to be calculated by the subnets $\mathcal{H}_i$ for all i. An interim value $\sigma_i(t+1)$ is an approximation of the target similarity $\sigma(X, Y_i)$ given the current state $x_i(t+1)$ of $\mathcal{H}_i$ at any one time $t+1$.
2. *Competition.* The interim values $\sigma_i(t+1)$ provide the basis for competition among the subnets via the WTA network. The classifier updates the current state $z(t)$ of the WTA network according to the dynamical rule (9) giving $z(t+1)$ as its new state vector. The external input $I_i(t)$ applied to unit c_i is of the form

$$I_i(t) = I(\sigma_i(t+1))$$

   where I is a function transforming the interim values in a suitable form for further processing by the WTA network.
3. *Adaption* The WTA classifier disables all subnets $\mathcal{H}_i$ for which $z_i(t+1) \leq 0$. Once a subnet is disabled it is excluded from the competition. A disabled subnet, which has not converged, aborts its computation and terminates before reaching a stable state.

Intertwined execution of the three basic steps, similarity matching, competition, and adaption implements the principle *elimination of competition*. This principle is based on the policy that a large similarity $\sigma(X, Y_i)$ emerges from a series of increasing large interim values $\sigma_i(t)$ at an early stage of the computation of $\mathcal{H}_i$. In detail the WTA classifier proceeds as follows:

ALGORITHM: Let X be a data graph and $\mathcal{Y} = \{Y_1, \ldots, Y_N\}$ a set of model graphs. Choose a value for the inhibition $w \in]0, \frac{1}{N-1}[$.

1. Enable all subnets $\mathcal{H}_i$. Each subnet can be *enabled* or *disabled*. Only if $\mathcal{H}_i$ is enabled it may proceed with its computation. Once $\mathcal{H}_i$ is disabled it resigns from the competition and is no longer in use.
2. Set $t = 0$.

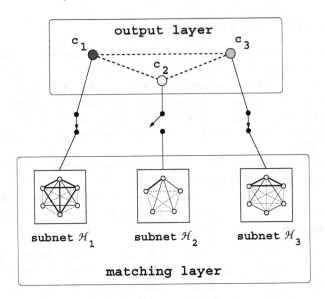

Fig. 1. Architecture of a WTA classifier.

3. **Repeat**
   a) **For** each enabled subnet $\mathcal{H}_i$ **do**
      i. Update $\mathcal{H}_i$ according to the dynamical rule (4) giving $x_i(t+1)$.
      ii. Compute an interim value $\sigma_i(t+1)$ of $\sigma(X, Y_i)$.
      iii. Compute the external input $I_i(t) = \sigma_i(t+1) + \varepsilon_i(t)$ of output unit c_i.
   b) Update the WTA network according to the dynamical rule (9) giving $z(t+1)$.
   c) Disable and terminate all subnets $\mathcal{H}_i$ for which the activation $z_i(t+1) \leq 0$.
   d) Set $t = t + 1$. steps
   until $z_i(t) > 0$ for only one output unit c_i.
4. Let c_{i^*} be the winning unit. Assign X to the category represented by model Y_{i^*}.

Figure 1 depicts a functional diagram of the WTA network for $N = 3$. Excitatory connections are represented by solid lines and inhibitory connections by dotted lines. The subnets $\mathcal{H}_1$, $\mathcal{H}_2$, and $\mathcal{H}_3$ in the matching layer are excitatory connected to the output units c_1, c_2, and c_3 via on-off switches. The switches enable or disable a subnet for or from further computation. The subnets $\mathcal{H}_1$ and $\mathcal{H}_3$ are enabled, while $\mathcal{H}_2$ is disabled. The shading of an output unit c_i indicates its activation $z_i(t)$. Brighter shading corresponds to lower activation and vice versa. A white shading represents output units c_i with activation $z_i(t) \leq 0$. Thus unit c_1 has the largest activation and unit c_2 the lowest. The white shading of unit c_2 indicates that $z_2(t) \leq 0$. In this case subnet $\mathcal{H}_2$ is disabled as illustrated by the off-position of the corresponding switch. In contrast the switches connecting $\mathcal{H}_1$ and $\mathcal{H}_3$ with c_1 and c_3, resp., are in an on-position. Thus the subnets $\mathcal{H}_1$ and $\mathcal{H}_3$ are enabled and further participate in the competition. This instantaneous state of the WTA classifier is consistent with the results we would expect if the subnets $\mathcal{H}_i$ converge to optimal solutions corresponding to maximum cliques. For each subnet a maximum clique of the associated graph is highlighted.

Computing Interim Values

The task of each subnet $\mathcal{H}_i$ is to approximate the similarity

$$\sigma_i = \sigma(X, Y_i) = \max\left\{ f(|\phi|) \ : \ \phi \in \mathcal{O}(X, Y_i)\right\}.$$

between X and Y_i as defined in (1) where the particular choice of a monotonously increasing function f does not effect the class decision. An interim value $\sigma_i(t)$ reports the current degree of similarity between X and Y_i as perceived by $\mathcal{H}_i$ at time step t. In other words, $\sigma_i(t)$ can be regarded as a preliminary estimate of σ_i given the state of $\mathcal{H}_i$ at time step t, where congruency with σ_i gradually emerges with the time spent on the matching process. We define an interim value $\sigma_i(t)$ of the target similarity σ given in (1) by

$$\sigma_i(t) = \frac{1}{n_i^2 - n_i} \sum_{\{k,l\} \in C_i^+} \mathrm{sgn}(w_{kl}) \left(o_k^i(t) + o_l^i(t) \right)$$

where $n_i = |X \diamond Y_i|$ is the number of units of $\mathcal{H}_i$, and $C_i^+ \subseteq V_i^{\{2\}}$ is the set of all connections $\{i, j\}$ between active units k, l of $\mathcal{H}_i$. An unit k of $\mathcal{H}_i$ is an *active unit* if $o_k^i(t) > 0$. The function $sgn(.)$ with

$$sgn(w_{ij}) = \begin{cases} 1 & : \quad w_{ij} = w_E \\ -1 & : \quad w_{ij} = w_I \end{cases}$$

denotes the signum function.

If a stable state $\mathcal{H}_i$ corresponds to a maximum clique in $X \diamond Y_i$, then

$$\sigma_i(t) = \sigma(X, Y_i) = \max_\phi \left\{ f_2(|\phi|) \right\} = \max_\phi \left\{ \frac{|\phi|^2 - |\phi|}{n_i^2 - n_i} \right\}$$

where $n_i^2 - n_i$ is the scaling factor of f_2 in (3).

Computing the External Input

The external input $I_i(t) = I(\sigma_i(t + 1))$ applied to output unit c_i wraps the interim value $\sigma_i(t + 1)$ in a form suitable to the WTA network in the output layer. If $\sigma_i(t)$ is directly applied to output unit c_i as external input, two problems may occur: (1) The WTA classifier disables subnets $\mathcal{H}_i$ with $\sigma_i(1) \leq 0$ in the first iteration step without giving them a *fair* chance to participate in the competition. (2) In order to assign input X to a class the winning unit c_{i*} must be unique. In general there is no guarantee that the interim values determine a unique winning unit.

To guard against premature disabling and ambiguities we define the external input $I_i(t)$ applied to the respective output unit c_i by

$$I_i(t) = [\sigma_i(t + 1)]_\theta^1 + \varepsilon_i(t) \tag{10}$$

where $[\,.\,]_\theta^1$ is the piecewise linear limiter function with lower saturation $\theta > 0$ and upper saturation 1, respectively. The function $\varepsilon_i(t) \in [-\varepsilon, +\varepsilon]$ with $\varepsilon > 0$ is a small random noise with mean zero. The positive constant θ with $1 \gg \theta > \varepsilon$ prevents a premature disabling. Imposing noise on the output units dissolves ambiguities almost surely.

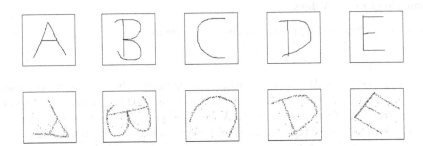

Fig. 2. Handwritten character models (top) and sample data characters (bottom)

4   Experimental Results

In our experiments we compared the performance of the WTA and MS classifier. To improve the performance of both classifiers we used a powerful technique based on local invariants to form the ϵ-association graphs [13]. Both classifier use a mean-field annealing algorithm described in [11] as HCN. The algorithms are implemented in Java using JDK 1.2. All experiments were run on a Sparc SUNW Ultra-4.

We used synthetic data to emulate handwriting recognition of alphanumeric characters as it typically occurs in pen technology of small hand-held devices, for example PDAs. We do not apply additional on-line information. To simulate handwriting recognition as a classification task for structured objects we draw five handwritten character models $\mathcal{Y} = \{$'A', 'B', 'C', 'D', 'E'$\}$ using an X windows interface. The contours of each image were discretized and expressed as a set of points in the 2D plane. The model images are shown in row 1 of Figure 2. For each model $Y_i \in \mathcal{Y}$ we generated k corrupted data characters X_{ik} as follows: First we randomly rotated the model image by an angle $\alpha \in [0, 360]$. Then to each point we added $N(0, 1)$ Gaussian noise. Each point had 10% probability to be deleted and 5% probability of generating spurious points. Row 2 of Figure 2 shows a sample of data images. From each point set we randomly selected 100 points to transform the image to a weighted 100-vertex graph. The vertices represent the points and the edges the normalized distances between two points. Thus each graph is a representation of a character, which is invariant to rotation, translation, and scaling. For each model Y_i we created 100 corrupted data characters giving a total of 500 classification tasks. Due to its higher computational effort, we applied the MS classifier only on 10 out of the 100 data characters of each model giving a total of 50 trials.

Table 1 shows the results. The column identifiers $\mathcal{E}_A$–$\mathcal{E}_E$ and $\mathcal{E}$ refer to the error rates for each model character and the total error rate, respectively. All error rates are given in percentage points. The following columns show the average number $\mathcal{I}$ of iteration steps of a single HCN and the average time $\mathcal{T}$ measured in *sec* to classify a single data character.

The results show that the WTA classifier can be successfully applied to classification and pattern recognition tasks of structures represented by weighted graphs. As can be seen from Table 1 the WTA classifier has a smaller error rate than the MS classifier. This unexpected result is due to the dynamics of the subnets. The error rate of the WTA

Table 1. Results of the MS and WTA classifier.

classifier	trials	$\mathcal{E}_A$	$\mathcal{E}_B$	$\mathcal{E}_C$	$\mathcal{E}_D$	$\mathcal{E}_E$	$\mathcal{E}$	$\mathcal{I}$ [steps]	$\mathcal{T}$ [sec]
MS	50	0.0	4.0	6.0	0.0	4.0	14.0	3979.8	61.7
WTA	500	0.0	0.4	0.0	0.8	0.0	1.2	22.9	1.2

classifier is small, if the correct subnet S_c is the most promising among all subnets at an early stage of the computation. In contrast the error rate of the MS classifier solely depends on the final state of all subnets. In average the correct subnets more often evolved most promising at the beginning than giving the best match after convergence.

As expected the WTA classifier outperforms the MS classifier with respect to speed. Due to [13] both classifiers are extremely fast. For example the graduated assignment algorithm [7], which is one of the fastest graph matching algorithm for weighted graphs reported in the literature, requires about 1 *min* to match two 100-vertex graphs, which would give a total of about 5 *min* to make a MS class decision. In contrast our MS classifier requires about 1 *min* for a class decision and WTA classifier about 1 *sec*.

5   Conclusion

We proposed a WTA classifier for structures. A distinguishing feature of WTA classification is elimination of competition. The winner of the competition determines the category the input structure is assigned to. In contrast to the commonly used MS classifier no complete structural similarity must be computed. In experiments on synthetic data mimicking handwritten characters we have shown that the WTA classifier is capable to significantly outperforms the MS classifier. In forthcoming papers we extend the WTA classifier to the k-nearest neighbor classifier and to competitive learning in the domain of structures.

References

1. A.P. Ambler, H.G. Barrow, C.M. Brown, R.M. Burstall, and R. J. Popplestone. A versatile computer-controlled assembly system. In *International Joint Conference on Artificial Intelligence*, pages 298–307. Stanford University, California, 1973.
2. H. Barrow and R. Burstall. Subgraph isomorphism, matching relational structures and maximal cliques. *Information Processing Letters*, 4:83–84, 1976.
3. I.M. Bomze, M. Budinich, P.M. Pardalos, and M. Pelillo. The maximum clique problem. In D.-Z. Du and P.M. Pardalos, editors, *Handbook of Combinatorial Optimization*, volume 4, pages 1–74. Kluwer Academic Publishers, Boston, MA, 1999.
4. W.J. Christmas, J. Kittler, and M. Petrou. Structural matching in computer vision using probabilistic relaxation. *IEEE Transactions on Pattern Analysis and Machine Intelligence*, 17(8):749–764, 1995.
5. M.A. Eshera and K.S Fu. An image understanding system using attributed symbolic representation and inexact graph-matching. *IEEE Transactions on Pattern Analysis and Machine Intelligence*, 8(5):604–618, 1986.

6. M. Garey and D. Johnson. *Computers and Intractability: A Guide to the Theory of NP-Completeness.* W.H. Freeman and Company, New York, 1979.
7. S. Gold and A. Rangarajan. A graduated assignment algorithm for graph matching. *IEEE Transactions on Pattern Analysis and Machine Intelligence,* 18(4):377–388, 1996.
8. S. Günter and H. Bunke. Adaptive self-organizing map in the graph domain. In H. Bunke and A. Kandel, editors, *Hybrid Methods in Pattern Recognition,* pages 61–74. World Scientific, 2002.
9. S. Günter and H. Bunke. Self-organizing map for clustering in the graph domain. *Pattern Recognition Letters,* 23:401–417, 2002.
10. J.J. Hopfield and D.W. Tank. Neural computation of decisions in optimization problems. *Biological Cybernetics,* 52:141–152, 1985.
11. A. Jagota. Approximating maximum clique with a Hopfield network. *IEEE Trans. Neural Networks,* 6:724–735, 1995.
12. B.J. Jain and F. Wysotzki. Fast winner-takes-all networks for the maximum clique problem. In *25th German Conference on Artificial Intelligence,* pages 163–173. Springer, 2002.
13. B.J. Jain and F. Wysotzki. A novel neural network approach to solve exact and inexact graph isomorphism problems. In *Proc. of the International Conference on Artificial Neural Networks (ICANN 2003).* Springer-Verlag, 2003.
14. K. Koutroumbas and N. Kalouptsidis. Qualitative analysis of the parallel and asynchronous modes of the hamming network. *IEEE Transactions on Neural Networks,* 15(3):380–391, 1994.
15. G. Levi. A note on the derivation of maximal common subgraphs of two directed or undirected graphs. *Calcolo,* 9:341–352, 1972.
16. R.P. Lippman. An introduction to computing with neural nets. *IEEE ASSP Magazine,* pages 4–22, April 1987.
17. A. Lumini, D. Maio, and D. Maltoni. Inexact graph matching for fingerprint classification. *Machine GRAPHICS and VISION, Special issue on Graph Transformations in Pattern Generation and CAD,* 8(2):231–248, 1999.
18. M. Pelillo, K. Siddiqi, and S.W. Zucker. Matching hierarchical structures using association graphs. *IEEE Transactions on Pattern Analysis and Machine Intelligence,* 21(11):1105–1120, 1999.
19. A. Rangarajan, S. Gold, and E. Mjolsness. A novel optimizing network architecture with applications. *Neural Computation,* 8(5):1041–1060, 1996.
20. J.W. Raymond, E.J. Gardiner, and P. Willett. Heuristics for similarity searching of chemical graphs using a maximum common edge subgraph algorithm. *Journal of Chemical Information and Computer Sciences,* 42(2):305–316, 2002.
21. K. Schädler and F. Wysotzki. Comparing structures using a Hopfield-style neural network. *Applied Intelligence,* 11:15–30, 1999.
22. L.G. Shapiro and R.M. Haralick. Structural descriptions and inexact matching. *IEEE Transaction on Pattern Analysis and Machine Intelligence,* 3:514–519, 1981.
23. K. Siddiqi, A. Shokoufandeh, S. Dickinson, and S. Zucker. Shock graphs and shape matching. *International Journal of Computer Vision,* 35:13–32, 1999.
24. A. Sperduti and A. Starita. Supervised neural networks for the classification of structures. *IEEE Transactions on Neural Networks,* 8(3):714–735, 1997.
25. P. Suganthan, E. Teoh, and D. Mital. Pattern recognition by graph matching using potts mft networks. *Pattern Recognition,* 28:997–1009, 1995.
26. A. Wong and M. You. Entropy and distance of random graphs with application to structural pattern recognition. *IEEE Transactions on Pattern Analysis and Machine Intelligence,* 7(5):599–609, 1985.
27. S.-S. Yu and W.-H. Tsai. Relaxation by the hopfield neural network. *Pattern Recognition,* 25(2):197–209, 1992.

Author Index

Lecture Notes in Computer Science

For information about Vols. 1–2609

please contact your bookseller or Springer-Verlag